T0141935

Advances in Intelligent Systems and Computing

Volume 859

The series "Advances in Intelligent Systems and Computing" contains publications on theory, applications, and design methods of Intelligent Systems and Intelligent Computing. Virtually all disciplines such as engineering, natural sciences, computer and information science, ICT, economics, business, e-commerce, environment, healthcare, life science are covered. The list of topics spans all the areas of modern intelligent systems and computing such as: computational intelligence, soft computing including neural networks, fuzzy systems, evolutionary computing and the fusion of these paradigms, social intelligence, ambient intelligence, computational neuroscience, artificial life, virtual worlds and society, cognitive science and systems, Perception and Vision, DNA and immune based systems, self-organizing and adaptive systems, e-Learning and teaching, human-centered and human-centric computing, recommender systems, intelligent control, robotics and mechatronics including human-machine teaming, knowledge-based paradigms, learning paradigms, machine ethics, intelligent data analysis, knowledge management, intelligent agents, intelligent decision making and support, intelligent network security, trust management, interactive entertainment, Web intelligence and multimedia.

The publications within "Advances in Intelligent Systems and Computing" are primarily proceedings of important conferences, symposia and congresses. They cover significant recent developments in the field, both of a foundational and applicable character. An important characteristic feature of the series is the short publication time and world-wide distribution. This permits a rapid and broad dissemination of research results.

More information about this series at http://www.springer.com/series/11156

Radek Silhavy · Petr Silhavy
Zdenka Prokopova
Editors

Computational and Statistical Methods in Intelligent Systems

Editors
Radek Silhavy
Faculty of Applied Informatics
Tomas Bata University in Zlin
Zlin, Czech Republic

Zdenka Prokopova
Faculty of Applied Informatics
Tomas Bata University in Zlin
Zlin, Czech Republic

Petr Silhavy
Faculty of Applied Informatics
Tomas Bata University in Zlin
Zlin, Czech Republic

ISSN 2194-5357 ISSN 2194-5365 (electronic)
Advances in Intelligent Systems and Computing
ISBN 978-3-030-00210-7 ISBN 978-3-030-00211-4 (eBook)
https://doi.org/10.1007/978-3-030-00211-4

Library of Congress Control Number: 2018953176

This Springer imprint is published by the registered company Springer Nature Switzerland AG
The registered company address is: Gewerbestrasse 11, 6330 Cham, Switzerland

Preface

This book constitutes the refereed proceedings of the Computational Methods in Systems and Software 2018 (CoMeSySo 2018), held in September 2018.

CoMeSySo 2018 conference intends to provide an international forum for the discussion of the latest high-quality research results in all areas related to cybernetics and intelligent systems. The addressed topics are the theoretical aspects and applications of software engineering in intelligent systems, cybernetics and automation control theory, econometrics, mathematical statistics in applied sciences and computational intelligence.

CoMeSySo 2018 has received (all sections) 117 submissions, and 63 of them were accepted for the publication.

The volume Computational and Statistical Methods in Intelligent Systems brings new approaches and methods to real-world problems and exploratory research that describes novel approaches in the computational statistics, econometrics, mathematical modelling and methods, software engineering. All these disciplines are studied in the scope of the intelligent systems.

The editors believe that readers will find the following proceedings interesting and useful for their research work.

July 2018

Radek Silhavy
Petr Silhavy
Zdenka Prokopova

Organization

Program Committee

Program Committee Chairs

Petr Silhavy	Department of Computers and Communication Systems, Faculty of Applied Informatics, Tomas Bata University in Zlin, Czech Republic
Radek Silhavy	Department of Computers and Communication Systems, Faculty of Applied Informatics, Tomas Bata University in Zlin, Czech Republic
Zdenka Prokopova	Department of Computers and Communication Systems, Tomas Bata University in Zlin, Czech Republic
Krzysztof Okarma	Faculty of Electrical Engineering, West Pomeranian University of Technology, Szczecin, Poland
Roman Prokop	Department of Mathematics, Tomas Bata University in Zlin, Czech Republic
Viacheslav Zelentsov	Doctor of Engineering Sciences, Chief Researcher of St. Petersburg Institute for Informatics and Automation of Russian Academy of Sciences (SPIIRAS), Russian Federation
Lipo Wang	School of Electrical and Electronic Engineering, Nanyang Technological University, Singapore
Silvie Belaskova	Head of Biostatistics, St. Anne's University Hospital Brno, International Clinical Research Center, Czech Republic

International Program Committee Members

Pasi Luukka
President of North European Society for Adaptive and Intelligent Systems & School of Business and School of Engineering Sciences Lappeenranta University of Technology, Finland

Ondrej Blaha
Louisiana State University Health Sciences Center New Orleans, New Orleans, USA

Izabela Jonek-Kowalska
Faculty of Organization and Management, The Silesian University of Technology, Poland

Maciej Majewski
Department of Engineering of Technical and Informatic Systems, Koszalin University of Technology, Koszalin, Poland

Alena Vagaska
Department of Mathematics, Informatics and Cybernetics, Faculty of Manufacturing Technologies, Technical University of Kosice, Slovak Republic

Boguslaw Cyganek
Department of Computer Science, University of Science and Technology, Krakow, Poland

Piotr Lech
Faculty of Electrical Engineering, West Pomeranian University of Technology, Szczecin, Poland

Monika Bakosova
Institute of Information Engineering, Automation and Mathematics, Slovak University of Technology, Bratislava, Slovak Republic

Pavel Vaclavek
Faculty of Electrical Engineering and Communication, Brno University of Technology, Brno, Czech Republic

Miroslaw Ochodek
Faculty of Computing, Poznan University of Technology, Poznan, Poland

Olga Brovkina
Global Change Research Centre Academy of Science of the Czech Republic, Brno, Czech Republic

Elarbi Badidi
College of Information Technology, United Arab Emirates University, Al Ain, United Arab Emirates

Gopal Sakarkar
Shri Ramdeobaba College of Engineering and Management, Republic of India

V. V. Krishna Maddinala
GD Rungta College of Engineering & Technology, Republic of India

Anand N. Khobragade — Maharashtra Remote Sensing Applications Centre, Republic of India

Abdallah Handoura — Computer and Communication Laboratory, Telecom Bretagne, France

Organizing Committee Chair

Radek Silhavy — Tomas Bata University in Zlin, Faculty of Applied Informatics
email: comesyso@openpublish.eu

Conference Organizer (Production)

OpenPublish.eu s.r.o.
Web: http://comesyso.openpublish.eu
Email: comesyso@openpublish.eu

Conference Website, Call for Papers

http://comesyso.openpublish.eu

Contents

An Analytical Modeling for Leveraging Scalable Communication in IoT for Inter-Domain Routing

A. Bhavana[1(✉)] and A. N. Nandha Kumar[2]

[1] Department of Computer Science and Engineering, VTU, Belagavi, Karnataka, India
bhavana.a.research@gmail.com

[2] Department of Computer Science and Engineering, GSSS, Mysuru, Karnataka, India

Abstract. Although Internet-of-Things (IoT) is one of the big name in the world of technology toward that integrates sophisticated system with cloud and sensors to construct a complete ubiquitous environment, but still it is not meant for offering massive connectivity with large number of heterogeneous sensors. This problem gives rise to scalability issue that is yet an open-end problem. Therefore, this paper introduces a unique approach of improving the accuracy by harnessing IoT ecosystem as well as various other resource-based processes that has never been found considered in prior research attempt. The system considers that if a correct planning of resources is carried out during inter-domain routing, scalability problems can be solved to large extent. The study outcome of this paper shows better communication performance in contrast to existing frequently used routing scheme in IoT.

Keywords: Internet-of-Things · Scalability · Resource utilization Uncertain traffic · Peak load · Heterogeneous nodes

1 Introduction

Internet-of-Things (IoT) aims for establishing a massive chain of connection among all forms of cyber physical devices also called as IoT nodes [1]. At present, the concept of smart city is developed on the basis of connection between installed sensors with cloud in order to develop a whole ubiquitous environment [2]. However, there are various challenges in doing so. The first challenge is for an effective gateway build up that works on distributed principle [3]. It is because without an effective gateway, it is not feasible for assisting in inter-domain routing operation. An effective inter-domain routing will require an effective gateway system in order to assist in connecting various heterogeneous IoT devices. At present, there are certain work carried out over inter-domain routing e.g. [4–6], but very less number of work has been actually offered as a solution although there are certain theoretical discussion on it [7]. Until and unless this problem is not solved, IoT will always encounter a scalability problem. Scalability feature offers incorporating a capability to the IoT devices for sustaining ever increasing connectivity of cyber-physical system. Although, there are some good

R. Silhavy et al. (Eds.): CoMeSySo 2018, AISC 859, pp. 1–11, 2019.
https://doi.org/10.1007/978-3-030-00211-4_1

initiative towards focusing on improving the scalability performance of IoT e.g. [8, 9], but they are not reported to improve the performance of inter-domain routing among heterogeneous devices. Therefore, there is a significant trade-off between inter-domain routing approaches towards minimizing the scalability problems in IoT.

Therefore, the proposed work introduces a unique implementation scheme using analytical methodology that targets to improve scalability with maximum communication performance in IoT. Section 2 discusses about the existing research work followed by problem identification in Sect. 3. Section 4 discusses about proposed methodology followed by elaborated discussion of algorithm implementation in Sect. 5. Comparative analysis of accomplished result is discussed under Sect. 6 followed by conclusion in Sect. 7.

2 Related Work

Our prior review work has discussed various aspects and ongoing research towards IoT while this section further upgrades it with briefings of some recent works [10]. In recent works, emphasis is found towards energy efficiency of IoT devices by using low-power sensors (Roy et al. [11]). Banchuen et al. [12] have discussed about formation of inter-domain network using Software defined network emphasizing on the gateway construction. Similar research direction towards improving gateway communication is also found reported in work of Chen et al. [13]. Adoption of bio-inspired routing is also found to assist in establishing communication in IoT (Hamrioui and Lorenz [14]). According to study of Kaed et al. [15], usage of semantic rules is found to enrich the data quality from IoT gateways. A unique protocol development for inter-domain routing was presented by Villa et al. [16]. The work carried out by Zhao et al. [17] has presented a provisioning mechanism for heterogeneous and distributed system in IoT. Distributed Hash Table was used to construct architecture for exploring distributed system followed by access in IoT in highly distributed manner (Tanganelli et al. [18]). Usage of distributed approach for identifying a discrete event using mobile sink was reported in the work of Suescun and Cardei [19]. A unique routing scheme was discussed by Shen et al. [20] focusing on energy problems among the sensors in IoT. Morabito et al. [21] has assessed the performance of the virtualization environment of IoT. Energy-based upgradation research towards IoT was carried out by Chen et al. [22] using prototype-based approach of antenna. According to literature, it was also explored that middleware could be significantly assists in offering security features using fog computing (Elmsery et al. [23]). Principle of adaptive routing is proven to be assisting communication in IoT in non-conventional communication medium (Javaid et al. [24]). There are various literatures that have witnessed the usage of RPL protocol viz. Taghizadeh et al. [25], Liu et al. [26], Tahir et al. [27], Iova et al. [28], Kim et al. [29, 30]. All these studies are found to solving different set of problem related to communication in IoT using different forms of experimental challenges as well as methodologies. Although, these research work assists in offering various guidelines to handle problems in IoT but they are also associated with various technical pitfalls. The next section briefs about some of the research problems from the existing approaches in IoT.

3 Problem Description

From the previous section, it can be seen that there are diversified form of research work towards IoT development each emphasizing on communication. However, a closer look into an existing system will show that the solutions are based on only symptomatic problems in IoT that is directly linked with communication system. However, we argue that there are various connected components and process e.g. consideration of gateway system, communication protocol, resource dependencies, cost involvement, etc. that is also required to be considered while dealing with communication improvement in IoT. It will mean that very few of existing research work has actually focused on building scalable IoT platform which is yet an open-end problem. At the same time, there is also less amount of work towards developing inter-domain routing considering scalability issues in IoT. The next section discusses about the proposed solution to countermeasure such identified problems.

4 Proposed Methodology

The core idea of the proposed system is to develop a model that assists in ensuring a better form of heterogeneous routing scheme in IoT for the sole purpose of enhancing the scalability performance. Analytical methodology is adopted in order to implement this scheme shown in Fig. 1

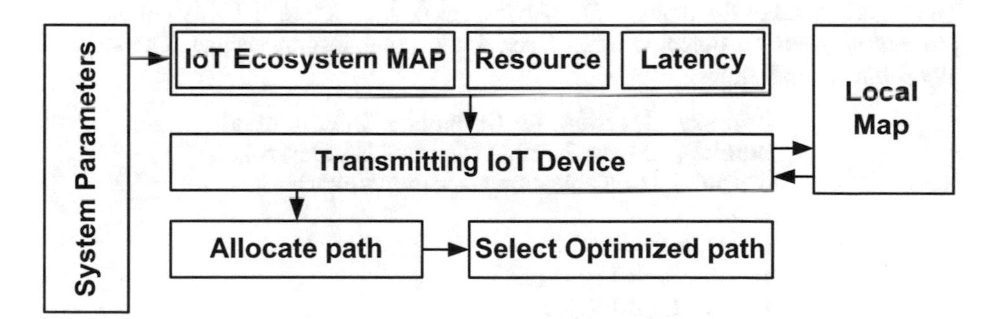

Fig. 1. Adopted research methodology

The proposed system considers various forms of system parameters (e.g. resources utilized, latency, etc.). Resource is an abstract term that will signify both network and computational resource for assisting in modeling. The proposed system also considers a new term IoT Ecosystem MAP that posses all the transactional information that can be accessed within a gateway system for obtaining information about the destination nodes. The local map retains the sequence of information about all the successful communication within each domain. Hence, accessing local map could assist in offering information about best IoT node availability. The core idea of proposed system is offer more number of information about the best alternative routes to the transmitting nodes in order to reduce the latency or any other form of communication degradation during increasing traffic condition. Therefore, scalability of each IoT nodes visibly increases and is proven with respect to cost involved during the process, packet delivery ratio during normal network and network with error condition. The next section illustrates about the algorithm designed for this purpose.

5 Algorithm Implementation

The prime purpose of the proposed algorithm is to address the core problems that lead to scalability issues among the IoT devices. This section presents two core algorithms that are responsible for leveraging the scalability performance among the IoT devices during peak traffic condition.

5.1 Primary Algorithm for Scalability Enhancement

This algorithm is basically responsible for selecting the most eligible IoT gateway node for assisting in heterogeneous communication among the IoT devices more efficiently. The design of the algorithm considers different practical constraints e.g. resource utilization, throughput, and latency. A virtual network with communication vector is formed as a result of this algorithm where the IoT gateway nodes are selected dynamically by the forwarding IoT device. In order to maintain better scalability, this algorithm is also responsible for allocating specific amount of resources towards establishing the communication between the IoT device and selected gateway node. The algorithm takes the input of D_s (Device Span), X (Selected IoT Device) that after processing yields an outcome of S_{cv} (scalable communication vector). The steps of algorithm are as follows:

Primary Algorithm for Scalability Enhancement
Input:D_s (Device Span), X (Selected IoT Device)
Output:S_{cv} (scalable communication vector)
Start
1. **For** n=1:m
2. $[R_{t1}, L_{t1}, c_1] \rightarrow f_1(D_s, X)^{\psi}$
3. $[R_{t2}, L_{t2}, c_2] \rightarrow f_2(D_s, X)^{\psi}$
4. Alloc $(R_{t1}, L_{t1}, R_{t2}, L_{t2}, R_0, L_0, c_1, c_2)$
5. $S_{cv} \rightarrow f_3(D_s, \gamma, \alpha)$
6. **End**
End

The algorithm considers its implementation for all the communicating nodes m (Line-1). The proposed system considers an IoT ecosystem map ψ that consists of all the updated information of all the communicating IoT devices with the gateway node.

This map system offers informative value about the IoT networking system, especially locally so that better form of gateway node could be selected. The algorithm than applies two significant functions f_1(x) (Line-2) and f_2(x) (Line-3) for carrying out two different form of transmission of data i.e. short and long transmission respectively. I also give estimations of cost c of two communicating nodes. Both the form of transmission considers similar input arguments i.e. D_s and X in order to give outcome of total amount of resources being used and cumulative latency. The computation of the cost is carried out mathematically as follows,

$$c_i = \frac{\Delta A.B}{y.\lambda_i}.[f_3(A, B)]^2 \tag{1}$$

In the above expression (1), the variable ΔA will represents effective signal i.e. (A-1), where A can be initialized to natural number Z. The variable B will represents resources required for performing processing within an IoT device, while λ represents channel error owing to its media or hardware. The algorithm also uses f_3(x) on similar parameters that acts as an enhanced Bessel function for further analyzing the heterogeneous communication system while y is a network constant (Line-3). The next step of the algorithm is basically to allocate all the necessary resources (Line-4) followed by applying function f_3(x) on D_s, γ (local maps), and α (signal). The proposed system constructs a Similarity Maps SM between two selected IoT gateway nodes as,

$$\mathrm{SM} \rightarrow \gamma^{\mathrm{u}} \tag{2}$$

Where γ (local maps) is the map constructed between two IoT device and power variable u represents distance ratio among the communicating IoT devices. The outcome gives scalable communication vector (Line-5).

5.2 Secondary Algorithm for Scalability Enhancement

The prior algorithm performs cost computation along with assurance that a communication vector is formulated dynamically to sustain the ever increasing load of networks in IoT. Although, it formulates path but there are many such path created in this way. Although, it is beneficial factor for the gateway nodes but still they have to spend more time in deciding the best path. Therefore, this algorithm is responsible for constructing the optimal path in such a way that it can withstand all forms of traffic scenario with least adverse effect on the IoT devices. Its steps are as follows:

Secondary Algorithm for Scalability Enhancement
Input: c (cost of route)
Output: opt_path (optimized path)
Start

```
1. init c, data
2. For n=1: size(data)
3.    For i=1:size(1)
4path=c(i,:);
5Fori=1:length(path)
6If path(i)~= Inf
7counter=counter+1;
8Ifpath(i)> 2
9          k=k+1;
10End
11.End
12.If counter==k
13.opt_path=path;
14.break;
15.End
16.End
17.End
End
```

The implementation of this algorithm is dynamically done on the basis of the incoming traffic and hence it starts its operation by considering the size of the incoming data (Line-2). It will also mean that the algorithm will act differently for different sizes of network traffic condition in IoT. A matrix *path* is constructed that will retain all the possible communication vectors linking to the IoT gateways node to all the connected IoT devices. For each path (Line-3/4), the algorithm sets its path, counter, and a variable k as zero. For the entire path that is found to be stabilized (Line-6), the algorithm records the path and increases its counter to read the similar type of path leading from the prior path itself (Line-7). A practical test case is presented in Line-8 to ensure that in order to select the optimal path, there should be atleast 2 minimum number of path (Line8). This condition is set to ensure that during the filtering process of selection of all unit path, the selection criteria must stop when there is only two or less than two path. The counter is increased accordingly to that maximum number of path is considered for analysis. All these paths are basically acting as the alternative routes when there is an increase in traffic condition. Only the path, whose counter matches with best path is selected as an optimal path i.e. *opt_path* (Line-12/13). It should be also noted from these that there are many number of optimal paths generated by this algorithms and proposed algorithms uses one-by-one selected optimal paths as an alternative paths for diverting the traffic when it is required during the bottleneck condition.

All this operation associated with cost will definitely require a media to be carried. The proposed system considered that such cost calculation is carried out by IoT gateway node against all the IoT devices that are connected to them or will be connected to them in future. This challenge of computing the cost of communication link to understand the traffic and controlling as necessary is carried out by developing two different forms of flag message that could forward request message for forwarding the data and another to forward acknowledgment message to process the data accordingly through a selected IoT gateway system. Another essential logic in the algorithm implementation is about IoT ecosystem MAP; basically it retains all the information associated with the network on the basis of this flag message. However, availability of precise and updated information from IoT ecosystem MAP cannot be guaranteed as it all depends upon the incoming traffic that cannot be predicted and is highly dynamic in nature. From the computational performance aspect, the proposed system applies probability theory in order to analyze the traffic performance. Moreover, the complete algorithm, combined together, to perform various serious of operation that involves exploring the best path (i.e. path where the resource cost involved is very less). It also perform repairing operation of certain intermittent route that doesn't assists in data delivery by consistently updating the IoT ecosystem MAP as much as possible. Hence, if there is any form of undesirable environmental condition, the proposed system has an inbuilt mechanism of communication system that assists in instantly generating all the possible forms of paths that could release lot of pressure of traffic and therefore offers successful scalability. By scalability viewpoint, the proposed system offers maximum performance of data delivery services in presence of highly uncertain traffic in IoT. The next section discusses about result obtained.

6 Results Discussion

As the proposed system claims of introducing a heterogeneous and scalable routing protocol for assisting communication among heterogeneous domains of nodes in IoT, therefore, it is essential to consider various number of performance parameters that are directly influenced by scalability. Considering probability theory, the study consider overall resource and latency to be 0.8 and 0.14 while circuit power of the IoT device is considered to be 250×10^{-3} J (Fig. 3).

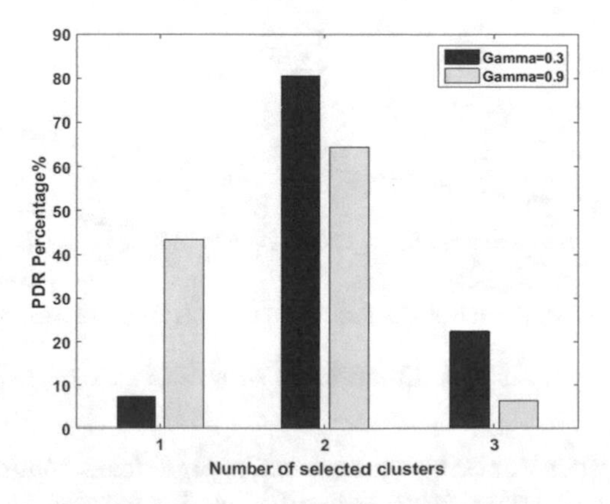

Fig. 2. Analysis of packet delivery ratio

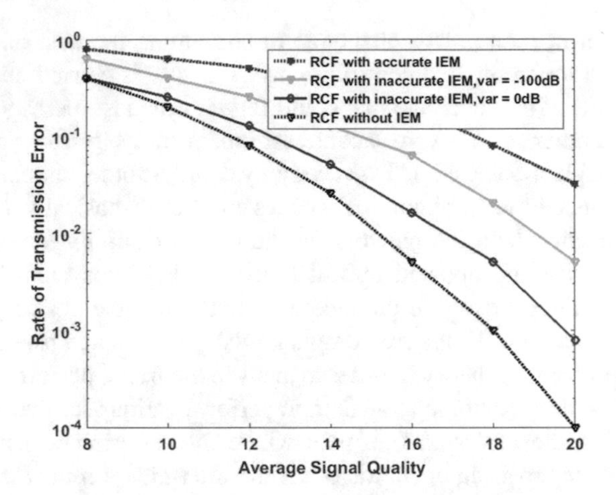

Fig. 3. Analysis of rate of transmission error

The study shows that proposed system is not dependent on obtaining local maps (γ) in order to offer more packet delivery ratio (Fig. 2). Hence, minimal value of local maps is good enough for data delivery in IoT. It was also found that there is a close connectivity between the transmission error and IoT Ecosystem MAP (IEM). Upon analysis, it shows that rate of transmission is absolutely not dependent on IEM. In fact it was found that Rise of Cost Factor (RCF) is found to be highly minimized without IEM value. By this analysis, it just proved that proposed system doesn't have much significant impact on the signal quality owing to the presence of its novel attributed in modeling; however, further it is subjected to comparison with Routing Protocol for Low-Power and Lossy network (RPL) [31].

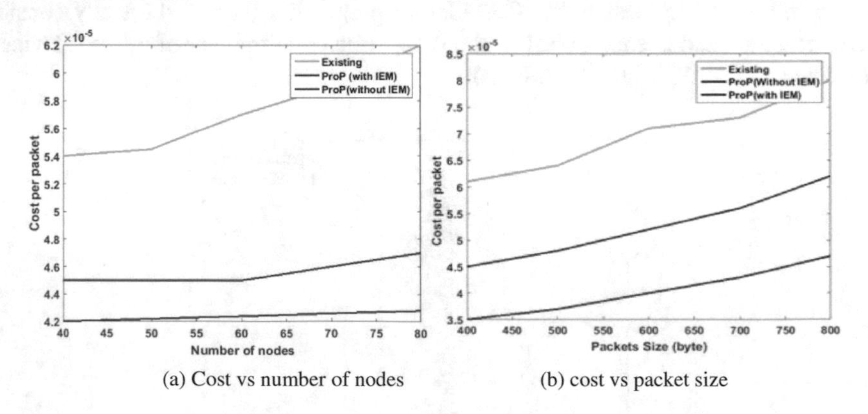

(a) Cost vs number of nodes (b) cost vs packet size

Fig. 4. Comparative analysis of cost

RPL routing system (existing system) offers significant routing among the constrained IoT nodes; however, it doesn't offer much supportability of point-to-point communication cost effectively. Hence, the cost per packet increases significantly with increase in number of nodes and packet (Fig. 4). This problem doesn't exist in proposed system. It also shows that it is not at all dependent on IEM.

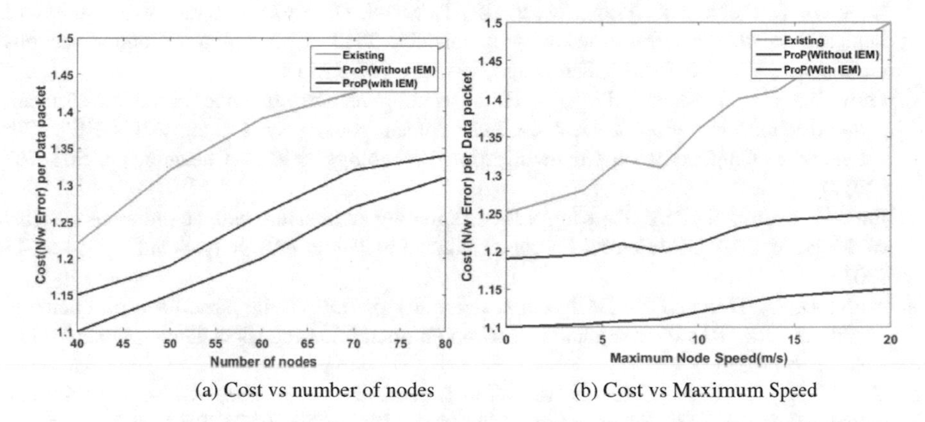

(a) Cost vs number of nodes (b) Cost vs Maximum Speed

Fig. 5. Comparative analysis of cost due to network error

The analysis is also carried out considering the scenario when the network is exposed to various forms of errors that cause extra consumptions of resources. In such condition, it was found that existing RPL routing offers increasing cost per data packet and is absolutely not scalable (Fig. 5). However, the proposed system offers reduced cost as the algorithm mainly runs within IoT gateway with faster and periodic exchange of message about the network. Hence, irrespective of any network condition, the proposed system could offer reduced cost with increasing nodes as well as packets.

7 Conclusion

This paper introduces a simplified approach towards inter-domain routing that emphasize on selection of best gateway node as a mean to control the uncertain or massive traffic system in IoT. The selection is carried on the basis of cost that directly represents resource utilization. The study outcome is analytically proven to offer reduced cost in increasing traffic condition and is found better than existing system. Our research will continue towards interoperation problem in IoT as an extension of this work.

References

1. Cravero, A.: Big data architectures and the Internet of Things: a systematic mapping study. IEEE Latin Am. Trans. **16**(4), 1219–1226 (2018)
2. Hassan, A.M., Awad, A.I.: Urban transition in the era of the Internet of Things: social implications and privacy challenges. In: IEEE Access

3. Hemmatpour, M., Ghazivakili, M., Montrucchio, B., Rebaudengo, M.: DIIG: a distributed industrial IoT gateway. In: 2017 IEEE 41st Annual Computer Software and Applications Conference (COMPSAC), Turin, pp. 755–759 (2017)
4. Al-Musawi, B., Branch, P., Armitage, G.: BGP anomaly detection techniques: a survey. IEEE Commun. Surv. Tutor. **19**(1), 377–396 (2017)
5. Gibbons, T., Hook, J.V., Wang, N., Shake, T., Street, D., Ramachandran, V.: A survey of tactically suitable exterior gateway protocols. In: 2013 IEEE Military Communications Conference, MILCOM 2013, San Diego, CA, pp. 487–493 (2013)
6. Guo, Y., Yan, J., Zhang, L., Qiu, H.: A routing recovery method based on structure properties and centralized control for inter domain routing system. In: 2017 IEEE 17th International Conference on Communication Technology (ICCT), Chengdu, pp. 563–567 (2017)
7. Park, S., Crespi, N., Park, H., Kim, S.H.: IoT routing architecture with autonomous systems of things. In: 2014 IEEE World Forum on Internet of Things (WF-IoT), Seoul, pp. 442–445 (2014)
8. Su-Seong, C., Dongjun, S.: Design of scalable IoT platform using hazard sensor, open and social data. In: 2018 International Workshop on Social Sensing (SocialSens), Orlando, FL, USA, p. 6 (2018)
9. Kim-Hung, L., Datta, S.K., Bonnet, C., Hamon, F., Boudonne, A.: A scalable IoT framework to design logical data flow using virtual sensor. In: 2017 IEEE 13th International Conference on Wireless and Mobile Computing, Networking and Communications (WiMob), Rome, pp. 1–7 (2017)
10. Bhavana, A.: Evaluating perception, characteristics and research directions for Internet of Things (IoT): an investigational survey. Int. J. Comput. Appl. **121**(4), 13–19 (2015)
11. Roy, S.S., Puthal, D., Sharma, S., Mohanty, S.P., Zomaya, A.Y.: Building a sustainable Internet of Things: energy-efficient routing using low-power sensors will meet the need. IEEE Consum. Electron. Mag. **7**(2), 42–49 (2018)
12. Banchuen, T., Kawila, K., Rojviboonchai, K.: An SDN framework for video conference in inter-domain network. In: 2018 20th International Conference on Advanced Communication Technology (ICACT), Chuncheon-si Gangwon-do, Korea (South), p. 1 (2018)
13. Chen, C.H., Lin, M.Y., Liu, C.C.: Edge computing gateway of the industrial Internet of Things using multiple collaborative microcontrollers. IEEE Netw. **32**(1), 24–32 (2018)
14. Hamrioui, S., Lorenz, P.: Bio inspired routing algorithm and efficient communications within IoT. IEEE Netw. **31**(5), 74–79 (2017)
15. Kaed, C.E., Khan, I., Van Den Berg, A., Hossayni, H., Saint-Marcel, C.: SRE: semantic rules engine for the industrial Internet-of-Things gateways. IEEE Trans. Ind. Inf. **14**(2), 715–724 (2018)
16. Villa, D., et al.: IDM: an inter-domain messaging protocol for IoT. In: IECON 2017 - 43rd Annual Conference of the IEEE Industrial Electronics Society, Beijing, pp. 8355–8360 (2017)
17. Zhao, S., Yu, L., Cheng, B.: An event-driven service provisioning mechanism for IoT (Internet of Things) system interaction. IEEE Access **4**, 5038–5051 (2016)
18. Tanganelli, G., Vallati, C., Mingozzi, E.: Edge-centric distributed discovery and access in the Internet of Things. IEEE Internet Things J. **5**(1), 425–438 (2018)
19. Aranzazu-Suescun, C., Cardei, M.: Distributed algorithms for event reporting in mobile-sink WSNs for Internet of Things. Tsinghua Sci. Technol. **22**(4), 413–426 (2017)
20. Shen, J., Wang, A., Wang, C., Hung, P.C.K., Lai, C.F.: An efficient centroid-based routing protocol for energy management in WSN-assisted IoT. IEEE Access **5**, 18469–18479 (2017)
21. Morabito, R.: Virtualization on Internet of Things edge devices with container technologies: a performance evaluation. IEEE Access **5**, 8835–8850 (2017)

22. Chen, Y., et al.: Energy-autonomous wireless communication for millimeter-scale Internet-of-Things sensor nodes. IEEE J. Sel. Areas Commun. **34**(12), 3962–3977 (2016)
23. Elmisery, A.M., Rho, S., Botvich, D.: A fog based middleware for automated compliance with OECD privacy principles in internet of healthcare things. IEEE Access **4**, 8418–8441 (2016)
24. Javaid, N., Cheema, S., Akbar, M., Alrajeh, N., Alabed, M.S., Guizani, N.: Balanced energy consumption based adaptive routing for IoT enabling underwater WSNs. IEEE Access **5**, 10040–10051 (2017)
25. Taghizadeh, S., Bobarshad, H., Elbiaze, H.: CLRPL: context-aware and load balancing RPL for Iot networks under heavy and highly dynamic load. IEEE Access **6**, 23277–23291 (2018)
26. Liu, X., Sheng, Z., Yin, C., Ali, F., Roggen, D.: Performance analysis of routing protocol for low power and lossy networks (RPL) in large scale networks. IEEE Internet Things J. **4**(6), 2172–2185 (2017)
27. Tahir, Y., Yang, S., McCann, J.: BRPL: backpressure RPL for high-throughput and mobile IoTs. IEEE Trans. Mob. Comput. **17**(1), 29–43 (2018)
28. Iova, O., Picco, P., Istomin, T., Kiraly, C.: RPL: the routing standard for the Internet of Things… or is it? IEEE Commun. Mag. **54**(12), 16–22 (2016)
29. Kim, H.S., Kim, H., Paek, J., Bahk, S.: Load balancing under heavy traffic in RPL routing protocol for low power and lossy networks. IEEE Trans. Mob. Comput. **16**(4), 964–979 (2017)
30. Kim, H.S., Im, H., Lee, M.S., Paek, J., Bahk, S.: A measurement study of TCP over RPL in low-power and lossy networks. J. Commun. Netw. **17**(6), 647–655 (2015)
31. The Routing Protocol for Low-Power and Lossy Networks (RPL) Option for Carrying RPL Information in Data-Plane Datagrams. https://tools.ietf.org/html/rfc6553. Accessed 05th June 2018

Cost-Sensitive Learner on Hybrid SMOTE-Ensemble Approach to Predict Software Defects

Inas Abuqaddom[1(✉)] and Amjad Hudaib[2]

[1] Department of Computer Science, King Abdullah II School for Information Technology, Jordan University, Amman, Jordan
inasqaddom@gmail.com

[2] Department of Computer Information System, King Abdullah II School for Information Technology, Jordan University, Amman, Jordan
ahudaib@ju.edu.jo

Abstract. A software defect is a mistake in a computer program or system that causes to have incorrect or unexpected results, or to behave in unintended ways. Machine learning methods are helpful in software defect prediction, even though with the challenge of imbalanced software defect distribution, such that the non-defect modules are much higher than defective modules. In this paper we introduce an enhancement for the most resent hybrid SMOTE-Ensemble approach to deal with software defects problem, utilizing the Cost-Sensitive Learner (CSL) to improve handling imbalanced distribution issue. This paper utilizes four public available datasets of software defects with different imbalanced ratio, and provides comparative performance analysis with the most resent powerful hybrid SMOTE-Ensemble approach to predict software defects. Experimental results show that utilizing multiple machine learning techniques to cope with imbalanced datasets will improve the prediction of software defects. Also, experimental results reveal that cost-sensitive learner performs very well with highly imbalanced datasets than with low imbalanced datasets.

Keywords: Cost matrix · Cost-sensitive learner · Data mining
Ensemble approaches · Imbalanced dataset · SMOTE
Software defect prediction · Software engineering

1 Introduction

The progression in software causes an expansion in the quantity of software products, what's more, their maintenance has become a challenging task. With increasing complexity in software systems, the likelihood of having defective modules in the software systems is getting higher [1]. It is basic to foresee and fix the defects previously of delivering software to clients, on the grounds that the software quality assurance is a tedious undertaking and once in a while does not take into account complete testing of the whole system because of budget

R. Silhavy et al. (Eds.): CoMeSySo 2018, AISC 859, pp. 12–21, 2019.
https://doi.org/10.1007/978-3-030-00211-4_2

issue. In this manner, identification of a defect software module can help us in allocating defects effectively [2,3].

Machine learning methods can be utilized to analyze data from alternate points of view and empower designers to retrieve helpful data. Literature utilizes many machine learning classifiers to distinguish defects in software datasets [4], such as; Multilayer perceptron [5], Naive Bayes classifier [6], Support vector machine [7], Random Forest [8] and Decision Trees [9]. However, all of them faceted the challenge of imbalanced distribution of software defects, where most instances are non-defective, which is called majority class, whereas few defective instances, which is called minority class. Standard classifier algorithms tend to only predict the majority class data. Thus, there is a high probability of misclassification of the minority class as compared to the majority class.

The rest of the paper is organized as follows: dealing with imbalanced dataset is discussed in next section, our proposed approach is described in Sect. 3. Datasets description and evaluation model are in Sects. 4 and 5, respectively. Finally, experimental results in Sect. 6 and conclusion in Sect. 7.

2 Dealing with Imbalanced Datasets

Dealing with imbalanced datasets, can be at both data level, where the distribution of classes is balanced possibly by increasing examples of minority class, and at the classifier level, where many techniques were introduced such as manipulating classifiers internally, one-class learning, ensemble learning, and Cost-Sensitive Learning (CSL) [10]. Some approaches to handle Imbalanced Datasets:

2.1 Resampling Techniques

Resample techniques are data level techniques for balancing classes in imbalanced dataset by either increasing number of examples for minority class or decreasing number of examples for majority class. Accordingly, to obtain almost the same number of examples for both classes. Some resampling techniques [10]:

- *Random Under-Sampling:* is randomly eliminating examples of majority class, in order to balance classes distribution. As a result, the examples of majority and minority classes are almost balanced out.
- *Random Over-Sampling:* is randomly replicating examples of the minority class, in order to present a higher representation of the minority class in dataset.
- *Cluster-Based Over Sampling:* in this case, the K-means clustering algorithm is independently applied to examples of minority and majority classes. Subsequently, each cluster is oversampled such that all clusters of the same class (either majority or minority), have an equal number of examples. Also the total number of examples for both classes are equal, too.
- *Informed Over Sampling:* is used to avoid overfitting which occurs when duplicating some examples of minority class. Such techniques as; SMOTE which

takes a subset of examples from minority class and create new synthetic similar examples from the selected subset. These synthetic examples are added to the original dataset [11].

2.2 Ensemble Techniques

Ensemble techniques are modifying classifiers to produce better prediction and more stable model. Hence, it is preferable to be used with imbalanced data sets to get better predictions [12]. Ensemble approach involves constructing model from training several classifiers, which are called base classifiers/weak classifiers. Accordingly, the final prediction is aggregated from their predictions, using various methods such as voting for classification problems or averaging for regression problems [3,13]. Some ensemble techniques:

- Bagging-Based: stands for Bootstrap Aggregating, which is a way to decrease the variance of prediction by selecting randomly n sets of examples from the original dataset with replacement, such that examples can be duplicated. Then each base classifier builds a model independently based on one of these n sets. Thus, Bagging decreases the variance, narrowly tuning the prediction to be expected correctly [14].
- Boosting-Based: combines weak classifiers to create a strong classifier that can make accurate classification. Boosting starts out with a weak classifier that is trained on training examples. In the next iteration, a new weak classifier focuses on those examples, which were incorrectly classified in the previous iteration. Such boosting-based techniques as adaptive boosting (AdaBoost), which creates a highly accurate classifier by combining many weak and inaccurate classifiers. Each classifier is sequentially trained examples with the goal of correctly classifying them, whom were incorrectly classified in the previous iteration. Furthermore, after each iteration, classifier gives more focus to examples that are harder to be classified. The quantity of focus is measured by a weight, which is initially the same for all examples, after each iteration, weights of misclassified examples increase, while weights of correctly classified examples decrease [14].
- Random Forests: are a collection of decision trees based on a modified version of Bagging. They build large decision trees and take the average of their predictions to obtain the final prediction. Random Forests are popular method because they work efficiently well with hyperparameters' default values [15,16].

2.3 Cost-Sensitive Learning (CSL)

CSL is a type of learning that takes the misclassification costs (that is either false negative or false positive prediction) into consideration, whereas Most classifiers utilize the same misclassification costs. In most real applications, this assumption isn't valid. For example, cancer diagnosis: misclassifying a tumor is significantly more genuine than the false terror since the patients could lose their life as a result of a late treatment. Cost is not always financial; it can be a waste of time or even seriousness of an ailment. In order to explain CSL for software defects, we illustrate CSL for binary classification problem; where the positive class (+) is the minority or the important class, and the negative class (−) is the majority or less important class, which represent software with defects and software without defects, respectively. Let C (i, j) be the cost of predicating an instance in class i when it is actually in class j; the cost matrix is shown in Table 1 [17–19].

Table 1. Cost matrix

	Positive prediction	Negative prediction
Positive class	C(+, +) or TP	C(−, +) or FN
Negative class	C(+, −) or FP	C(−, −) or TN

The main assumption that there is no cost for correct classifications. In other words, cost of C(+, +) and C(−, −) are zeros, hence the cost matrix can be described by the cost ratio as in Eq. (1). The purpose of CSL is to build a model that minimizes the total cost as in Eq. (2) :

$$\text{Cost ratio} = \frac{C(-,+)}{C(+,-)} \tag{1}$$

$$\text{Total cost} = \big(C(-,+) * \#FN\big) + \big(C(+,-) * \#FP\big) \tag{2}$$

where #FN and #FP are the number of false negative and false positive examples, respectively [17,20].

3 Proposed Approach

We propose to utilize multiple imbalanced techniques to get higher performance for classifying software defects, as shown in Fig. 1. Our approach deals with imbalanced datasets in two levels:

1. Data level: utilizing resampling algorithm such as SMOTE.
2. Classifier level: utilizing CSL and AdaBoost such as an ensemble technique.

On other words, we apply CSL to the most recently proposed and evaluated approach [12], which utilizes SMOTE as a resample technique and AdaBoost classifier. Hybrid SMOTE-ensemble approach emphasizes the outstanding of AdaBoost among bagging and random forest as ensemble technique, also the outstanding of decision tree among Naïve Bayes and multilayer perceptron as the base classifier, to predict software defects. Thus, cost-sensitive learner on hybrid SMOTE-ensemble model utilizes AdaBoost with decision tree as its base classifier, as shown in Fig. 1. Furthermore, CSL supposes that there is a cost matrix available for each application. Unfortunately, software defects have no given cost matrix, hence, we dealt with cost ratio as a hyper-parameter and use experiment to select the best cost matrix for each dataset.

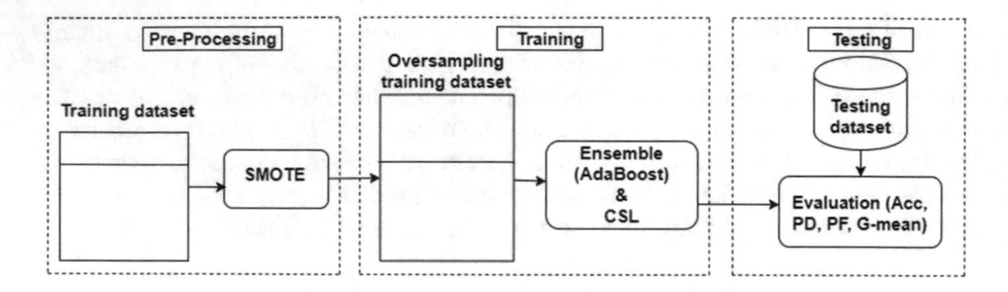

Fig. 1. Cost-sensitive learner on hybrid SMOTE-ensemble approach.

4 Datasets Description

This paper utilizes four public available datasets of software defects; CM1, PC3, KC1 and JM1 [21] from http://openscience.us/repo/ PROMISE[1,2] versions. These datasets were originated from McCabe and Halstead features extractors of source code. These features were characterized trying to equitably describe code features that are related to programming quality. The McCabe and Halstead measures are "module"- based, where a module is the smallest unit of usefulness. In C, modules would be called functions or methods [22,23]. However, CM1 dataset is the NASA space instrument written in C, and PC3 is the flight software for earth circle satellite metadata. On the other hand, KC1 dataset is storage management system that is receiving and processing ground data, whereas, JM1 is a real-time predictive ground system that uses simulations to generate predictions. These four datasets are imbalanced with different ratios between their majority and minority classes, as shown in Table 2.

Table 2. Datasets description

Datasets	Language	#Features	#modules	#Non-defects	#Defects	%Non-defects	%Defects
CM1	C	22	498	449	49	90.16%	9.84%
KC1	C++	22	2109	1783	326	84.54%	15.45%
JM1	C	22	10885	8779	2106	80.65%	19.35%
PC3	C	38	1563	1403	160	89.76%	10.24%

5 Model Evaluation

The evaluation of the proposed approach is measured by set of equations based on confusion matrix, which contains information about the actual and the predicted classes, as shown in Table 3 [24]. Literature shows many equations; that each of them is fit for a specific goal. For our proposed approach, we use Eqs. 3, 4 and 5.

1 http://promise.site.uottawa.ca/SERepository/datasets-page.html.
2 http://tunedit.org/repo/PROMISE/DefectPrediction.

Table 3. Confusion matrix

	Positive prediction	Negative prediction
Positive class	True positive (TP)	False negative (FN)
Negative class	False positive (FP)	True negative (TN)

$$\text{Model Accuracy: } Acc = \frac{TP + TN}{TP + FN + FP + TN} \tag{3}$$

$$\text{True positive rate: } TP_{rate} = \frac{TP}{TP + FN} \tag{4}$$

$$\text{False positive rate: } FP_{rate} = \frac{FP}{FP + TN} \tag{5}$$

Additionally, in software defects true positive rate is called probability of detection (PD), whereas, false positive rate is called probability of false alarm (PF). Best classifier generates high PDs and low PFs [22]. However, while working on imbalanced dataset, accuracy is not an appropriate measure to evaluate model performance [25]. For example, a binary classification dataset with distribution ratio as 98:2 then any classifier that classifies all examples as the majority class will achieve an accuracy of 98%, even though it does not predict correctly any example of minority class [26]. And so on, we need more accurate measure that combines both majority and minority classes, which is G-mean measure in Eq. 6:

$$G\_mean = \sqrt{PD * (1 - PF)} \tag{6}$$

Such that higher g_mean means higher PD and lower PF, which indicates better evaluation for the classifier. In other words, better classifier predicts more defect modules correctly with little chance of misclassification for non-defect modules [22].

6 Experimental Results

The experiments are applied using 10 folds cross-validation as training and testing methodology. However, the four datasets are normalized to be in the interval of [0, 1] to put all features in the same scale. In our experiments, the Min-Max scaling is applied on each feature separately according to Eq. 7.

$$X_{norm} = \frac{X_{actual} - X_{\min}}{X_{\max} - X_{\min}}(New_{\max} - New_{\min}) + New_{\min} \tag{7}$$

where $X_{actual}$is the value to be normalized to X_{norm} which will be in the range of [New_{max}, New_{min}]. Also X_{max} and X_{min} are the actual max and min values of the feature, respectively [27].

SMOTE parameters and AdaBoost number of iterations are chosen according to the recommendations of hybrid SMOTE-ensenble approach [12] as shown in Table 4. To estimate the best value for cost ratio, we experimented different values for cost ratio; 2, 3, 4, 6, 8, 10, 16, 32, 64 and 128. Finally, the best cost ratio for CM1 and JM1 is 10, but for KC1 and PC3 are 6 and 3, respectively.

The proposed approach is compared with decision tree (DT), AdaBoost with DT as its base classifier, and hybrid SMOTE-AdaBoost approach, which are chosen as the best classifiers for software defects prediction among Naive Bayes, multilayer perceptron and decision tree as base classifiers for the most widespread ensemble methods; AdaBoost, Bagging and random forest [12]. The experimental results show the outstanding of CSL on hybrid SMOTE-AdaBoost approach among DT, AdaBoost and hybrid SMOTE-AdaBoost approach, in terms of maximizing g_mean that is the possible maximum PD and minimum PF rates for software defects, as shown in Fig. 2, and Tables 5 and 6.

Table 4. Experiment setup

Dataset	SMOTE		AdaBoost	Cost ratio
	#nearest neighbors	Percentage	#iteration	
CM1	5	120%	20	10
KC1	5	120%	20	6
JM1	5	180%	10	10
PC3	5	200%	10	3

Table 5. CM1 and KC1 results

Classifier	CM1 results				KC1 results			
	PD	PF	G_mean	Acc	PD	PF	G_mean	Acc
DT	0.061	0.031	0.2431	87.95%	0.331	0.061	0.5575	84.54%
AdaBoost	0.102	0.038	0.3132	87.75%	0.34	0.047	0.5692	85.87%
SMOTE-AdaBoost	0.523	0.069	0.6978	85.25%	0.71	0.07	0.8126	86.72%
CSL-SMOTE-AdaBoost	**0.617**	0.085	**0.7514**	85.79%	**0.756**	0.08	**0.8340**	87.32%

It is obvious that the performance metrics improve, as the used imbalanced techniques increase. Such as CM1 dataset, the g-means are 31.32%, 69.78% and 75.14% for AdaBoost classifier, SMOTE-AdaBoost classifier and CSL on SMOTE-AdaBoost classifier, respectively. Also, the ratio of correct prediction for defect modules (PD) increases, but with little increasing of the misclassification ratio for non-defect modules (PF). Nevertheless, the best improvement achieved, that of CM1 dataset, whereas the lowest improvement, that of JM1 dataset. By comparing this note with Table 2, CM1 has the highest imbalanced ratio (90.16:9.84), as JM1 has lowest ratio (80.65:19.35) among the experimented datasets. As a result, we conclude that cost-sensitive learner performs very well with highly imbalanced dataset than with low imbalanced dataset.

Table 6. JM1 and PC3 results

Classifier	JM1 results				PC3 results			
	PD	PF	G_mean	Acc	PD	PF	G_mean	Acc
DT	0.235	0.069	0.4677	79.66%	0.206	0.041	0.4445	88.23%
AdaBoost	0.271	0.066	0.5031	80.59%	0.188	0.03	0.4270	89.00%
SMOTE-AdaBoost	0.753	0.11	0.8186	83.52%	0.723	0.06	0.8244	88.48%
CSL-SMOTE-AdaBoost	**0.789**	0.142	**0.8228**	83.03%	**0.76**	0.064	**0.8434**	89.11%

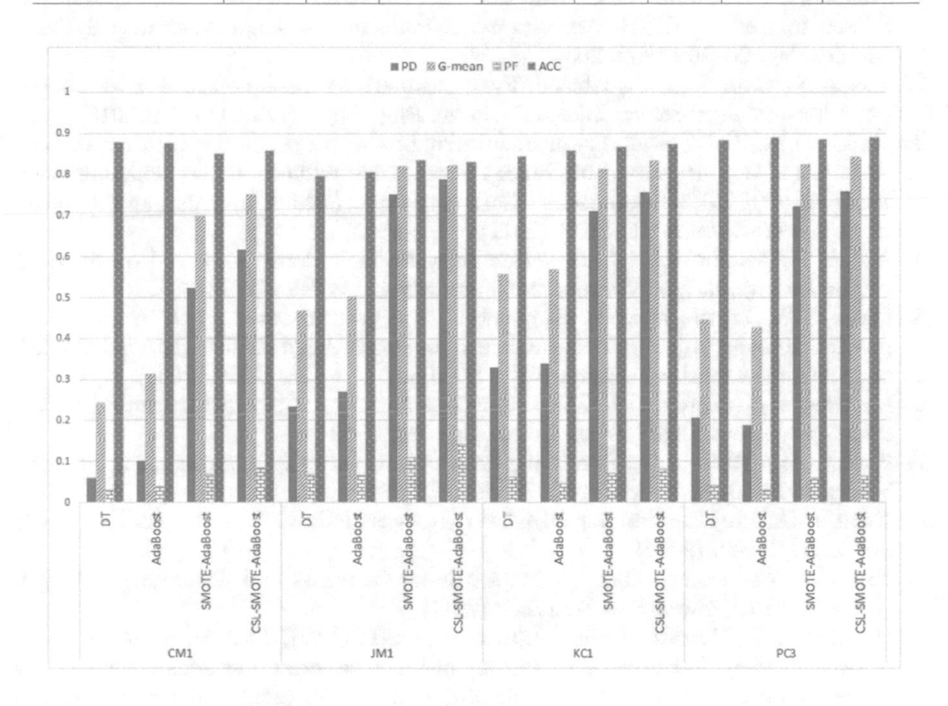

Fig. 2. Aggregated experimental results.

7 Conclusion

Machine learning methods are helpful in software defect prediction, even though with the challenge of imbalanced software defect distribution, such that the non-defect modules are much higher than defective modules. In this paper we introduce the cost-sensitive learner on hybrid SMOTE-ensemble approach to deal with software defects problem, utilizing three techniques to handle imbalanced distribution issue, which are CSL, SMOTE and ensemble classifier. This paper utilizes four public available datasets of software defects and provides comparative performance analysis with the most resent approaches for software defects. Experimental results show that utilizing multiple machine learning techniques to cope with imbalanced datasets will improve the prediction of software defects. Additionally, the best achieved improvement is 75.14% for CM1 dataset, which has the highest imbalanced ratio. Furthermore, we can generalize the results to improve prediction of imbalanced datasets by utilizing multiple techniques together. Also, we conclude that cost-sensitive learner performs very well with highly imbalanced datasets than with low imbalanced datasets.

References

1. Menzies, T., Di Stefano, J.S.: How good is your blind spot sampling policy? Submitted to the 8th IEEE International Symposium on High Assurance Systems Engineering, 25–26 March 2004
2. Aleem, S., Capretz, L.F., Ahmed, F.: Benchmarking machine learning techniques for software defect detection. Int. J. Softw. Eng. Appl. (IJSEA) **6**(3) (2015)
3. Chitraranjan, C.D., et al.: Frequent substring-based sequence classification with an ensemble of support vector machines trained using reduced amino acid alphabets. In: IEEE 2011 10th International Conference on Machine Learning and Applications and Workshops (ICMLA) (2011)
4. Abaei, G., Selamat, A.: A survey on software fault detection based on different prediction approaches. Vietnam J. Comput. Sci. **1**, 79–95 (2014)
5. Quah, T.S., Thwin, M.M.T.: Application of neural networks for software quality prediction using object-oriented metrics. In: Proceedings on International Conference on Software Maintenance, ICSM 2003, pp. 116–125. IEEE (2003)
6. Menzies, T., Greenwald, J., Frank, A.: Data mining static code attributes to learn defect predictors. IEEE Trans. Softw. Eng. **33**, 2–13 (2007)
7. Elish, K.O., Elish, M.O.: Predicting defect-prone software modules using support. Vector Mach. J. Syst. Softw. **81**, 649–660 (2008)
8. Koru, A.G., Liu, H.: Building effective defect-prediction models in practice. IEEE Softw. **22**, 23–29 (2005)
9. Han, J., Kamber, M., Jian, P.: Data Mining Concepts and Techniques. Morgan Kaufmann Publishers, San Francisco (2011)
10. Hernandez, J., Carrasco-Ochoa, J.A., Martinez-Trinidad, J.F.: An empirical study of oversampling and undersampling for instance selection methods on imbalance datasets. In: Ruiz-Shulcloper, J., Sanniti di Baja, G. (eds) Progress in Pattern Recognition, Image Analysis, Computer Vision, and Applications. CIARP. Lecture Notes in Computer Science, vol. 8258. Springer, Heidelberg (2013)
11. Jiang, K., Lu, J., Xia, K.: A novel algorithm for imbalance data classification based on genetic algorithm improved SMOTE. King Fahd University of Petroleum & Minerals. Springer (2016)
12. Alsawalqah, H., Faris, H., Aljarah, I., Alnemer, L., Alhindawi, N.: Hybrid SMOTE-ensemble approach for software defect prediction. In: Silhavy, R., Silhavy, P., Prokopova, Z., Senkerik, R., Kominkova Oplatkova, Z. (eds) Software Engineering Trends and Techniques in Intelligent Systems. CSOC 2017. Advances in Intelligent Systems and Computing, vol. 575. Springer, Cham (2017)
13. Amrieh, E.A., Hamtini, T., Aljarah, I.: Mining educational data to predict student's academic performance using ensemble methods. Int. J. Database Theory Appl. **9**(8), 119–136 (2016). https://doi.org/10.14257/ijdta.2016.9.8.13
14. Buhlmann, P.: Bagging, Boosting and Ensemble Methods. Seminar fur Statistik, ETH Zurich, Zurich

15. Breiman, L.: Random Forests. Kluwer Academic Publishers, Dordrecht (2001). Manufactured in The Netherlands
16. Suleiman, D., Al-Naymata, G.: SMS spam detection using H2O framework. In: The 8th International Conference on Emerging Ubiquitous Systems and Pervasive Networks (EUSPN 2017). Published by Elsevier (2017)
17. Thai-Nghe, N., Gantner, Z., Schmidt-Thieme, L.: Cost-Sensitive Learning Methods for Imbalanced Data
18. Weiss, G.M., McCarthy, K., Zabar, B.: Cost-sensitive learning vs. sampling: which is best for handling unbalanced classes with unequal error costs? Copyright by Foxit software company (2007)
19. Ling, C.X., Sheng, S.: Cost-sensitive learning. In: Sammut, C., Webb, G.I. (eds.) Encyclopedia of Machine Learning. Springer, Boston (2011)
20. Lopez, V., Fernandez, A., Moreno-Torres, J.G., Herrera, F.: Analysis of preprocessing vs. cost-sensitive learning for imbalanced classification. Open problems on intrinsic data characteristics. Expert Syst. Appl. **39**, 6585–6608 (2012)
21. Shepperd, M., Song, Q., Sun, Z., Mair, C.: Data quality: some comments on the NASA software defect datasets. IEEE Trans. Softw. Eng. **39**, 1208–1215 (2013)
22. Sayyad Shirabad, J., Menzies, T.J.: The PROMISE Repository of Software Engineering Databases, School of Information Technology and Engineering, University of Ottawa, Canada. http://promise.sitc.uottawa.ca/SERepository (2005)
23. Hudaib, A., AL zaghoul, F., AL Widian, J.: Investigation of software defects prediction based on classifiers (NB, SVM, KNN and decision tree). J. Am. Sci. **9**(12) (2013). http://www.jofamericanscience.org
24. Menzies, T., DiStefano, J., Orrego, A., Chapman (Mike), R.: Assessing predictors of software defects. Workshop on Predictive Software Models, Chicago, USA Co-located with ICSM (2004)
25. Alqatawna, J., Faris, H., Jaradat, K., Al-Zewairi, M., Adwan, O.: Improving knowledge based spam detection methods: the effect of malicious related features in imbalance data distribution. Int. J. Commun. Netw. Syst. Sci. **8**, 118–129 (2015)
26. Niu, N., Mahmoud, A.: Enhancing Candidate Link Generation for Requirements Tracing: The Cluster Hypothesis Revisited. 978-1-4673-2785-5/12/$31.00 c 2012 IEEE (2012)
27. Patro, S.G.K., Sahu, K.K.: Normalization: a preprocessing stage. arXiv: 1503.06462, March 2015. https://doi.org/10.17148/IARJSET.2015.2305

A Compressive Sensing Based Channel Estimator and Detection System for MIMO-OFDMA System

B. Archana[1(✉)] and T. P. Surekha[2]

[1] Department of Electronics and Communication Engineering, GSSS Institute of Engineering and Technology for Women, Mysore, India
archanab.research@gmail.com

[2] Department of Electronics and Communication Engineering, Vidyavardhaka College of Engineering, Mysore, India

Abstract. The wireless communication system with MIMO-OFDMA has gained popularity in current generation. The communication through MIMO-OFDMA integrated with three element based steering antenna can offer the significant improvement in the performance with respect to bit error rate (BER) and provides the extra diversity gain without altering the radio frequency (RF) of front end circuits. This system uses a channel estimator i.e., mean square error (MSer) to get the information of the channel state but lags with accuracy in channel estimation and complexity issue on computation due to calculation of inverse matrix. This paper, compressive sensing based low complexity detection is introduced to have less complex channel estimation and detection for MIMO-OFDMA receiver with same antenna. The final outcome of the proposed system indicates the improvement in BER which further allows minimizing the computational cost.

Keywords: MIMO-OFDMA · BER · Channel estimation · Computational cost Complexity detection · Radio frequency

1 Introduction

The combinational advantage of MIMO and OFDMA brings the significant performance with high data rate in wireless communication. A three element based steerable antenna is the small sized antenna which have the capability of steering the directivity pattern with low power utilization [1–3]. In Reinoso and Okada [4], a 2 × 2 MIMO-OFDMA is presented with the implementation of this kind of antenna where two receiver elements are used with changing directivity at same frequency of subcarrier frequency spacing. In comparison with the traditional MIMO-OFDMA, the [4] scheme obtains the diversity gain and bit-error-rate (BER) performance improvement in frequency selective fading channel with no increment in the radio frequency (RF) front end circuits. In order to perform the channel estimation, a minimum mean square error (MSer) is used. Due to this mechanism, the BER performance will be decreased due to insufficient accuracy in estimation. Also, the minimum MSer leads to the higher complexity due to inverse matrix computation and other multiplication processes

R. Silhavy et al. (Eds.): CoMeSySo 2018, AISC 859, pp. 22–31, 2019.
https://doi.org/10.1007/978-3-030-00211-4_3

needed for system operation. The works of Wu et al. [5] and Gao et al. [6] performed the investigation on complexity minimization of matrix inversion for the large scale MIMO systems. But, these systems have failed to achieve the complete channel estimation with reduced complexity in MIMO-OFDMA system with minimum MSE mechanism. Compressed sensing is a set of new algorithms that allows the reconstruction of sparse signals from much fewer measurements. These algorithms can be applied in different areas including wireless communications.

This paper introduces a channel estimation technique based on compressive sensing for the receiver, which expresses the issues of channel estimation as compressive sensing problem and then estimates the impulse response of channel. Here a low complexity detection scheme is introduced on the basis of channel matrix sub division. The overall paper is organized as: Sect. 2 discusses the review of existing researches, Sect. 3 describes the system design, Sect. 4 explains the algorithm implementation, Sect. 5 illustrates the results analysis and Sect. 6 gives the conclusion.

2 Review of Existing Techniques

The most relevant researches evolved with an intension of energy minimization and proper resource allocations in MIMO-OFDMA are discussed here. A review work on resource allocation for futuristic LTE is addressed in Archana and Surekha [7], which expels various challenges, methods evolved in MIMO-OFDMA with future scope. A significant research survey on wireless communication for energy efficiency is performed by Feng et al. [8] which includes discussion on physical layer techniques like MIMO and OFDM, cognitive radio, distributive antennas, heterogeneous etc. A survey work on OFDMA concept for future developments in wireless communication is discussed in Maeder and Zein [9] with current state of art in OFDMA networks. In Yin and Alamouti [10] have expressed the review work on design philosophies of OFDMA systems along with the significance of OFDMA over traditional techniques like CDMA and TDMA. The analysis research towards the issues of design parameters for upcoming OFDMA communication system is done in Boddu and Kamili [11]. The author has considered the inter-cell-interference problem in the OFDMA cellular networks. The cell capacity performance subjected to the number of users by selecting proper design parameters. Here a soft frequency reuse mechanism is presented by considering the above problem and design parameters and obtained better capacity.

Focusing on the energy efficiency in MIMO-OFDMA and resource allocation, Mao et al. [12] have introduced an optimization technique involving user selection, subspace selection of receive signal and power allocation for energy efficiency enhancement. Similarly, the issues associated with the distributed resource allocation in MIMO is extensively considered in Zappone et al. [13] and proposed a novel approach by considering the same user selection, subspace selection of receive signal and power allocation factors with the implementation of Dinkelbach's algorithm to maximize the energy efficiency. The focus on the energy efficiency with proper scheduling and resource allocation in MIMO-OFDMA is found in Feminias and Riera-Palou [14]. In this, a channel and queue aware scheduling and resource allocation framework is introduced and achieves significant performance.

To tackle the issues of power consumption or power efficiency and resource allocation in MIMO-OFDMA systems, author Moretti et al. [15] have utilized a layered architecture where the users are divided into different groups on channel quality and assignment basis to achieve better QoS.

The research towards the investigation on the trade-off in energy & spectral efficiency is found in Tang et al. [16]. In this, the practical consideration on power consumption by more number of users and active transmit antennas are examined for implementation of two layer resource allocation in the proposed power model for energy efficiency optimization. For further improvement in the energy efficiency optimization a mechanism of Frobenius norm based dynamic selection was designed. Also, to minimize the computational complexity, fixed user set was developed for all the subcarriers to admit the users.

From the various survey analysis it is found that to get the information of the channel state but lags with accuracy in channel estimation and complexity issue on computation due to calculation of inverse matrix. Hence compressive sensing based low complexity detection is introduced to have less complex channel estimation and detection for MIMO-OFDMA.

3 System Methodology

This section evolves with the discussion of research methodology employed. In this, two transmitters and receiver each are considered with pilot frequency and energy as 4 and 1 respectively. In the following Fig. 1 indicates the research methodology employed. An input image is selected for experimentation. For the same image modulation is applied which brings the changes in the resolution and contrast of the image is then transferred over the OFDMA channel. Later, OFDMA modulation includes addition of cyclic prefix and IFFT operation which is applied for the modulated data which helps to split the signals into channels and minimize the crosstalk in wireless communication. The inputs SNR are considered to apply the channel. Further at receiver end demodulation is applied which includes removal of cyclic prefix and FFT operation. Then the channel estimation is performed with channel estimation (No_ch), Estimating is performed using 3 parameters such as, Least Square (LS), minimum MSer and compressive sensing (CS) methods. Later, demodulation is applied to extract the original information from carrier wave. Then, the data will be received with different channel estimation methods and then analysis is performed with signal to noise ratio (SNR) and BER.

The three element based steering antenna is smaller in size and it consumes low power. The antenna consists of a radiator element subjected to the radio frequency (RF) front end and a single or multiple parasitic or passive elements are terminated with variable capacitors. The directivity of the beam is controlled by introducing the variation in bias voltage supplied to the capacitor. The separation among the parasitic elements and radiator elements can be small. i.e., $0.1 \times \lambda$, where λ is meant for wavelength.

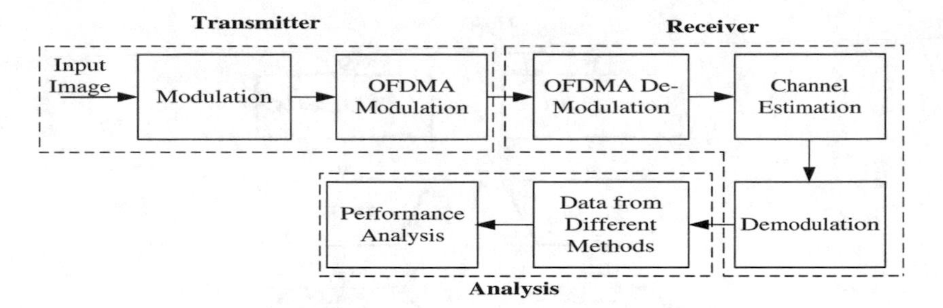

Fig. 1. Research methodology

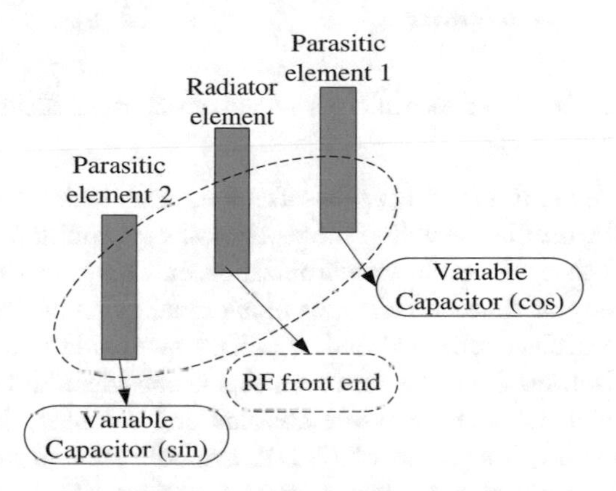

Fig. 2. Three element steering antenna

The Fig. 2 indicates the model of 3 element antenna with periodic change in directivity. This antenna consists of one radiator and two parasitic elements with sinusoidal waves as bias voltage. The sinusoidal wave's frequency is same as spacing of OFDM subcarrier frequency. The effect of periodic variation (directivity) is a signal frequency modulation received by parasitic elements. The analysis performed in frequency domain, where received signal is the addition of all three elements.

A modified system of MIMO-OFDMA receiver is introduced by considering the periodically changing along with three electronic elements based steering antenna (ESA) which considers electronic elements of received signal. The mechanisms achieve the diversity gain within the frequency fading selective channel and obtain improved BER. The proposed system model consists of two receivers and two transmitters. The transmitter is based on the wireless LAN standard i.e., IEEE 802.11n. The following Fig. 3 indicates the block representation of the 3 element antenna model. In this system, every transmitter performs the modulation process, inverse fast Fourier transforms (IFFT) operation and guard interval operation.

For OFDMA receiver, synchronization is must and the absence of synchronization may leads to degradation in BER performance. Hence, synchronization help to know

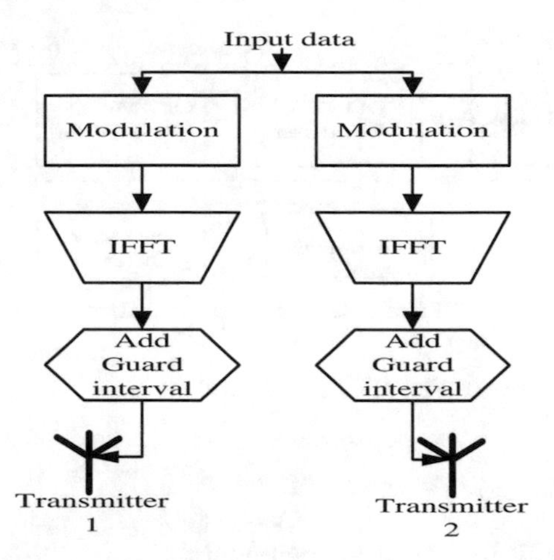

Fig. 3. Block representation of system model of transmitter unit

the initial position of the frame and carrier offset frequency which is occurred due to the mismatch among the oscillators within the receiver and transmitter. Various techniques have been evolved to achieve the synchronization and they may utilize the received symbols in frequency or time domain. For time domain signals, the common characteristics among the cyclic prefix and end of OFDM symbol is utilized to achieve less complex synchronization. Thus, the above concept is considered for proposed detection and estimation model. In order to assure constant and common characteristic among cyclic prefix of size (*Scp*) and end of OFDM symbol. Thus, the signal $m(k)$ at the output of three electronic element based steering antenna (ESA) can be represented with following Eq. 1.

$$m(k) = m_r(k) + R_1 + R_2 \tag{1}$$

In Eq. (1),
k - refers to discrete time index
N - means window size of FFT
$m_r(k)$ - signal received by radiator element

$$R_1 = e^{j2\Pi k/N} \times m_{pe_1}(k)$$

$$R_2 = e^{-j2\Pi k/N} \times m_{pe_2}(k)$$

Where, $m_{pe_1}(k)$ and $m_{pe_2}(k)$ indicates the signals received by parasitic element 1 and 2 respectively.

Next, to bring the simplicity in the proposed model, the weighing factors like x, y, z are considered as 1. The separation among the element within cyclic prefix and OFDM symbol tail is equal to N. Under this equality condition, if m(k) indicates an element in cyclic prefix (Cp) then the element in tail can be referred with Eq. (2).

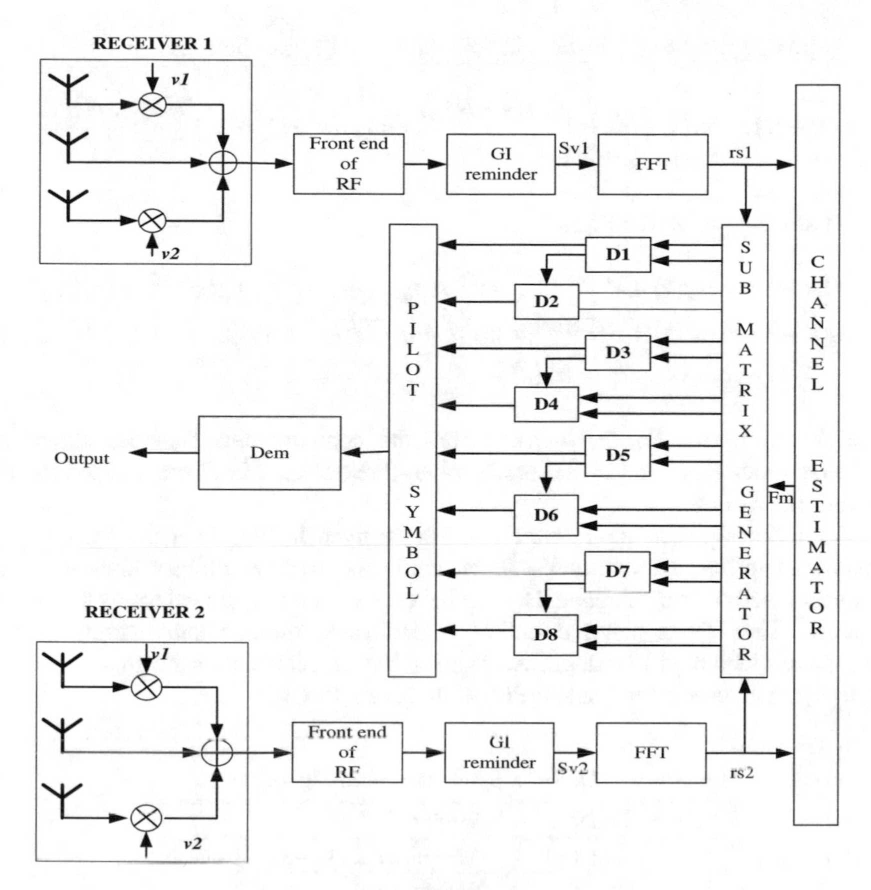

Fig. 4. Architecture of receiver model

$$m(k+N) = m_r(k+N) + R_3 + R_4 \tag{2}$$

In above Eq. (2),

$$R_3 = e^{j2\Pi(k+N)/N} \times m_{pe_1}(k+N)$$

$$R_4 = e^{-j2\Pi(k+N)/N} \times m_{pe_2}(k+N)$$

With the consideration of negligible channel effect and $k \in (-S_{cp}, -1)$, the following assumptions are made,

$$\begin{aligned} m_r(k+N) &= m_r(k) \\ m_{pe_1}(k+N) &= m_{pe_1}(k) \\ m_{pe_2}(k+N) &= m_{pe_2}(k) \end{aligned} \tag{3}$$

On putting the Eq. (3) in Eq. (2) following Eq. (4) can be obtained,

$$m(k+N) = m_r(k+N) + e^{j2\Pi(k+N)/N} \times m_{pe_1}(k+N) + e^{-j2\Pi(k+N)/N} \times m_{pe_2}(k+N) \quad (4)$$

On solving the above Eq. (4),

$$\begin{aligned} m(k+N) &= m_r(k) + e^{j2\Pi(k)/N} \times e^{j2\Pi} \times m_{pe_1}k + e^{-j2\Pi(k)/N} \times e^{-j2\Pi} \times m_{pe_2}k \\ m(k+N) &= m_r(k) + e^{j2\Pi(k)/N} \times m_{pe_1}k + e^{-j2\Pi(k)/N} \times m_{pe_2}k \\ m_r(k+N) &= m_r(k) \end{aligned} \quad (5)$$

Thus the above Eq. (5) indicates that the common characteristics among the OFDMA symbol tail and cyclic prefix when three electronic element based steering antenna (ESA) is used.

The block representation of the receiver is given in Fig. 4. In this system, each transmitter utilizes a traditional antenna while the receiver utilizes three electronic elements based steering antenna. Here a channel estimator is utilized to get the channel response. Then the channel detection is performed through sub matrix detection mechanisms which will be described more in the implementation section.

In symbols used in the below Fig. 4 are given Table 1,

Table 1. Symbols used in Fig. 4.

SI. No	Symbols	Meaning
1	r1 and r2	n size observation vector 1 and 2
2	D1, D2…D8	Detections
3	Fm	Channel response matrix
4	rs1 and rs2	Received symbol

4 Algorithm Implementation

In the transmission end, the input image (In) is selected from the desired folder and resized with the scale of 0.1 and obtained the rescaled image (Io). The same Io image is converted into unsigned 8 bit integer of size $Nr \times Nc$ where, Nr is number of rows and Nc is number of columns. Then, the modulation is applied to the Io image to get the modulated image (Im) with improved resolution and contrast then it's converted from decimal to binary of 8 bit image.

Further to apply the OFDMA modulation on Im which involves addition of cyclic prefix (α) and inverse FFT. Then, the signals will be split as channels. Later, the input SNR is provided to the channel to transmit the data via wireless communication. Then at receiver end the OFMA demodulation is applied which includes removal of cyclic prefix and FFT operation. Then the channel estimation (β) is performed by with channel estimation method, Least Square (LS) method, minimum MSer method and

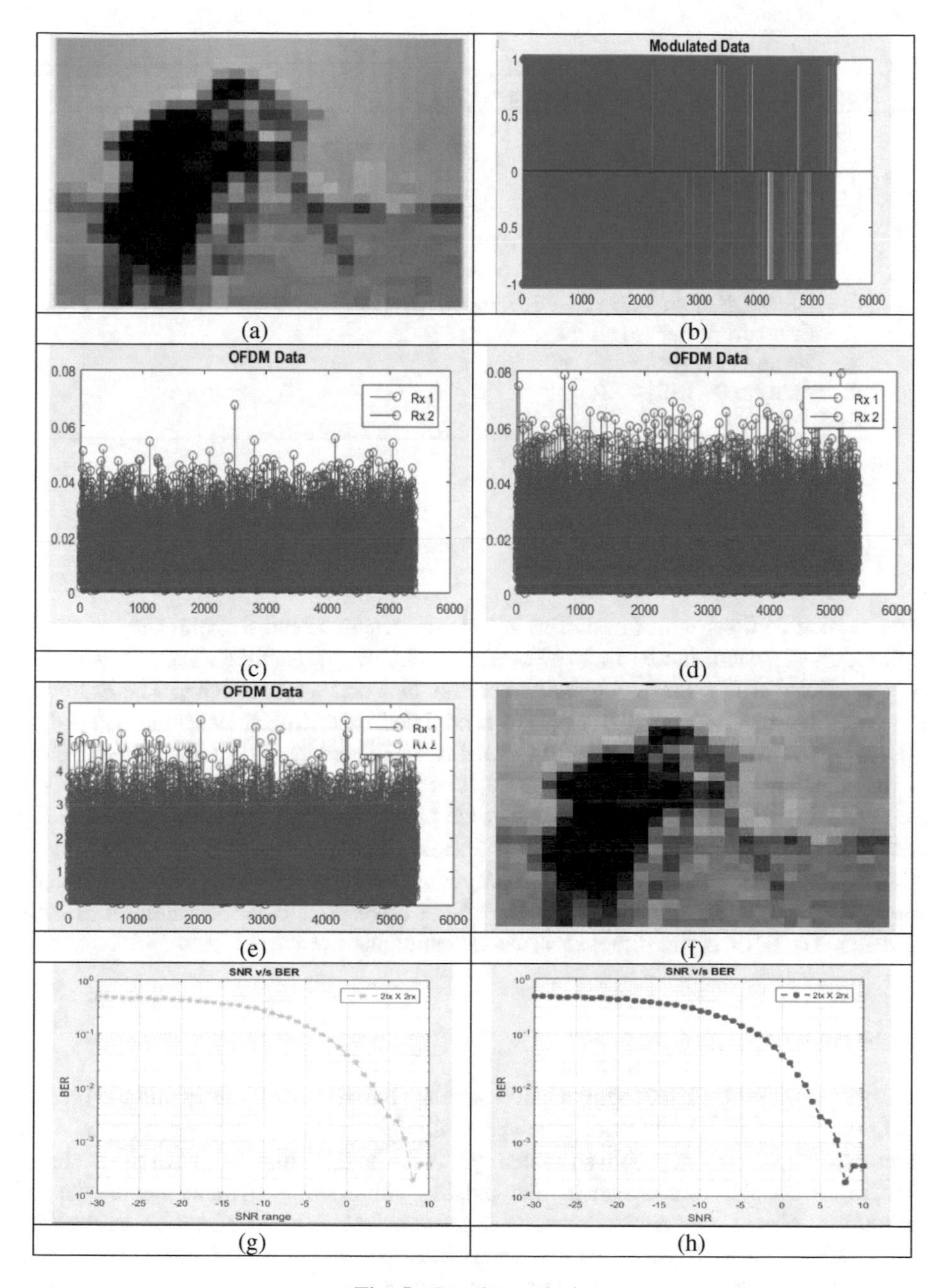

Fig. 5. Results analysis

compressive sensing (CS) method. Demodulation is applied to extract the original information from carrier wave. Then, the data will be received with different channel estimation methods and then analysis is performed with signal to noise ratio (SNR) and BER. The entire algorithm is given as below:

Algorithm for system design and analysis

Start
1. Select → In
2. Io← rescale (In)
3. Im←modulate (Io)
4. OFDM modulation←Inverse FFT, α (Im)
5. Apply channel → SNR
6. OFDM demodulation←FFT, remove α (Im)
7. Estimate → β from all the methods
8. Demodulate (Im)
9. Analyze→ BER, SNR

End

5 Result Analysis

This section evolves with the discussion on the results obtained from the successful simulation of system model in MATLAB. In this, two transmitters and receiver each are considered with pilot frequency and energy as 4 and 1 respectively. The following Fig. 5(a) indicates the input image, (b) modulated data, (c) OFDM modulated data, (d) data after channel application, (e) OFDM demodulated data, (f) demodulated image. In Fig. 5(g) and (h) the performancc analysis of SNR Vs BER is illustrated. The SNR is the amount of unwanted voltage levels which have crept into transmitted signal as it is transmitted via the medium. The BER is the rate where the error is generated due to interpretation of transmitted message comparing between SNR and BER, the inverse relation exist i.e., BER increases if the SNR decreases and BER decreases if SNR increase. The SNR is the direct measure of the relative power of noise.

6 Conclusion

This paper introduced the compressive sensing based channel estimation with less complexity in MIMO-OFDMA receiver with three element steering antenna. The outcomes obtained from proposed system gives the better improvement in the accuracy and improved BER performance than existing minimum MSE scheme of channel estimation. Also, a sub matrix detection a mechanism is introduced which reduces the computation cost with least degradation in BER performance. This proposed system can be considered for futuristic improvement in the BER performance with computational cost minimization.

References

1. Stuber, G.L., Barry, J.R., Mclaughlin, S.W., Li, Y., Ingram, M.A., Pratt, T.G.: Broadband MIMO-OFDM wireless communications. Proc. IEEE **92**(2), 271–294 (2004)
2. Ohira, T., Iigusa, K.: Electronically steerable parasitic array radiator antenna. Electron. Commun. Jpn. (Part II: Electron.) **87**(10), 2545 (2004)
3. Ohira, T., Gyoda, K.: Electronically steerable passive array radiator antennas for low-cost analog adaptive beamforming. In: Proceedings of Phased Array Systems and Technology, Dana Point, CA, USA (2000)
4. Reinoso, D.J., Okada, M.: MIMO-OFDM receiver using a modified ESPAR antenna with periodically changed directivity. In: 13th International Symposium on Communications and Information Technologies, Samui Island, Thailand (2013)
5. Wu, M., Yin, B., Vosoughi, A., Studer, C., Cavallaro, J.R., Dick, C.: Approximate matrix inversion for high-throughput data detection in the large-scale MIMO uplink. In: 2013 IEEE International Symposium on Circuits and Systems (ISCAS), pp. 2155–2158 (2013)
6. Gao, X., Dai, L., Yuen, C., Zhang, Y.: Low-complexity MMSE signal detection based on Richardson method for large-scale MIMO systems. In: 2014 IEEE Vehicular Technology Conference (VTC Fall), pp. 1–5 (2014)
7. Archana, B., Surekha, T.P.: Resource allocation in LTE: an extensive review on methods, challenges and future scope. Resource **3**(2) (2015)
8. Feng, D., Jiang, C., Lim, G., Cimini, L.J., Feng, G., Li, G.Y.: A survey of energy-efficient wireless communications. IEEE Commun. Surv. Tutor. **15**(1), 167–178 (2013)
9. Maeder, A., Zein, N.: OFDMA in the field: current and future challenges. ACM SIGCOMM Comput. Commun. Rev. **40**(5), 71–76 (2010)
10. Yin, H., Alamouti, S.: OFDMA: a broadband wireless access technology. In: 2006 IEEE Sarnoff Symposium. IEEE (2006)
11. Boddu, S., Kamili, J.B.: Analysis of design parameter issues for next generation OFDMA downlinks. In: 2017 Progress in Electromagnetics Research Symposium - Fall (PIERS - FALL), Singapore, pp. 2707–2717 (2017)
12. Mao, J., Chen, C., Cheng, X., Xiang, H.: Energy efficiency optimisation in MIMO-OFDMA systems with block diagonalisation. IET Commun. **11**(18), 2681–2690 (2017)
13. Zappone, A., Jorswieck, E., Leshem, A.: Distributed resource allocation for energy efficiency in MIMO OFDMA wireless networks. IEEE J. Sel. Areas Commun. **34**(12), 3451–3465 (2016)
14. Femenias, G., Riera-Palou, F.: Scheduling and resource allocation in downlink multiuser MIMO-OFDMA systems. IEEE Trans. Commun. **64**(5), 2019–2034 (2016)
15. Moretti, M., Sanguinetti, L., Wang, X.: Resource allocation for power minimization in the downlink of THP-based spatial multiplexing MIMO-OFDMA systems. IEEE Trans. Veh. Technol. **64**(1), 405–411 (2015)
16. Tang, J., So, D.K.C., Alsusa, E., Hamdi, K.A., Shojaeifard, A.: On the energy efficiency-spectral efficiency tradeoff in MIMO-OFDMA broadcast channels. IEEE Trans. Veh. Technol. **65**(7), 5185–5199 (2016)

A Novel Approach to the Potentially Hazardous Text Identification Under Theme Uncertainty Based on Intelligent Data Analysis

Vladislav Babutskiy(✉) and Igor Sidorov

South Federal University, Taganrog, Russia
vbabutskiy@gmail.com

Abstract. The problem of potentially hazardous text identification is an important one in the intelligent data analysis area. As usual, this problem is solved by methods and techniques, which are of a low efficiency in conditions of theme uncertainty.

Within this paper, a novel approach to the potentially hazardous text identification under theme uncertainty is presented. The main idea of data processing approach proposed is based on the user and automatically extracted keywords comparison. This paper contains the brief overview of the text identification methods, the description of the approach presented, some statistical experimental results, discussion and conclusion.

Keywords: Data analysis · Intelligent data analysis · Text identification Theme uncertainty

1 Introduction

The problem of potentially hazardous text identification is one of the important issues in contemporary data processing and analysis research area [1]. Potentially hazardous texts in the Internet contain a wide range of information, which biases the audience to the suicide, extremism, terrorism, etc. [2].

Nowadays text identification methods allow to identify such potentially hazardous text in conditions of theme certainty. These methods are:

- Statistical methods [3, 4];
- Geometrical methods [5, 6];
- Methods based on neural networks [7, 8];

However, within social networks we deal with a huge amount of posts; lots of them appear daily, and the themes of them can be various. Therefore, we cannot presuppose which hashtags can be used in "special" communities, which involve the Internet users to the various terroristic, extremist, fanatical, and suicidal organizations, due to the usage of special cypher. Obviously, as the time passed, those new communities become popular and known, but the time is gone. Therefore, the problem of potentially hazardous text identification in conditions of theme uncertainty is an important issue of contemporary intelligent data analysis.

R. Silhavy et al. (Eds.): CoMeSySo 2018, AISC 859, pp. 32–38, 2019.
https://doi.org/10.1007/978-3-030-00211-4_4

Within this paper, we present a new approach to the potentially hazardous text identification. It is based on a preliminary assumption that potentially hazardous text (PHT) published in a social network can use hashtags which differ from keywords automatically extracted from the post body.

The following sections of the current paper contain:

- The brief overview of well-known and widespread PHT identification methods and techniques;
- A novel approach of the PHT identification presentation and description;
- Some statistic research results, which are the preliminary proof of assumptions made;
- Conclusion.

2 Text Identification Methods and Techniques

2.1 Statistical Methods [9]

Statistical methods are based on the frequency of word usage in texts. The main idea of these methods is that each subject domain has some relevant words. If the processed text has more words from any subject domain, then we can consider the text as a part of it. The well-known algorithms used by statistical methods are Naïve Bayes classifier and Max Entropy Classifier [4].

Statistical method advantages are listed below:

1. These methods are language-independent.
2. They are very fast.
3. They require less training data at the learning stage.

Statistical method disadvantages are:

1. If keyword occurs in a text only once, the statistical methods will not recognize this word.
2. If the distribution of classes within the training data is non uniform, the results of learning stage will be of a poor quality.
3. These methods require training data.
4. The semantic word meaning is not considered.
5. The structure of the text is ignored.

2.2 Geometrical Methods

Each document is described by a set of attributes and can be represented as a point in multidimensional space where the number of dimensions is equal to number of attributes in a vector. Algorithms used by geometrical methods are k Nearest Neighbors classifier and Support Vector Machines [5, 6].

The advantages of geometrical methods are:

1. These methods may be more accurate than previous ones.
2. These methods are simple to implement.

Geometrical methods disadvantages are:

1. The semantic word meaning is not considered.
2. The structure of the text is ignored.
3. Computational complexity.

2.3 Neural Networks Based Methods

A neural network is a graph, with patterns represented in terms of numerical values attached to the nodes of the graph and transformations between patterns achieved via simple message-passing algorithms. Some nodes in the graph are input nodes or output nodes, and the graph as a whole can be viewed as a representation of a multivariate function linking inputs to outputs. Numerical values (weights) are attached to the links of the graph, parameterizing the input/output function and allowing it to be adjusted via a learning algorithm [10].

Neural networks based methods have the following advantages:

1. These methods are language-independent.
2. These approaches may be more accurate than previous.

The disadvantages of neural networks based methods are:

1. It is necessary to create big dataset for neural network learning stage.
2. There can be a possibility of data labeling errors.
3. The decision about neural network architecture has to be made.
4. Learning stage is a very long and resource consuming operation.

The methods considered can be more or less accurate, more or less time consuming due to the learning stage, but the key issue (and the key disadvantage) is that they can give acceptable results in conditions of a-priori known themes or keywords only.

3 A Novel Approach to the PHT Identification Under Theme Uncertainty

We assume that potentially hazardous text keywords are probably not the same as user hashtags. Based on this assumption, the novel approach to the potentially hazardous text identification under thematic uncertainty is presented below.

We assume that neutral texts have three features:

1. User hashtags coincide with the automatically extracted keywords.
2. The subject domain of the user hashtags coincides with automatically extracted keywords.
3. The user hashtags have the same subject domain (Fig. 1).

The method proposed includes three main components:

1. Keyword extraction;
2. Hashtag analysis;
3. The comparison of hashtag and keyword subject domain.

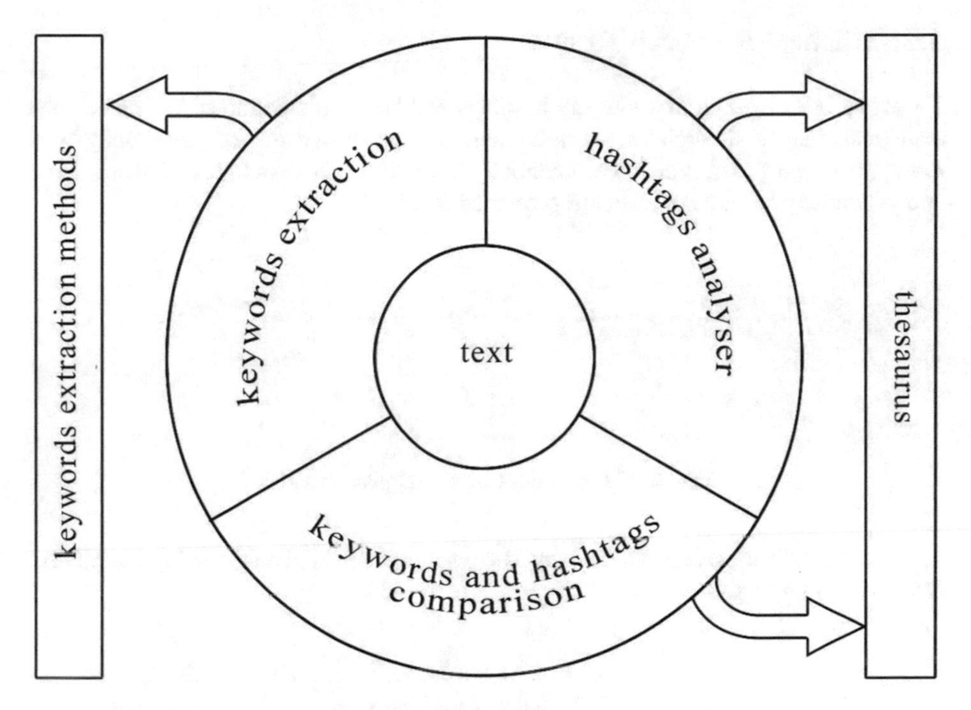

Fig. 1. The scheme of the novel approach to the PHT identification under theme uncertainty

Automatic keyword extraction methods are divided into two types:

1. Statistical methods;
2. Hybrid methods.

The advantages of the statistical methods are the universality and simplicity in terms of implementation. However, these methods accuracy is unsatisfactory. Hybrid methods has a greater potential. In hybrid methods statistical methods are supplemented by linguistic databases.

For the description of any subject domain a certain set of terms is used. Each term designates or describes any concept from this subject domain. The set of the terms describing this subject domain with the indication of the semantic relations (communications) between them is the thesaurus. Such relations in the thesaurus always indicate existence of semantic communication between terms.

The component named “hashtag analyzer” tightly coupled with thesaurus because of necessity to analyze the hashtags in terms of their relevance to the same subject domain.

The last component of the method is the comparison of keywords extracted and hashtags (or comparison of their subject domains). Obviously, it can be easier to compare keywords and hashtags in terms of computational complexity, so, the comparison between related subject domains can be presented as an additional facility to make a conclusion about the text to be analyzed.

4 Statistical Research Results

To prove our assumption about the hashtags and keywords extracted we made some statistical analysis of social network content. We chose two sets of posts: daily "positive" posts and posts, which are potentially hazardous. The examples of neutral posts and potentially hazardous ones are presented in Fig. 2.

Fig. 2. The examples of neutral post and PHT.

The size of both sets is 2000 posts. The result of statistical analysis for neutral texts is presented on Fig. 3.

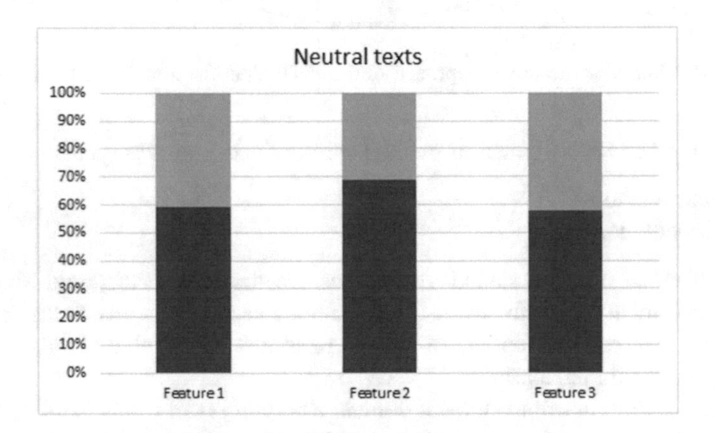

Fig. 3. The result of statistical analysis for neutral texts.

In previous section, some assumptions were made:

1. User hashtags coincide with the automatically extracted keywords.
2. The subject domain of the user hashtags coincides with automatically extracted keywords.
3. The user hashtags have the same subject domain.

According to the analysis results, we can see that feature 1 occurs in neutral texts in 59% of the cases. Feature 2 occurs more often that feature 1. It is caused by the fact that the second feature includes texts, which are characterized by the first feature. Feature 3 occurs in neutral texts in 57% of the cases.

The result of the same statistical analysis for PHT is presented in Fig. 4.

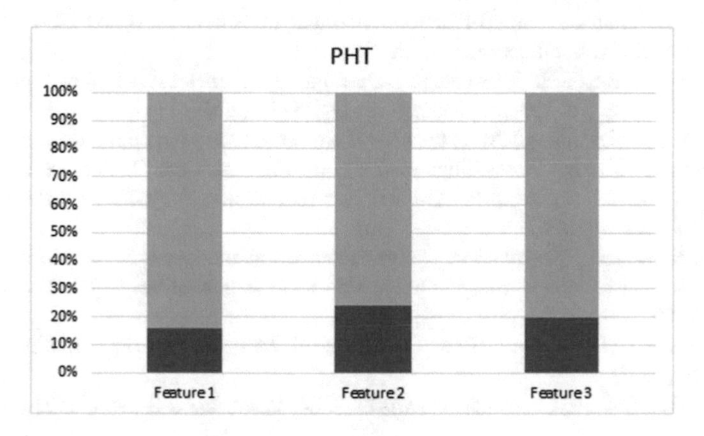

Fig. 4. The result of statistical analysis for PHT.

In the PHT feature mentioned above occurs much rarer. Feature 1 is detected in the 15% of PHT. Feature 2 occurs in the 23% of PHT. Feature 3 is observed in the 20% of PHT.

The results of the statistical research showed that our assumptions are quite correct. Based on these features, texts at least preliminary can be identified as potentially dangerous.

5 Conclusion

Social networks are the important part of life of any person nowadays. The users' knowledge level in such areas as information security and psychology is very low, whereas the same skills of the terrorists or extremists are opposite. That is why the problem of PHT identification is important nowadays.

To solve this problem several methods are used. The methods are statistical, geometrical and methods based on neural networks. These methods have their advantages and disadvantages. The main disadvantages is that they are useless under theme uncertainty.

According to the statistic research results a novel approach to the potentially hazardous text identification seems to be quite promising. It allows to identify hazardous text under thematic uncertainty.

References

1. Nigam, K., McCallum, A.D., Thrun, S., Mitchell, T.M.: Text classification from labelled and unlabeled documents using EM. Mach. Learn. **39**, 103–134 (2007)
2. Omer, E.: Using machine learning to identify jihadist messages on Twitter. M.S. theses, Department of Information Technology, Uppsala University Sweden (2015)

3. Zhang, L., Zhu, J., Yao, T.: An evaluation of statistical spam filtering techniques. ACM Trans. Asian Lang. Inf. Process. (TALIP) **3**(4), 243–269 (2004)
4. Rish, I.: An empirical study of the Naïve Bayes classifier. In: IJCAI 2001 Work Empire Methods Artificial Intelligence, vol. 3 (2001)
5. Kwon, O.-W., Lee, J.-H.: Text categorization based on k-nearest neighbor approach for web site classification. Inf. Process. Manag. **39**(1), 25–44 (2003)
6. Tresch, M., Luniewski, A.: An extensible classifier for semi-structured documents. In: Park, E.K., Makki, K. (eds.) Proceedings of the Fourth International Conference on Information and Knowledge Management (CIKM 1995), Niki Pissinou, Avi Silber-schatz, pp. 226–233. ACM, New York (1995)
7. Haykin, S.: Neural Networks - A Comprehensive Foundation. Canada (1999)
8. Sebastiani, F.: Machine learning in automated text categorization. ACM Comput. Surv. **34**, 1–47 (2002)
9. Mahinovs, A., Tiwari, A.: Text Classification Method Review. Cranfield University, Cranfield (2007)
10. Jordan, M.I., Bishop, C.: Neural Networks. CRC Press, Boca Raton (1997)

Towards a Message Broker Based Platform for Real-Time Streaming of Urban IoT Data

Elarbi Badidi(✉)

College of Information Technology, United Arab Emirates University, Al Ain, United Arab Emirates
ebadidi@uaeu.ac.ae

Abstract. Modern cities deploy a variety of sensors, electric and water meters, or other devices used to guarantee efficient provisioning of services. Internet of Things (IoT) devices and sensors monitor and report on weather conditions, parking space availability, the structural integrity of bridges, historical monuments, and buildings, trash levels in waste containers, night activity and traffic in the streets of the city, and many more. The massive volumes of data produced by sensors and devices need to be harnessed to help smart city applications make informed decisions on the fly. This requires the ability to stream sensed data efficiently, process data streams in real-time, and utilize big data analytics. This paper describes a platform that aims to deal with the streaming of IoT data across the various systems of the city to permit conducting urban data analytics and creation of value-added services. The framework relies on a powerful message broker that can deal with the heterogeneity of IoT devices and adequately scale with the data volumes.

Keywords: Internet of Things · Smart cities · Real-time processing Batch processing · IoT interoperability · Middleware

1 Introduction

Several issues such as ensuring a steady supply of water and energy, reliable transportation, public safety, city planning, sustainable growth of cities, and effective management of natural and financial resources are the heart of many local government agendas. Many cities worldwide are now deploying IoT technologies to reach the above goals, improve the efficiency of their operations, and improve the quality of services offered to their citizens [12].

As a result of the increasing deployment of sensors and meters in cities, tremendous amounts of data are generated by different city systems. Cities will be able to improve city and citizen services and reduce costs when insight skills from analytics power them. For example, cities can improve traffic flow in real-time, improve waste management solutions by picking-up bins only when they are full, detect water leaks, enhance car parking discovery, and offer smart streets lighting. IoT data provides municipalities with information about how much water has been used, where and when. So, they can accurately predict future needs and plan for continuous service. Communities can ensure that water is clean, transportation is reliable, and neighborhoods are safe. As safety is the

R. Silhavy et al. (Eds.): CoMeSySo 2018, AISC 859, pp. 39–49, 2019.
https://doi.org/10.1007/978-3-030-00211-4_5

highest priority for most citizens, analytics informs police officers to ensure they are equipped and prepared for calls to keep themselves and their citizens safe.

A look at the literature on smart cities and at some pilot projects in several countries shows that IoT deployments in cities are still in early stages. Despite the increasing usage of IoT devices and sensors in cities, a comprehensive digital and communication infrastructure to support IoT is still missing. Hence, the increased intelligence in or at the edge of many city systems remains under-exploited. Sensors and physical equipment need a connectivity infrastructure to unlock all their full potential. Urban IoT's success highly depends on the availability of middleware software to link IoT devices and applications. However, most IoT deployments are somewhat localized and limited in the number of involved parties. For example, sensors might be deployed in city parking spaces, but they are rarely accessible by drivers looking for available spots to park their cars without the hassle of going around several times before being able to locate free places. Similarly, a road-side sensor might be accessible by the traffic authority, but by no other devices or parties.

The connectivity between IoT devices and higher-level applications and services is a pressing issue that needs to be addressed to unlock the potential of IoT. Limited connectivity is mainly due to the non-interoperability of the various and heterogeneous IoT devices and technologies in addition to the security concerns [5, 11]. Effective communication with IoT devices, secure transmission and storage of massive amounts of data, data streams processing, and privacy protection, are the main challenges that need to be addressed. Enabling technologies include edge/cloud computing, streams processing, and big data analytics.

In the current state of IoT technologies, dealing with the non-interoperability issue requires the deployment of intermediary components to address the above challenges. In this paper, we propose a message broker component to extend the capabilities of IoT gateways by allowing the streaming of large amounts of data and providing the basis for developing smart city applications and services.

The remainder of the paper is organized as follows: Sect. 2 briefly discusses the issue of IoT interoperability. Section 3 describes related work on middleware to tackle some of the challenges of IoT. Section 4 describes our proposed message broker-based architecture that will serve for the streaming of IoT data. Section 5 illustrates an implementation scenario. Finally, Sect. 6 concludes the chapter.

2 IoT Interoperability

IoT is transforming the way organizations connect and interact with customers, suppliers, and partners. Also, IoT is changing the way cities connect with citizens and businesses. IoT involves connecting IoT devices, sensors and actuators to the organization's network, enabling the collection, exchange, and analysis of sensed data.

Figure 1 depicts the essential elements of a typical IoT solution. Low-power devices use a gateway to communicate through a network (or the Internet) with an enterprise back-end server running an IoT software platform.

The initial IoT vision considers a global ecosystem where devices are hyper-connected and able to communicate with other devices to deliver various services as

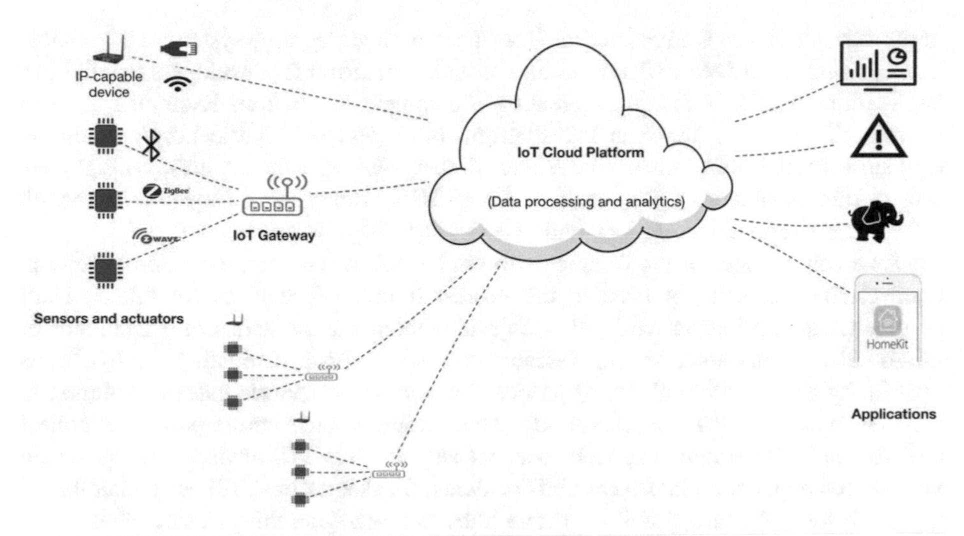

Fig. 1. Typical components of an IoT solution

needed by users. However, with the lack of standards in this emerging field, each manufacturer of IoT products and solutions ends up having proprietary solutions, which are incompatible with solutions from other vendors. The next step in the IoT arena is to interconnect heterogeneous things (devices, sensors, actuators, etc.), manufactured by different vendors with different data formats and semantics, both between themselves and with enterprise back-end servers in a uniform manner to permit the creation of value-added and interoperable services.

Several issues arise from the lack of interoperability of IoT-enabled devices. These issues include:

- the difficulty of integrating and deploying devices manufactured by different manufacturers because devices might have different types of connectors and support different communication protocol standards,
- the difficulty of using common management and monitoring systems to manage and monitor devices,
- the inability to use the same interfaces for pulling and pushing information from the devices,
- the inability to use common mechanisms and approaches to test the devices APIs, and
- the inability to use third-party security software to secure the devices.

As IoT devices increasingly flood the market, interoperability will be the challenge to moving IoT forward. Today, every IoT device is deployed in its manufacturer platform and ecosystem. Tech giants are still flooding the market with their proprietary solutions leading to the fragmentation in IoT. However, many companies and products are emerging to enable interoperability via open source development. Several enterprises are working on IoT cloud platforms, which provide open source frameworks on which to build IoT solutions. These frameworks help setting up and managing the organization's Internet-connected devices from a single control system. They enable interoperability between IoT devices and provide a platform for developers to build IoT applications quickly rather than creating a complete integration. Examples of open source IoT platforms are: Kaa IoT Platform (https://www.kaaproject.org/), Mainflux (https://www.mainflux.com/), SiteWhere (https://www.sitewhere.com/), ThingSpeak (https://thingspeak.com/), DeviceHive, Zetta, DSA, Thingsboard (https://thingsboard.io/), Thinger (https://thinger.io/), and WSo2 (https://wso2.com/iot).

As a consequence of the fragmentation in the IoT market and the heterogeneity in connectivity protocols, at least in the nearest future, IoT devices (or things) can't communicate and interact with other objects without the assistance of a facilitator or intermediate component. Several research works are also investigating the IoT interoperability issue. Aloi et al. [4, 5] proposed a smartphone-centric gateway solution to provide support for IoT interoperability. Their proposed architecture permits to collect and forward data originating from sensors and wireless IoT devices that are using various communication interfaces and standards. Blackstock et al. [8] used a hub-based approach for IoT interoperability, where hubs can aggregate things using Web protocols. They proposed a staged base method to interoperability. Asensio et al. [6] suggested an IoT gateway to provide seamless connectivity and interoperability among things. Perera et al. [15] suggested a plug-in based IoT middleware on mobile devices that aims to allow collecting and processing sensor data before its transmission to the cloud for storage or further processing. Desai et al. [11] proposed a gateway and Semantic Web-based IoT architecture to handle the interoperability between IoT systems. The Semantic Gateway as Service (SGS) integrates sensor and services standards with semantic Web technologies to translate messaging protocols like the Message Queue Telemetry Transport (MQTT), the Constrained Applications Protocol (CoAP), and the Extensible Messaging and Presence Protocol (XMPP) using a multi-protocol proxy architecture.

3 Related Work

There have been several research efforts into middleware to tackle some of the challenges of the IoT. Blackstock et al. [7] proposed a middleware platform, called magic broker 2 (MB2), which provides an interface for implementing IoT applications in which things or smart object can cooperate. MB2 is a low-level event broker that permits to decouple devices, data, and users to support robust services. MB2 uses a publish/subscribe model and leverages OSGi to allow direct communication with devices without the need of using gateways. Brokerage of services is performed via a remote method call. Da Silva [10] proposed a system for automating the deployment of IoT applications based on the standardized MQTT messaging protocol, the Mosquitto message broker, and the runtime environment OpenTOSCA, which is an implementation of the TOSCA standard [14]. They describe a small IoT scenario and how it can be deployed automatically using their proposed system.

Collina et al. [9] proposed a broker, called QEST, which can bridge the world of the physical devices and the Web, used mainly by end-uses, using the MQTT and the Representational State Transfer (REST) protocols. They exposed MQTT topics as

REST resources. The two ecosystems have different semantic models of communication. MQTT implements pub/sub, while the HyperText Transfer Protocol (HTTP) which is the basis of the REST pattern is just a request/response protocol.

Almeida et al. [3] described The Thing Broker, a core platform for Web of Things that provides interfaces to things using a Twitter-like set of abstractions and communication model. The Thing Broker evolved from previous work on IoT platforms, mainly from the Magic Broker 2 [7]. The communication model uses the notions of "following" and "followed" links, to represent the relationships between things. These relationships are dynamic as things can follow and unfollow other things. Events generated by a thing are made available to the followers of the thing. The Thing Broker was designed into layers to make it possible to easily add security layers on the top of its architecture to provide the desired security and privacy levels needed by WoT applications. The architecture is composed of three layers. The Data Access and Object (DAO) layer offers required interfaces for data manipulation, access, and storage using a SQL or No/SQL database. The core layer provides abstractions of things and events. The front-end layer supports different protocols and communication schemes, such as RESTful and Web services.

Abdelwahab et al. [1, 2] proposed FogMQ, a message broker and device cloning system to provide scalable geographically distributed edge analytics support. FogMQ provides each device with a cloning service, which subscribes to device messages. FogMQ uses Flock, an algorithm that mimics flocking behavior, to allow device clones to migrate autonomously across heterogeneous cloud/edge platforms. FogMQ aims to solve the problem of controlling messaging latency of IoT devices in Cloud/Edge platforms. Dynamic changes in computing and networking infrastructure lead to such latency instability. The authors show that, given the current infrastructure conditions, simple migration of IoT device clones between cloud/edge computing resources can minimize latency and keep it at a stable point.

Guner et al. [13] proposed a middleware architecture that uses a database (DB) as a data repository, a data and context orchestrator, and a message broker to manage the communication between requestors and providers in an IoT environment. The Data and Context Orchestrator (DCO) monitors the traffic between requesters and providers and assist applications exchange data privately and securely. The middleware interfaces with the providers and requestors via a control channel, for connection management, authentication and authorization, and a data channel for data transfer based on MQTT.

In this paper, we propose a message-broker based architecture to facilitate interactivity amongst IoT devices and enterprise applications in a smart city environment. The primary goal of the architecture is to promote the streaming of urban IoT data and its management and processing in a scalable fashion while respecting any given security and privacy policies.

4 A Message Broker-Based Urban IoT Architecture

Figure 2 depicts our proposed architecture to address the IoT data streaming issue in smart cities.

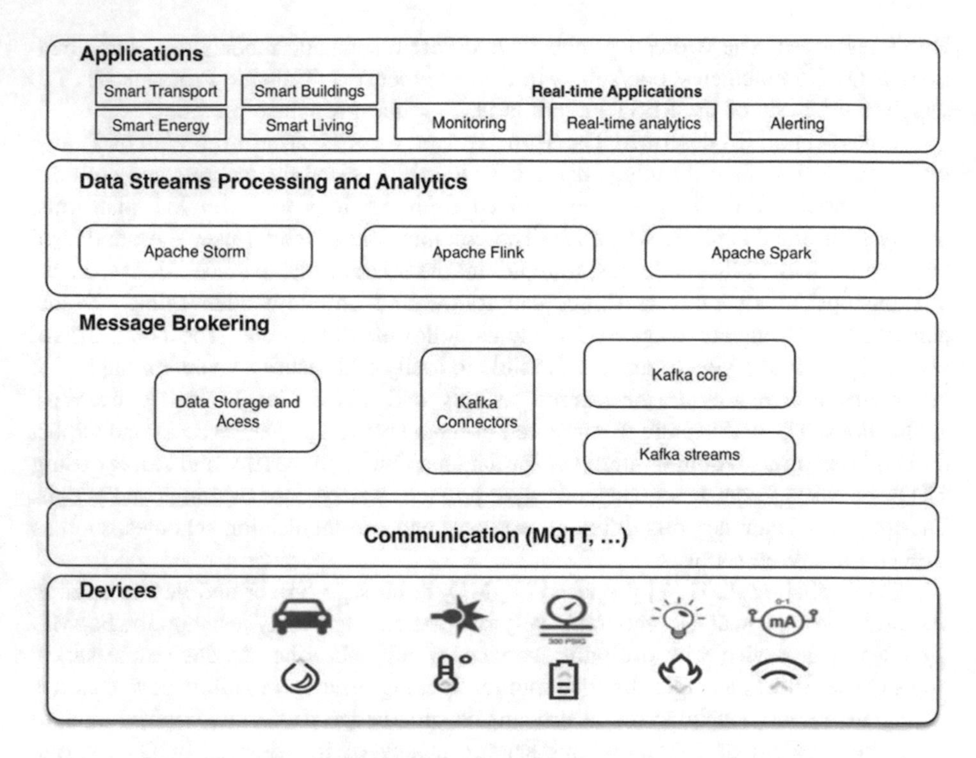

Fig. 2. A message broker-based architecture for urban data streaming

4.1 IoT Devices Layer

The realization of smart energy, smart transport, smart health, smart agriculture, etc. will be permitted by IoT technologies, which require the deployment of a vast number of IoT devices and sensors. This layer is made up of various smart IoT devices. An IoT device detects some input from its surrounding environment and responds to it. The particular input could be light, motion, speed, vibration, pressure, water level, heat, or any other environmental phenomenon. The device reading is then converted into a human-readable form or sent over a network to a gateway for further processing. An IoT device, with typically an IP address, can connect to a network to exchange data. Smart IoT devices enable automating operations of a city by collecting data on various physical assets (equipment, vehicles, buildings, facilities, etc.) to monitor their behavior and status, and using collected data to optimize resources and processes.

To be considered as an IoT device, the device must attach to the Internet either directly or indirectly. Examples of devices with direct connections to the Internet are:

- Arduino with Ethernet or Wi-Fi connection
- Raspberry Pi with Ethernet or Wi-Fi connection
- Intel Galileo with Ethernet or Wi-Fi connection.

Devices and actuators, which do not have operating systems, connect to edge devices or edge gateways using Wi-Fi or Ethernet connections of a Local Area Network (LAN) or using Bluetooth, ZigBee, and Ultra-Wide-band (UWB) of Personal Area Network (PAN). Examples of devices with indirect connections to the Internet are:

- devices connecting via a ZigBee gateway
- low energy devices connecting via Bluetooth
- Devices connecting via low power radios to a Raspberry Pi.

4.2 Communication Layer

Communication between IoT devices and edge gateways and the Internet includes several models:

- Bluetooth Low Energy
- Near Field Communication (NFC)
- Zigbee or other mesh radio networks
- SRF and point-to-point radio links
- UART or serial lines
- SPI or I2C wired buses
- Direct Ethernet or Wi-Fi connectivity using TCP or UDP.

4.3 Message Brokering Layer

The interconnection of IoT devices and sensors is often centered on the infrastructure required for communication protocols to work. Today, the broker-based and bus-based architectures dominate the architectures for data exchange. In the broker-based architecture, a central component, called the broker, controls the dissemination of information between the publishers (source of information) and the subscribers (consumers of data). Decoupling between publishers and subscribers is a fundamental quality attribute of this architecture. Examples of broker-based protocols include MQTT, Advanced Message Queuing Protocol (AMQP), CoAP, and Java Message Service API (JMS). In the bus-based architecture, there is no centralized broker, and the messages posted by publishers for a specific topic are directly delivered to the subscribers of that topic. Examples of bus-based protocols include REST, XMPP, and Data Distribution Service (DDS).

Thus, in a smart city context, a message broker is a physical component that handles the communication between IoT devices and the applications at the upper layers of the architecture. Instead of directly communicating with each other, in a point-to-point fashion, devices and applications communicate only with the message broker. A device sends a message to the message broker to report an event. The message broker looks up applications, which subscribed to the event and then passes the message to them. Before using the message broker, the applications must register their interest in receiving communications so that the message broker can dispatch messages to them.

Several implementations of the message broker are now available. However, the most famous are RabbitMQ, ActiveMQ, and Kafka. RabbitMQ (https://www.rabbitmq.com) is an open source message broker, which is lightweight and easy to deploy on premises as well as on the cloud. It initially implemented AMQP and had since been extended to MQTT and multiple other protocols. RabbitMQ can be deployed in various configurations to meet the scalability and availability requirements. RabbitMQ runs on several operating systems and cloud environments and provides developers with libraries for interfacing with the broker for all popular programming languages.

Apache ActiveMQ (http://activemq.apache.org/) is a message broker written in Java with JMS, WebSocket, and REST interfaces. It supports protocols like MQTT, AMQP, and OpenWire that can be used by applications in various programming languages.

Apache Kafka (http://kafka.apache.org) is recognized as the most advanced message broker that can handle data streams efficiently and in a scalable manner. Kafka typically runs on a cluster of one or many brokers (servers). Kafka immutably stores the messages coming from multiple sources ("producers") in queues ("topics"), which are organized into several partitions. Messages of a partition are indexed and saved with a timestamp. Other processes ("consumers") can query the messages stored in Kafka partitions. Kafka partitions are replicated across the cluster brokers to guarantee fault tolerance. Kafka has four APIs:

- Producer API – allows applications and services to publish streams of events into Kafka topics.
- Consumer API – allows applications and services to subscribe to Kafka topics of interest and process the stream of events.
- Streams API – permits to convert the input streams, stored in the topics, into other output streams that other applications can consume.
- Connector API – permits to connect Kafka cluster with external sources such as key-value stores, and relational databases.

4.4 Data Streams Processing Layer

After enabling the efficient storage and access to of data in the messaging layer, the data streams processing layer would allow city stakeholders to transform and analyze vast amounts of urban data streams efficiently. This layer is in charge of processing and analyzing data streams stored in the message broker to permit applications to use it and generate valuable insights. An extensible set of established open source and commercial solutions for data processing can power this layer. These solutions include Apache Storm, Apache Flink, Apache Samza, Amazon IoT, Google Cloud Dataflow, Amazon Elastic MapReduce that allows for processing both streaming and historical data, which is extremely important for smart cities.

5 Implementation Scenario

Apache Kafka is widely seen as the best distributed messaging system to build robust IoT platforms. In our proposed architecture, Kafka acts as a gateway to the data processing layer, which can be powered by Apache Storm, Apache Flink, Apache Spark, and Apache Hadoop clusters. Figure 3 depicts an implementation scenario and Fig. 4 shows a Kafka-based data pipeline to implement the scenario.

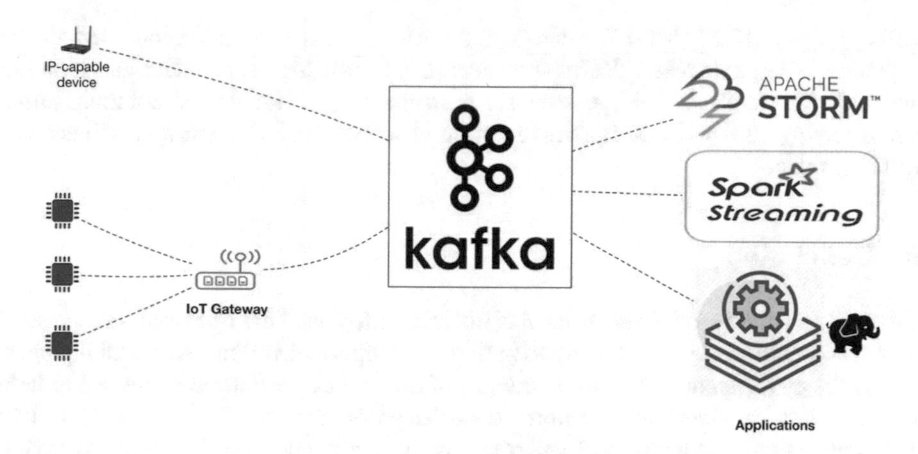

Fig. 3. Implementation scenario

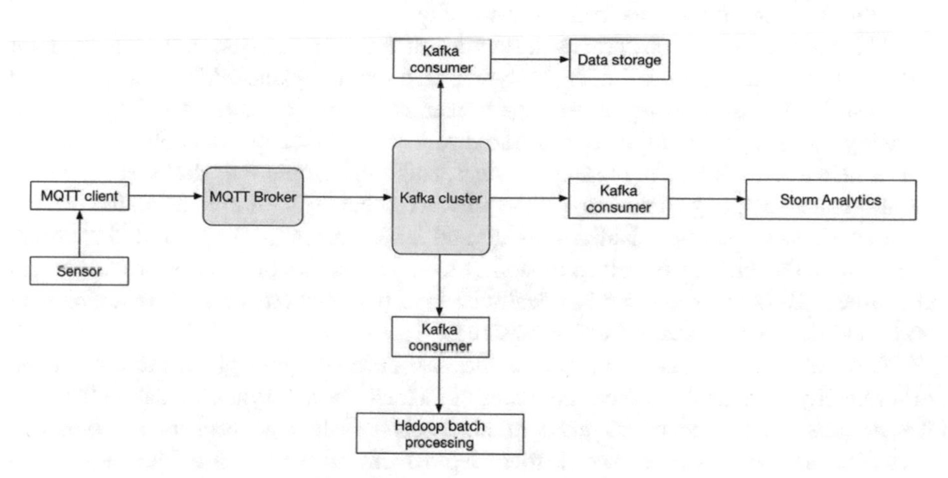

Fig. 4. A Kafka-based data pipeline

The IoT gateway pushes the data set, received from the devices, to a Kafka cluster, where it is stored in various topics. The gateway is typically an MQTT broker. The components of the data processing layer responsible for near real-time processing and batch processing and analytics are subscribers of Kafka topics. Apache Storm, Apache Flink, and Apache Spark provide the computational environment needed to perform near real-time processing; while a Hadoop cluster offers support for batch processing.

Metrics such as temperature, humidity, and air quality that may require taking real-time corrective actions require near real-time processing using Apache Storm or Apache Flink. Other metrics like energy and water consumption are usually collected over a period of time and then analyzed for averages. These data points are collected and analyzed through batch processing using a MapReduce job running in the Hadoop cluster.

Kafka does not replace an MQTT broker, which is a message broker typically used for Machine-to-Machine (M2M) communications. Kafka's design goals are very different from MQTT. MQTT is designed for low-power devices and cannot handle the ingestion of large data sets. Kafka, however, can handle high-speed data ingestion but not with M2M. Thus, a usage scenario is to use MQTT for device communication while relying on Apache Kafka for ingestion of sensor data. The gateway will act as a Kafka producer.

6 Conclusion

Urban data streams originate from the Internet of Things (IoT) devices and sensors, deployed in different parts of a city to monitor and report on various events taking place across the city systems. The large volumes of urban data need to be harnessed to help smart city stakeholders make informed decisions on the fly. Furthermore, effective management and governance of smart city components relies on the ability to stream data as events occur, process data streams in near real-time, and use data analytics to get more insights of what happening in the city.

This paper describes our proposed framework that aims to assist in streaming urban IoT data streams, perform data analytics, and create value-added services. The framework relies on a powerful message broker, at its core, capable of scaling with the growing amount of sensed data. The framework will allow city stakeholders to detect real-time events, implement alerting services, and build monitoring dashboards, which would enhance the governance of the city. A scenario of implementation of the framework was presented. Kafka is more and more becoming the preferred message broker in IoT environments given its ability to scale with the number of producers and consumers. Kafka was designed to deal with ingesting vast amounts of streaming data and handling data replication and persistence.

A federation of message brokers could also be considered given the size of the different city systems. However, the issues of interactions and synchronization between the brokers of the federation need to be resolved. Message brokers are typically deployed on-premises even though their deployment on the cloud is becoming common. IBM Bluemix has Message Hub, a fully managed, cloud-based messaging service based on Kafka. Cloud Karafka (https://www.cloudkarafka.com) is another Message streaming as a service platform in the public cloud, powered by Apache Kafka.

References

1. Abdelwahab, S., Hamdaoui, B.: FogMQ - a message broker system for enabling distributed, internet-scale IoT applications over heterogeneous cloud platforms. CoRR (2016)
2. Abdelwahab, S., Zhang, S., Greenacre, A., Ovesen, K., Bergman, K., Hamdaoui, B.: When clones flock near the fog. IEEE Internet Things J. **5**(3), 1914–1923 (2018)
3. Almeida, R.A.P., Blackstock, M., Lea, R., Calderon, R., Prado, A.F., Guardia, H.C.: Thing broker: a twitter for things. In: The Proceedings of the UbiComp 2013 Adjunct - Adjunct Publication of the 2013 ACM Conference on Ubiquitous Computing, New York, NY, USA, pp. 1545–1554 (2013)

4. Aloi, G., Caliciuri, G., Fortino, G., Gravina, R., Pace, P., Russo, W., Savaglio, C.: A mobile multi-technology gateway to enable IoT interoperability. In: 2016 IEEE First International Conference on Internet-of-Things Design and Implementation (IoTDI), Berlin, pp. 259–264 (2016)
5. Aloi, G., Caliciuri, G., Fortino, G., Gravina, R., Pace, P., Russo, W., Savaglio, C.: Enabling IoT interoperability through opportunistic smartphone-based mobile gateways. J. Netw. Comput. Appl. **81**, 73–83 (2017)
6. Asensio, A., Marco, A., Blasco, R., Casas, R.: Protocol and architecture to bring things into internet of things. Int. J. Distrib. Sens. Netw. 18 p. (2014). Article ID 158252
7. Blackstock, M., Kaviani, N., Lea, R., Friday, A.: MAGIC Broker 2: an open and extensible platform for the internet of things. In: The Proceedings of the 2010 International Conference on Internet of Things (IOT), pp. 1–8 (2010)
8. Blackstock, M., Lea, R.: IoT interoperability: a hub-based approach. In: The Proceedings of the 2010 IEEE International Conference on the Internet of Things (IOT), pp. 79–84 (2014)
9. Collina, M., Corazza, G.E., Vanelli-Coralli, A.: Introducing the QEST broker: scaling the IoT by bridging MQTT and REST. In: The Proceedings of the 2012 IEEE 23rd International Symposium on Personal, Indoor and Mobile Radio Communications (PIMRC 2012), pp. 36–41 (2012)
10. da Silva, A.C.F., Breitenbucher, U., Képes, K., Kopp, O., Leymann, F.: OpenTOSCA for IoT. Presented at the 6th International Conference, New York, NY, USA, pp. 181–182 (2016)
11. Desai, P., Sheth, A., Anantharam, P.: Semantic gateway as a service architecture for IoT interoperability. In: The Proceedings of the 2015 IEEE 3rd International Conference on Mobile Services, MS 2015, pp. 313–319 (2015)
12. Diginomica.com: Bright Lights. Smart City. San Diego's pioneering IoT platform (2017). https://diginomica.com/2017/10/31/bright-lights-smart-city-san-diegos-pioneering-iot-platform/. Accessed 05 May 2018
13. Guner, A., Kurtel, K., Celikkan, U.: A message broker based architecture for context aware IoT application development. In: The Proceedings of the 2nd International Conference on Computer Science and Engineering, UBMK 2017, pp. 233–238 (2017)
14. OASIS: Topology and Orchestration Specification for Cloud Applications (TOSCA) Version 1.0 (2013)
15. Perera, C., Jayaraman, P.P., Zaslavsky, A., Georgakopoulos, D., Christen, P.: MOSDEN: an internet of things middleware for resource constrained mobile devices. In: The Proceedings of the 2014 47th Hawaii International Conference on System Sciences (HICSS), pp. 1053–1062 (2013)

Alternative Approach to Fisher's Exact Test with Application in Pedagogical Research

Tomas Barot(✉) and Radek Krpec

Department of Mathematics with Didactics, Faculty of Education, University of Ostrava, Fr. Sramka 3, 709 00 Ostrava, Czech Republic
{Tomas.Barot, Radek.Krpec}@osu.cz

Abstract. Mathematical statistical methods have been frequently appeared in practical implementations of intelligent systems and as a tool for processing research data. In favor of these applications or for purposes of a reduction of a computational complexity, modifications of established approaches can be advantageous. In this paper, a statistical approach to practical providing hypotheses testing with two categorical variables is simplified. This presented proposal can be appropriate for purposes of unfulfilling necessary input conditions of Chi-squared test instead a more complex Fisher's exact test. The proposed approach can be concretely an alternative to this Fisher's exact test and brings a simplification of considered problem. The presented recommendation is demonstrated on a practical example of a statistical research; especially, in case of a concrete statistical quantitative research with pedagogical aspects. In this utilization, two categorical variables are considered, both with two items, corresponding to a case of four-field contingency table.

Keywords: Hypotheses testing · Contingency table · Chi-squared test
Fisher's exact test · Quantitative research

1 Introduction

Mathematical statistics [1] is a significantly important part of mathematics with a wide spectrum of particular applications, e.g. [2–6]. In these current researches, statistical approaches can be even considered as a part of intelligent systems. Therefore, research of statistical approaches can bring suitable advantages, which can be implemented in practice. Especially, concrete statistical methods can be further modified with aim of decreasing the computational complexity [7] or improving in favor of concrete practical implementations.

In practice, a major trend of utilization of mathematical statistics became bound on processing particular methods in statistical software with results based on *p*-value [8]. However, an advanced approach to understanding concrete principles of methods could be still considered as essential in favor of developments focused on a specific plane of an application.

In field of pedagogy, researches are frequently based on quantitative methods [1], as can be seen e.g. in [9–13]. Using software possibilities can be advantageous for academics; however, inappropriate situations can occur. Typical example of this

R. Silhavy et al. (Eds.): CoMeSySo 2018, AISC 859, pp. 50–59, 2019.
https://doi.org/10.1007/978-3-030-00211-4_6

situation can be seen by hypotheses testing with two categorical variables. Chi-squared method [14] can be applied; however, its necessary conditions have to be fulfilled.

Necessary conditions for Chi-squared test depend on a structure of values in a contingency table with empiric frequencies [14]. All empiric frequencies must be greater than or equal to 1. Only 20% of empiric frequencies may be lower than 5. These empiric frequencies are computed from basic contingency table [14] with absolute frequencies [14] of data in a considered research. In case of fulfilling these conditions, both contingency tables are used for purposes of Chi-squared test.

In other case, Chi-squared test can not be applied. In literature, Fisher's exact test [15] is recommended for this case. Although Fisher's exact test can be advantageous instead of Chi-squared test, its numerical algorithm can be for academics difficult. Therefore, an alternative approach is presented in this paper. This modified proposal transforms a previous contingency tables in form useful for further hypotheses testing by generally known T-test [16] & F-test [17] or Mann-Whitney test [18] depending on normality of data [19]. Results of practical utilization of this presented approach is compared with conclusions of hypotheses testing by Fisher's exact test. Practical application is bound on a particular pedagogical research.

2 Hypotheses Testing Based on Contingency Table

In hypotheses testing with two categorical variables X^i (1) [1], a significant dependence (Table 1) can be tested on significance level α. For this purpose, statistical method Chi-squared [14], based on contingency table [14], can be suitable used.

Random variables X^1 and X^2 (1) can have only non-numerical form. Where n is a size of population and sets Γ^i include all possible non-numerical items of each i-th categorical random variable.

$$\exists! n \in \mathrm{N}, i \in \langle 1;\, 2 \rangle, j \in \langle 1; n \rangle : X^i = [x_1^i, \cdots, x_j^i, \cdots, x_n^i]^T;\; x_j^i \in \Gamma^i \tag{1}$$

Table 1. Statistical hypotheses testing with two categorical variables

Hypothesis	Definition
H_0	There are not significant dependences between categorical variables X^1 and X^2
H_1	There are significant dependences between categorical variables X^1 and X^2

In a frame of a quantitative research [1], n pairs, which are subsets of Cartesian product (2), can be found e.g. in data (Table 2) in questionnaires with nominal scaled items [1].

$$j \in \langle 1; n \rangle : [x_j^1, x_j^2] \subset \left(\Gamma^1 \times \Gamma^2\right) \tag{2}$$

In a contingency table, absolute frequencies n_{rs} (3) [1] of occurrence of paired items (2) are determined from Table 2. Where indices r and s express order of a categorical item in set Γ^1 and Γ^2.

Table 2. Structure of pairs of categorical random variables

j-th record	Data of X^1 & X^2
1	$[x_1^1, x_1^2]$
$\vdots$	$[x_j^1, x_j^2]$
n	$[x_n^1, x_n^2]$

$$n_{rs} = n_{rs}([\gamma_r^1, \gamma_s^2]);\ [\gamma_r^1, \gamma_s^2] \subset \left(\Gamma^1 \times \Gamma^2\right) \tag{3}$$

Frequencies (3) express main part of contingency table (Table 3) which is complemented by horizontal sums s_h and by vertical sums s_v. Variable s should be equal to both these sums and to range n. In this paper, four-pole contingency table (Table 3) [14] is further considered with sums (4).

Table 3. Structure of four-pole contingency table

Categorical variable X^2	Categorical variable X^1		Sums
	Item γ_1^1	Item γ_2^1	
Item γ_1^2	$n_{11} = n_{11}([\gamma_1^1, \gamma_1^2])$	$n_{21} = n_{21}([\gamma_2^1, \gamma_1^2])$	$\mathbf{s}_v$
Item γ_2^2	$n_{12} = n_{12}([\gamma_1^1, \gamma_2^2])$	$n_{22} = n_{22}([\gamma_2^1, \gamma_2^2])$	
Sums	$\mathbf{s}_h$		s

$$\left.\begin{array}{r}\mathbf{s}_h = \begin{bmatrix} s_{h1} & s_{h2} \end{bmatrix};\ \mathbf{s}_v = \begin{bmatrix} s_{v1} & s_{v2} \end{bmatrix}^T; \\ \mathbf{s}_h = \begin{bmatrix} (n_{11} + n_{12}) & (n_{21} + n_{22}) \end{bmatrix}; \\ \mathbf{s}_v = \begin{bmatrix} (n_{11} + n_{21}) & (n_{12} + n_{22}) \end{bmatrix}^T \end{array}\right\} \tag{4}$$

Although research data (Table 2) can be presented by contingency table (Table 3), the second four-pole table (Table 4) has to be constructed for using Chi-squared test. In the second table, empirical frequencies are computed for values from Table 3 by equations in (5) with results in field of positive integer numbers.

$$\left.\begin{array}{r} i \in \langle 1;2 \rangle;\ j \in \langle 1;2 \rangle;\ \tilde{n}_{ij} \in \mathrm{I}_0^+ : \\ \left(\frac{s_{vi}s_{hj}}{s} - 1\right) \leq \tilde{n}_{ij} \leq \frac{s_{vi}s_{hj}}{s} \end{array}\right\} \tag{5}$$

Table 4. Construction of four-pole table of empiric frequencies

Categorical variable X^2	Categorical variable X^1		Sums
	Item γ_1^1	Item γ_2^1	
Item γ_1^2	$\tilde{n}_{11}$	$\tilde{n}_{21}$	$\mathbf{s}_v$
Item γ_2^2	$\tilde{n}_{12}$	$\tilde{n}_{22}$	
Sums	$\mathbf{s}_h$		s

Necessary conditions for using Chi-squared test depend on properties of empiric frequencies computed in Table 4. 80% of empirical frequencies have to be greater than or equal to 5. Each empirical frequency has to be greater than or equal to 1. In case of fulfilling these both conditions, Chi-squared test can be used for hypotheses testing on significant dependences (Table 1). Results of this testing, in form of p-value [8], can be obtained by statistical software, e.g. [20].

However, unfulfilling these conditions can frequently occur in case of four-pole contingency table. Therefore, Fisher's exact test [15] can be appropriate in this situation. Final p-value (6) for hypotheses testing (Table 1) is computed in a numerical algorithm with ω steps and exit condition (7). Where k is index of step of this algorithm.

$$k \in \mathrm{N}; \omega \in \mathrm{N}; \exists \omega \, . \, k = \langle 1; \omega \rangle : p = \sum_{k=1}^{\omega} p(k) \tag{6}$$

$$k \in \mathrm{N}; i, j \in \langle 1; 2 \rangle; \ \exists! \omega \in \mathrm{N}; \ \exists n_{ij}(k = \omega) \colon n_{ij}(k = \omega) = 0 \tag{7}$$

Procedure of each step of Fisher's exact test consists of construction of new modified contingency tables. As initial contingency table, Table 3 is considered. End of procedure occurs, if some of frequency $n_{ij}(k)$ of k-th step appears with value 0. Each contingency table is built by following rules. A minimal frequency is found and decreased by one. All other frequencies are recomputed with aim to get the same horizontal and vertical sums. In k-th step of the procedure, a partial k-th p-value is given by rule (8).

$$p(k) = \frac{s_{v1}! s_{v2}! s_{h1}! s_{h2}!}{n! \prod_{r=1}^{2} \prod_{s=1}^{2} n_{rs}(k)!} \tag{8}$$

3 Proposal of Alternative Approach to Fisher's Exact Test

Fisher's exact test can be significantly suitable in favor of hypothesis testing in case of two categorical variables (Table 1) in situation of unfulfilling of necessary conditions for Chi-squared test. In this situation, this numerical method can be applied with implementation of Eqs. (6)–(8). For reasons of a modular utilization of statistical software solutions, this described algorithm can have a disadvantageous property, which is bound on a manual solution of the algorithm. An alternative to Fisher's exact test is presented in this paper with respect to using generally used statistical methods, which can be suitably processed in statistical software.

In this paper, proposal of alternative approach is defined by following practical rules, which are focused on transformations (9) of the second categorical variable to forms of integer numbers. As can be seen in Table 5, contingency table (Table 2) with frequencies of values of both categorical variables is rewritten in form with one category variable X^1 and with one numbered variable X^2.

$$\gamma_1^2, \gamma_2^2 \in \Gamma^2; \beta_1, \beta_2 \in \mathrm{I}; (\gamma_1^2 \rightarrow \beta_1) \wedge (\gamma_2^2 \rightarrow \beta_2) \tag{9}$$

Table 5. Structure of transformed variables in proposed alternative approach

	Categorical Variable X^1		
	Item γ_1^1	Item γ_2^1	
n_{11}	β_1	β_1	n_{12}
	$\vdots$	$\vdots$	
	β_1	β_1	
n_{21}	β_2	β_2	n_{22}
	$\vdots$	$\vdots$	
	β_2	β_2	

The proposal of transformation can be followed by definition of modified hypotheses, as can be seen in Table 6.

Table 6. Definition of hypotheses of transformed approach

Hypothesis	Definition
H_0	There are not significant dependences between categorical variables X^1 and numerical variable X^2
H_1	There are significant dependences between categorical variables X^1 and numerical variable X^2

This modified definition of hypotheses can be then tested using method T-test [16] with F-test [17] or using statistical method Mann-Whitney [18]. Selection between these two options depends on normality of numerical data in Table 5. Normality can be analyzed by method Shapiro-Wilk [19], e.g. in a free-available software PAST [20], with results in form of p-value.

4 Results

For purposes of application of the proposed approach, a practical quantitative research was realized in this paper. Field of implementation of the presented alternative approach is based on educational approaches in connection with internationalization. Opinions of students about foreign study possibilities were main topic of this research. Number of considered responds n was equal to 37.

Structure of items was proposed with respect to categorical character of variables in topic of this paper. An on-line questionnaire was consisted of nominal scaled items bound on identification of facts about respondents and on questions based on research problems. Defined hypotheses were focused on following topics, which can be seen in Table 7. Significance level α was declared as 0.05 related on type of field of application [1].

Table 7. Hypotheses defined for purposes of practical research

Definition of Hypotheses
$1H_0$: There are not significant dependences $1H_1$: There are significant dependences between knowledge about existence of departmental coordinators and type of study.
$2H_0$: There are not significant dependences $2H_1$: There are significant dependences between knowledge about existence of Internationalization at home and type of study.
$3H_0$: There are not significant dependences $3H_1$: There are significant dependences between knowledge about existence of foreign language educational software and type of study.
$4H_0$: There are not significant dependences $4H_1$: There are significant dependences between attending foreign-study and type of study.

In a frame of the practical example of a quantitative research, 37 pairs of data were obtained. Categorical items for each hypothesis can be seen in Table 8.

Table 8. Categorical items of categorical variables in hypothesis testing

Hypotheses	γ_1^1	γ_2^1	γ_1^2	γ_2^2
$1H_0$, $1H_1$	Known	Unknown	Bachelor's Degree	Master's Degree
$2H_0$, $2H_1$	Known	Unknown		
$3H_0$, $3H_1$	Known	Unknown		
$4H_0$, $4H_1$	Complete	Not yet		

Source of data can be rewritten into form of contingency table for each hypothesis, as can be seen in Table 9. Second contingency table with empiric frequencies for each hypothesis, is displayed in Table 10.

In case of empiric frequencies in Table 10, necessary conditions for realization of Chi-squared test are not fulfilled, because there are not included occurrence 3 of 4 frequencies greater than or equal to 5.

Table 9. Four-pole contingency table for each hypothesis

Hypotheses	Categorical Variable X^2	Categorical Variable X^1 Item γ_1^1	Item γ_2^1	Sums
$1H_0, 1H_1$	Item γ_1^2	5	0	5
	Item γ_2^2	12	20	32
	Sums	17	20	37
$2H_0, 2H_1$	Item γ_1^2	3	2	5
	Item γ_2^2	6	26	32
	Sums	9	28	37
$3H_0, 3H_1$	Item γ_1^2	4	1	5
	Item γ_2^2	14	18	32
	Sums	18	19	37
$4H_0, 4H_1$	Item γ_1^2	5	0	5
	Item γ_2^2	3	29	32
	Sums	8	29	37

Table 10. Contingency table with empiric frequencies for each hypothesis

Hypotheses	Categorical Variable X^2	Categorical Variable X^1 Item γ_1^1	Item γ_2^1	Sums
$1H_0, 1H_1$	Item γ_1^2	2	3	5
	Item γ_2^2	15	17	32
	Sums	17	20	37
$2H_0, 2H_1$	Item γ_1^2	1	4	5
	Item γ_2^2	8	24	32
	Sums	9	28	37
$3H_0, 3H_1$	Item γ_1^2	2	3	5
	Item γ_2^2	16	16	32
	Sums	18	19	37
$4H_0, 4H_1$	Item γ_1^2	1	4	5
	Item γ_2^2	7	25	32
	Sums	8	29	37

In Table 11, results obtained by Fisher's exact test can be seen with described inter-results of this numerical algorithm, which is based on values in Table 9.

Table 11. Results of hypothesis testing obtained by fisher's exact test

Hypotheses	Iteration $k = 1 \ldots \omega$	$p(k)$	Final p-value	Conclusion of Hypotheses Testing
$1H_0, 1H_1$	1	0.0142	0.0142<0.05	Reject $1H_0$ in Favor of $1H_1$ on Sign. Level $\alpha = 0.05$
$2H_0, 2H_1$	1	0.0728	0.0812>0.05	Fail to Reject $2H_0$ on Sign. Level $\alpha = 0.05$
	2	0.0081		
	3	0.0003		
$3H_0, 3H_1$	1	0.1334	0.1530>0.05	Fail to Reject $3H_0$ on Sign. Level $\alpha = 0.05$
	2	0.0197		
$4H_0, 4H_1$	1	0.0001	0.0001<0.05	Reject $4H_0$ in Favor of $4H_1$ on Sign. Level $\alpha = 0.05$

For purposes of simplification of described hypotheses testing procedure, following transformation (10) of categorical variables is provided. Type of study was substituted by an appropriate year of study. 37 pairs of data were rewritten into a suitable modified form with one category variable X^1 and with one numbered variable X^2.

Using proposed approach, hypotheses testing was provided using Mann-Whitney method in software PAST [20] with results in Table 12. Mann-Whitney test was used for non-normal character of values verified by Shapiro-Wilk test on normality.

In Table 12, consistence of conclusions of hypotheses testing in comparison with Table 11 can be seen. Therefore, presented approach can be suitable applied in this case of the pedagogical research.

$$\left.\begin{array}{c}\gamma_1^2, \gamma_2^2 \in \Gamma^2; \beta_1, \beta_2 \in \mathrm{I}: \\ \left(\gamma_1^2 \rightarrow \beta_1 = 2\right) \wedge \left(\gamma_2^2 \rightarrow \beta_2 = 4\right)\end{array}\right\} \tag{10}$$

Table 12. Results of hypotheses testing achieved by proposed approach

Hypotheses	Normality tested for values under items γ_1^1, γ_2^1	p-value of hypotheses testing	Conclusion of hypotheses testing
$1H_0$, $1H_1$	Unconfirmed on Sign. Level $\alpha = 0.05$	$0.0109 < 0.05$	Reject $1H_0$ in Favor of $1H_1$ on Sign. Level $\alpha = 0.05$
$2H_0$, $2H_1$	Unconfirmed on Sign. Level $\alpha = 0.05$	$0.0521 > 0.05$	Fail to Reject $2H_0$ on Sign. Level $\alpha = 0.05$
$3H_0$, $3H_1$	Unconfirmed on Sign. Level $\alpha = 0.05$	$0.1437 > 0.05$	Fail to Reject $3H_0$ on Sign. Level $\alpha = 0.05$
$4H_0$, $4H_1$	Unconfirmed on Sign. Level $\alpha = 0.05$	$7.31 \times 10^{-6} < 0.05$	Reject $4H_0$ in Favor of $4H_1$ on Sign. Level $\alpha = 0.05$

5 Conclusion

In an applied statistical research based on quantitative research principles, the proposal aimed for unfulfilling of input necessary conditions of Chi-squared test was compared with Fisher's exact test. Number of items of both considered variables was considered as two. The proposed approach brings a computational simplification based on research data transformation in comparison of case of Fisher's exact test. Both tests are appropriate for solving the unusable Chi-squared test; however, the presented proposal does not use so complicated analytical procedure with computations of factorials of variables of a contingency table. Proposed approach can be practically applied in the presented analyzed situation by hypotheses testing on significant dependences between categorical variables after unfulfilling necessary input conditions for Chi-squared test.

References

1. Teo, T.: Handbook of Quantitative Methods for Educational Research. Sense Publishers (2013). ISBN 978-946209404-8. https://doi.org/10.1007/978-94-6209-404-8
2. Lech, P., Wlodarski, P.: Analysis of the IoT WiFi mesh network. In: 6th Computer Science On-line Conference: Cybernetics and Mathematics Applications in Intelligent Systems, pp. 272–280. Springer (2017). ISBN 978-3-319-57263-5. https://doi.org/10.1007/978-3-319-57264-2_28
3. Kularbphettong, K.: Analysis of students' behavior based on educational data mining. In: 1st Conference on Computational Methods in Systems and Software: Applied Computational Intelligence and Mathematical Methods, pp. 167–172. Springer (2017). ISBN 978-331967620-3. https://doi.org/10.1007/978-3-319-67621-0_15
4. Achuthan, K., Murali, S.S.: Virtual lab: an adequate multi-modality learning channel for enhancing students' perception in chemistry. In: 6th Computer Science On-line Conference: Cybernetics and Mathematics Applications in Intelligent Systems, pp. 419–433. Springer (2017). ISBN 978-3-319-57263-5. https://doi.org/10.1007/978-3-319-57264-2_42
5. Meghanathan, N.: Correlation Analysis of decay centrality. In: 6th Computer Science On-line Conference: Cybernetics and Mathematics Applications in Intelligent Systems, pp. 407–418. Springer (2017). ISBN 978-3-319-57263-5. https://doi.org/10.1007/978-3-319-57264-2_41

6. Sulovska, K., Belaskova, S., Adamek, M.: Gait patterns for crime fighting: statistical evaluation. In: Proceedings of SPIE - The International Society for Optical Engineering, vol. 8901. SPIE (2013). ISBN 978-081949770-3. https://doi.org/10.1117/12.2033323
7. Agrawal, M., Arvind, V.: Perspectives in Computational Complexity. Springer, Berlin (2014). https://doi.org/10.1007/978-3-319-05446-9. ISBN 978-3-319-05445-2
8. Cortes, J., Casals, M., Langohr, M., et al.: Importance of statistical power and hypothesis in P value. Med. Cl. **146**(4), 178–181 (2016). ISSN 0025-7753. https://doi.org/10.1016/j.medcle.2016.04.057
9. Krpec, R.: The factors influencing the education value-added models. In: 9th Annual International Conference of Education, Research and Innovation: ICERI Proceedings, pp. 3600–3608. IATED Academy (2016). ISBN 978-84-617-5895-1. https://doi.org/10.21125/iceri.2016.1852
10. Zemanova, R.: Geometry taught in 1st to 5th year of study of primary schools in Czech Republic—comparison of educational content in selected primary schools. In: 10th Annual International Conference of Education, Research and Innovation: ICERI Proceedings, pp. 694–698. IATED Academy (2017). ISBN 978-84-697-6957-7. https://doi.org/10.21125/iceri.2017.0263
11. Barot, T.: Complemented adaptive control strategy with application in pedagogical cybernetics. In: 7th Computer Science On-line Conference: Cybernetics and Algorithms in Intelligent Systems. Advances in Intelligent Systems and Computing, vol. 765, pp. 53–62. Springer (2018). ISBN 978-3-319-91191-5. https://doi.org/10.1007/978-3-319-91192-2_6
12. Schoftner, T., Traxler, P., Prieschl, W., Atzwanger, M.: E-learning introduction for students of the first semester in the form of an online seminar. In: Pre-conference Workshop of the 14th E-learning Conference for Computer Science, pp. 125–129. CEUR-WS (2016). ISSN 1613-0073
13. Horsley, M., Sikorova, Z.: Classroom teaching and learning resources: international comparisons from TIMSS—a preliminary review. Orbis Scholae. **8**(2), 43–60 (2017). ISSN 1802-4637. https://doi.org/10.14712/23363177.2015.65
14. Andres, A.M., Quevedo, M.S., Garcia, J.T., Silva-Mato, A.: On the validity condition of the chi-squared test in 2×2 tables. Test **14**(1), 99–128 (2005). ISSN 1133-0686. https://doi.org/10.1007/bf02595399
15. Camilli, G.: The relationship between Fisher's exact test and Pearson's chi-square test: a bayesian perspective. Psychometrika **60**(2), 305–312 (1995). ISSN 0033-3123. https://doi.org/10.1007/bf02301418
16. Rasch, D., Kubinger, K.D., Moder, K.: The two-sample t test: pre-testing its assumptions does not pay off. Stat. Pap. **52**(1), 219–231 (2011). ISSN 1613-9798. https://doi.org/10.1007/s00362-009-0224-x
17. Mewhort, D.J.K.: A comparison of the randomization test with the F test when error is skewed. Behav. Res. Methods **37**(3), 426–435 (2005). ISSN 1554-3528. https://doi.org/10.3758/bf03192711
18. Fischer, D., Oja, H.: Mann-Whitney type tests for microarray experiments: the R package gMWT. J. Stat. Softw. **65**(1), 1–19 (2015). ISSN 1548-7660. https://doi.org/10.18637/jss.v065.i09
19. Alizadeh Noughabi, H.: Two powerful tests for normality. Ann. Data Sci. **3**(2), 225–234 (2016). ISSN 2198-5812. https://doi.org/10.1007/s40745-016-0083-y
20. Hammer, O., Harper, D.A.T., Ryan, P. D.: PAST: paleontological statistics software package for education and data analysis. Palaeontol. Electron. **4**(1) (2001). http://palaeo-electronica.org/2001_1/past/issue1_01.htm

Development of the Forecasting Component of the Decision Support System for the Regulation of Inter-regional Migration Processes

V. V. Bystrov[1(✉)], M. G. Shishaev[1,2], S. N. Malygina[1,3], and D. N. Khaliullina[1]

[1] Institute for Informatics and Mathematical Modeling—Subdivision of the Federal Research Centre «Kola Science Centre of the Russian Academy of Sciences», Apatity, Murmansk Region, Russia
{bystrov,shishaev,malygina,khaliullina}@iimm.ru
[2] Murmansk Arctic State University, Murmansk, Russia
[3] Apatity Branch, Murmansk Arctic State University, Apatity, Murmansk Region, Russia

Abstract. The article is devoted to the development of the forecasting component of the decision support system in the management of interregional migration. The architecture of the forecasting component is considered. Its program modules with indication of their functions are described. Special attention is paid to the development of the prototype of a poly-model complex. It serves as a tool for the formation of analytical information for the development of recommendations in the field of regulation of migration flows.

Keywords: Decision support system · Forecasting component Simulation · Mathematical modeling · Inter-regional migration

1 Introduction

Migration processes have a direct impact on the transformation of social spatial relations, which leads to the formation of a new labor market. The pace of development these relations varies and is not always consistent with the adaptive capacity of both migrants and indigenous people in the regions. These inconsistencies can cause increased social tensions in some regions.

The development of special information systems and technologies, particularly decision support systems, is needed to manage the flow of migrants across different regions. The basis of the operation of such systems includes a variety of models. They serve as the main means of forecasting the development of the situation under different conditions scenarios. At the same time, the development of tools for forecasting inter-regional migration flows is a non-trivial task [1, 2]. The main reason for the complexity simulation of migration is the lack of complete information about the structure of migration flows. In this regard, in solving the problem of forecasting migration, it is

R. Silhavy et al. (Eds.): CoMeSySo 2018, AISC 859, pp. 60–71, 2019.
https://doi.org/10.1007/978-3-030-00211-4_7

almost impossible to achieve the same accuracy as in calculating the natural movement of the population.

This article describes conception of decision support system in the field of migration flows regulation. Particular attention is given to the problem of creating components for forecasting interregional migration in such a system.

2 Background

Today, there are research works that analyze the genesis of the development of various approaches to modeling and forecasting of migration processes [3]. This creates certain grounds for studying migration processes on the basis of modern information technologies using mathematical and simulation modeling.

In recent years, the most interesting approaches to the study of migration are presented in [4–6].

Schmertmann [4] proposed a model of population change in conditions when the stable birth rate is below the reproduction level and there is a constant migration flow. The results of the theoretical analysis show that the population in such a country eventually becomes constant. Numerical calculations performed using statistical data show that, in the limit, the ratio between migrants and the indigenous population will be minimized.

In the work of Fejtinger, Prskavec and Velov [5] are considered the tasks of optimal management of demographic policy, related to the distribution of migration by age groups. For the purpose on the based of a combination the Lotka model of population dynamics and the Solow economic model the authors developed the macromodel. The results of the work are illustrated by numerical experiments on statistical data of Austria.

In the work of Simon, Skritek and Veliov [6] the results of simulation modeling of demographic policy are presented. The developed model is a problem of optimal control of the age density of the population with restrictions. The behavior of the control object is described by the first-order partial differential equation Kendrick-von Foerster on an infinite period of time.

Many mathematical models associated with the study of population dynamics reflect the relationship of demographic and economic processes. The approach to studying of such kind of socio-economic systems is offered by academician Popkov [7]. The main idea of the proposed approach is to identify certain states of demo-economic systems. These states appear as a result of the interaction of the processes of the space-time evolution of the population and the economy in terms of macro indicators. At the same time, the basic assumption is that the population of the region is economically motivated, i.e. the behavior of individuals is mainly determined by a variety of economic indicators. In an aggregated form, the demo-economic system is represented in the form of an appropriate model structurally consisting of two main subsystems - "Population" and "Economy", as well as an auxiliary subsystem "Interaction". The subsystems of the demoeconomical model constructed in this way are included in the closed circuit, and therefore the designation of direct and reverse connections in it depends on the priority given to the subsystem “Population” in relation to the subsystem “Economy”.

It should be separately noted that taking steps to increase the level of macro-modeling, usually the accuracy of forecasting is reduced, but the horizon of forecasting and its qualitative characteristics are preserved. One of the possible ways to eliminate this disadvantage is the development of poly-model complexes that integrate different modeling approaches. For example, analytical modeling in terms of economic and demographic macro-indicators can be used to identify (calculate) the basic trends of migration processes, and simulation – as a means to improve the adequacy of the forecast based on collective expertise. It should be noted that this article focuses on the development of the mentioned concept for the study of migration processes.

3 Architecture of Forecasting Component

To improve the quality of modeling of migration processes, it is proposed to supplement the analytical models with simulation models. The result of such integration is the creation of a whole multi-model complex. It will be developed using the information technology of the synthesis of randomized complex migration models proposed by the authors. The technology is based on the approaches, methods and means of research of complex socio-economic systems developed by the author’s team in recent decades. This information technology will allow to generate complex models of migration processes in an automated mode, based both on the results of a computational experiment of mathematical models, and on formalized expert knowledge about the macro system “population-economy”.

Figure 1 shows the general architecture of the developed information technology for the synthesis of migration processes models. In the final implementation, the technology will be represented by a forecasting component of the decision support system (DSS), including a set of interacting relatively independent program modules. Each module performs a specific set of functions according to the proposed scheme. From the above structure, processes associated with the organization of interaction between the end user and the forecasting component are deliberately omitted. Randomized complex migration models will be part of the decision support tool that the user will interact with directly.

According to the scheme (Fig. 1) the information technology being developed includes the following components:

- Analytical model. It is a randomized parametric migration model configured using the entropy randomized machine learning algorithm. At the output of this component are formed parameters and regularities between them in a formalized form.
- Computing component. Designed to conduct a computational experiment for the developed analytical migration models. It is implemented as a software module that performs calculations of mathematical models in numerical form. At the output it forms ensembles of trajectories of the dynamics of migration flows.
- Knowledge extraction (elicitation) component. It is used for organization of interaction with experts with the purpose of formation and subsequent replenishment of

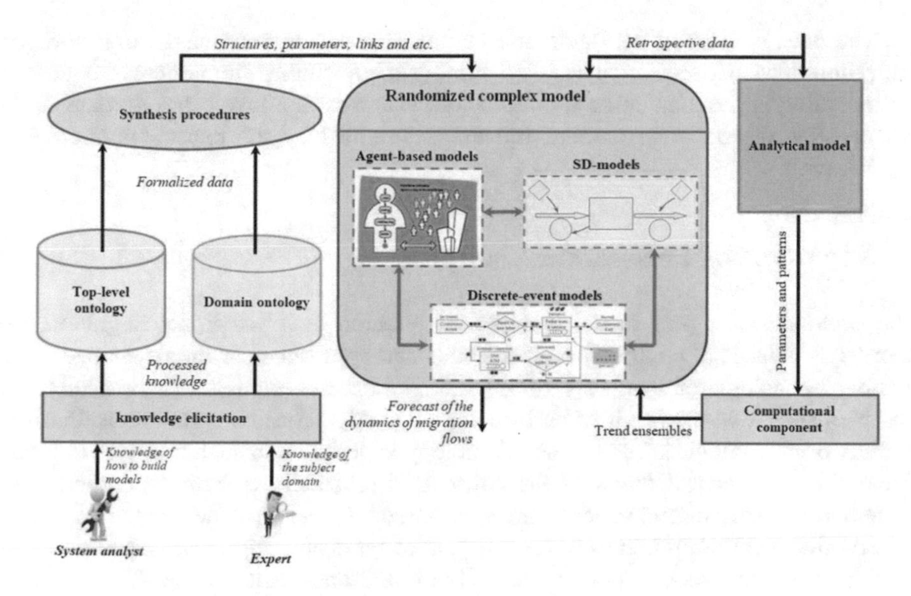

Fig. 1. Architecture of forecasting component

knowledge base and data on simulated processes. Two categories of users for this component are identified: experts and system analyst. The expert is the carrier of knowledge about the phenomena and processes of the subject area. System analyst is an expert in the field of system analysis and designing of simulation models of complex systems. At the output of the component formed expert knowledge, presented in a convenient for entering into ontology form.

- Domain ontology. It serves as a knowledge base and data on the phenomena and processes under study. It is implemented as an applied ontology in OWL format.
- Top-level ontology. It contains information about the terms, relations, rules and mechanisms used in the field of simulation modeling. Acts as a knowledge base used to synthesize system-dynamic, agent-based and discrete-event models and their combinations in a specific simulation tool.
- Synthesis procedures. They are algorithms for generating simulation models, formed on the basis of rules stored in the top-level ontology. They allow forming the basic construction of a model complex as a result of establishing the correspondence between the concepts from the domain ontology and the objects of a certain notation of the modeling method, between the relations and relations in the simulation model. To determine the procedures for the synthesis of simulation models used developments of the research team of the Institute of Informatics and Mathematical modeling CSC RAS.
- Randomized complex model. It is a poly-model complex consisting of a set of interconnected system-dynamic, agent-based and discrete-event models, synthesized on the basis of information presented in the domain ontology and top-level ontology, as well as on the parameters of the entropy randomized migration model. The simulation tool Anylogic 7 is used as a software platform.

The main purpose of the randomized complex model is to refine the trajectories of migration flows between territorial and administrative entities (in particular, countries) by including in the simulation model factors that were not taken into account in the construction of analytical models, but are determined by the expert community as significant.

4 Prototype of Poly-Model Complex

The prototype of a poly-model complex of research of interregional migration was created as a result of approbation of methods and approaches of the developed information technology of synthesis of migration processes models. Theoretically, the complex should be the result of performing automated procedures for the synthesis of models from ontologies. The top-level ontology is designed in such a way as to reflect the basic concepts and terms of the concept of probabilistic demo- economics, the system dynamics, discrete-event and agent-based modeling. The domain ontology reflects the relationship between the concepts of regional socio-economic systems. In practice, the synthesis of models is carried out in a semi-manual mode. This is due to the imperfection and incompleteness of the developed ontologies and inaccuracies in the software implementation of synthesis procedures.

The preparatory stage for applying the concept of probabilistic demo-economics to study the migration process included several steps:

1. Processing and organization of statistical data storage by basic indicators: demographic (birth rate, death rate, migration rates, etc.), economic (pricing, number of economic sectors, their power, etc.), the share of influence of economic sectors within each zone on each other, etc.
2. Formation of a set of simulated objects taking into account the results of processing statistical data and expert knowledge. Zones (clusters) are formed by grouping of individual countries of the Eurasian continent on the basis of their characteristics.

Depending on the object of study, the type of model was determined:

(a) System-dynamic, which allow to study the behavior of complex systems in time, presenting them as a network of interacting levels and flows, characterized by the presence of cause-effect relationships, feedback loops, reaction delays, the influence of the environment, etc. System-dynamic modeling was used in the implementation of the agents of sectors of the economy, as it allows tracing the accumulation of certain parameters in the system.

(b) Agent-based models, the distinctive feature of which is the formation of the global behavior of the system as a result of reproduction of individual behavior of its parts. The following objects were implemented in the form of agent-based models: person, zone, sector of economy.

Analytical models (equations or systems of equations written in the form of algebraic, integral, differential, finite-difference and other relations and logical conditions) were used to determine individual parameters of simulation models. For example, to solve optimization problems to determine the values of specific indicators (parameters), entropy-robust modeling was used.

3. Organization of interaction of the above models to obtain a general idea of the development of the macrosystem migration model by means of system integration.
4. Conducting a computational experiment and verifying the complex of migration models.

According to the approach under consideration, the macro-system demo-economic model generally consists of two subsystems: “Population” and “Economy”, which are connected with each other by direct and reversed links. “Population” models natural population change (birth and mortality) and migration (internal and external). “Economy” reflects the economic condition of each zone and the distribution of economic sectors in it.

At the current stage of the study only internal migration (migration of the population between zones) is taken into consideration. One of the main criteria for the country’s classification to a specific zone is the economic characteristics and type of childbearing.

Each zone is represented by an agent and includes considered subsystems - “Population” and “Economy”. The zone is described by the following parameters: zone name; zone size; number of economic sectors; percentage distribution of men and women of different types.

The filling of the zone by people occurs in accordance with the percentage distribution of women and men, as well as the probability distribution of types of women (western and eastern). Western type is characterized by the birth of an average of 1-2 children, and the eastern - 4-6.

People, as the object of research, are also represented in the prototype by agents that are determined by the appropriate parameters: age (*p.age*), migration type (for economic reasons or for family reunion, *p.migration*), satisfaction with the zone of residence (*p.satisfaction*), probability death (*p.death*), gender (*p.Male*), number of children (*p.birth*), and others.

Implementation of the Subsystem “Population”

When creating an agent «person» (calling the function *people.add ((Person))*) of any type, regardless of gender, it is automatically attached to a certain birth zone, as well as to the parent (mother). At the birth of a girl, she is given the type of fertility of the mother (western or eastern).

At the death of a person (calling the function *people.remove (o)*) of any type, regardless of gender, it is automatically removed from the area of residence.

Programmatically internal migration of the population is organized by moving people from one zone to another. In this case, the corresponding function is called:

```
if ((zone!=null) && ((zone!=target_zone)))
{
get_Main().movesCounter[zone.getIndex()][target_zone.getI
ndex()]++;}
```

An important factor in the model is the consideration of the age of the created agent of any type, regardless of gender. This option allows the following:

- destroy the agent in the case of the alleged death of the person, as the age of a person is between 0 and 85 years;
- transfer the agent to the adult population for the organization of migratory flows (>20 years);
- in the case of female agents of any type, this parameter will be responsible for reproductive function. At the same time, the fertile age is in the interval [15; 45].

The age update happens in a year and changes in all agents of «person» type. When a woman gives birth to a child, the parameter «birth» (number of children) of this agent increases by one, with the probability of a boy or a girl being born is the same and is 50%. This makes it possible to monitor the birth rate of women in accordance with the restrictions imposed on agents of this type. The organization of migration flows takes into account the relationship "mother and child". This link is broken when a child reaches the age of 20.

Implementation of the Subsystem "Economy"

When the model is initialized in each zone, in addition to the population subsystem, the economy subsystem is also formed - economic sectors related to this zone are created. Each sector of the economy is represented by an agent, for which the following parameters are set: name of the sector of the economy (*se.name*); economy sector index (*se.sector_index*); zone to which the economy sector belongs (*se.zone*); the equilibrium price characteristic of this sector of the economy of each zone (*se. (p)*).

Each sector of the economy is a system-dynamic model, the template of which is presented in Fig. 2.

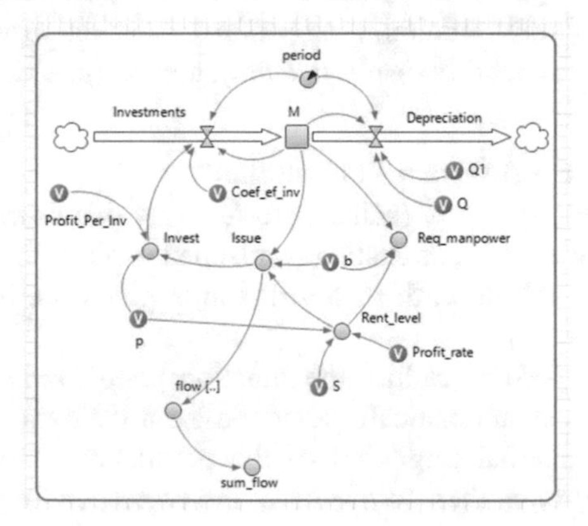

Fig. 2. Template of model of the economic sector

One indicator of the economic sectors is "Production Capacity" (M), represented in the system-dynamic model level. Its change is characterized by the incoming update flow (due to investments – «Investments») and the outflow of depreciation (due to natural aging – «Depreciation»):

$$\frac{dM}{dt} = Investments - Depreciation \tag{1}$$

The incoming update flow is described by the following equation:

$$Investments = period * M * Coef\_ef\_inv * Invest, \tag{2}$$

With: $period$ - the period of relaxation; M - the production capacity; $Coef\_ef\_inv$ - the coefficient of efficiency of investments; $Invest$ - investment.

The outflow of depreciation is described by the following equation:

$$Depreciation = period * M * (Q + Q1 * M), \tag{3}$$

With: $period$ - the period of relaxation; M - the production capacity; Q, $Q1$ - depreciation rates.

At the same time, investments are calculated according to the formula:

$$Invest = \Pr ofit\_Per\_Inv * p * Issue, \tag{4}$$

With: $\Pr ofit\_Per\_Inv$ - share of profit allocated to investments; p - equilibrium price; $Issue$ - output of this sector of the economy.

The threshold of profitability for this sector of the economy:

$$Rent\_level = \frac{1}{S} * [(1 - Profit\_rate) * p - process\_costs], \tag{5}$$

With: S - wages; $\Pr ofit\_rate$ - rate of return; p - equilibrium price; $process\_costs$ - production costs, the value of which is calculated according to the formula:

$$process\_costs = \sum_{j=1, j \neq k}^{K} a^{k,j}(n,t) * p^{j}(n,t), \tag{6}$$

With $a^{k,j}(n,t)$ - the share of production of the j-th sector in the production of the k-th sector (industry technological coefficients); p^{j} - equilibrium prices of the j-th sector of economy. The values of the corresponding variables are taken from tables containing information on the distribution of influence. The main fields of the table contain the names of the sectors of the economy, and at the intersection of these values, the share of the impact of sector j on sector k is indicated.

The necessary information is obtained during the call of the corresponding function *yearlySec()*:

```
for(int asd = 2;
asd<(int)get_Main().data.getLastRowNum("Sec_Sec"); asd++)
{ double ak =
get_Main().data.getCellNumericValue("Sec_Sec",asd, stol-
bec);
  if ( ak > 0)
  { for (int gh = 0; gh<fgh.sectors.size(); gh++)
    {  if (asd == fgh.sectors.elementAt(gh).sector_index)
       {summa = summa+ak*fgh.sectors.elementAt(gh).p;} } }
}
```

It is worth noting that the interaction between the subsystems of the prototype is carried out by means of information transmission through a set of specific model parameters. At the output of the subsystem "Population" a spatial sex and age distribution of the population is formed by zones. It is used to calculate the supply of labour in each of the zones required for the production of goods and services. This information is transferred to the subsystem "Economy", at the output of which the parameters influencing the economic attractiveness of the considered zone for potential migrants are determined.

5 Results

For the purpose of preliminary analysis of the working capacity of the methods and approaches proposed in the study, a number of computational experiments were carried out. The objects of computational experiments were the countries of the continent Eurasia. Demographic and economic indicators of official statistics from 2012 to 2016 [8, 9] were used to set up a prototype of a poly-model complex.

Unfortunately, due to the lack of detailed information on migration processes in some areas (clusters), the results of modeling were compare with the statistical data for one «EU» zone to assess the adequacy of models. Table 1 presents the quantitative values of such aggregate indicator as «total emigration», in person. Average relative error for EU countries ranges from 2.7% (Slovakia) to 6.6% (Sweden).

Table 2 contains information on the simulation results and statistical data on the aggregated "general immigration" in persons. Average relative error on general immigration by EU countries lies in the range from 2.7% (Netherlands) to 6.8% (Slovenia).

During the simulation of the developed poly-model complex, a forecast of total emigration and total immigration for the EU countries for 5 years (2018–2023) was build. The simulation results are presented in Figs. 3 and 4.

As can be seen from Figs. 3 and 4, the leader of the EU countries in terms of the number of leaving and people coming to the country during the year is Germany. Emigration in European countries is expected to be more evenly distributed across countries than immigration processes.

Table 1. Statistic data and simulation result of total emigration, person

GEO/TIME	Type	2012	2013	2014	2015	2016	Average relative error, %
Belgium	stat	93600	102657	94573	89794	92471	5.2
	sim	87823	99784	92444	82783	86147	
Bulgaria	stat	16615	19678	28727	29470	30570	3.8
	sim	17436	20507	29213	30682	29327	
Denmark	stat	43663	43310	44426	44625	52654	4.4
	sim	43154	42601	41875	47177	56698	
Germany	stat	240001	259328	324221	347162	533762	6.3
	sim	224548	241636	333163	319020	494831	
...							
United Kingdom	stat	321217	316934	319086	299183	340440	4.3
	sim	309533	303110	334163	288108	322543	
Iceland	stat	4758	4372	4052	4046	4159	3.4
	sim	4921	4360	3871	4339	4091	
Norway	stat	22693	26523	29308	29173	34694	3.9
	sim	21500	24566	30702	29323	35271	
Switzerland	stat	103881	106196	111103	116631	120653	5.2
	sim	109499	103601	100640	112944	127598	

Table 2. Statistic data and simulation result of total immigration, person

GEO/TIME	Type	2012	2013	2014	2015	2016	Average relative error, %
Belgium	stat	129477	120078	123158	146626	123702	6.4
	sim	141935	114230	133287	153250	129555	
Bulgaria	stat	14103	18570	26615	25223	21241	4.2
	sim	15442	18916	27563	23936	21432	
Denmark	stat	54409	60312	68388	78492	74383	5.8
	sim	58943	64459	66175	73939	70692	
Germany	stat	592175	692713	884893	1543848	1029852	4.8
	sim	575678	626096	962178	1542949	1058033	
...							
United Kingdom	stat	498040	526046	631991	631452	588993	3.3
	sim	503201	526412	616649	568393	572455	
Iceland	stat	4960	6406	5368	5635	8710	5.0
	sim	4925	6723	5160	5223	9403	
Norway	stat	69908	68313	66903	60816	61460	4.8
	sim	66962	64851	71450	58136	63766	
Switzerland	stat	149051	160157	156282	153627	149305	5.1
	sim	156888	165789	169183	141974	148354	

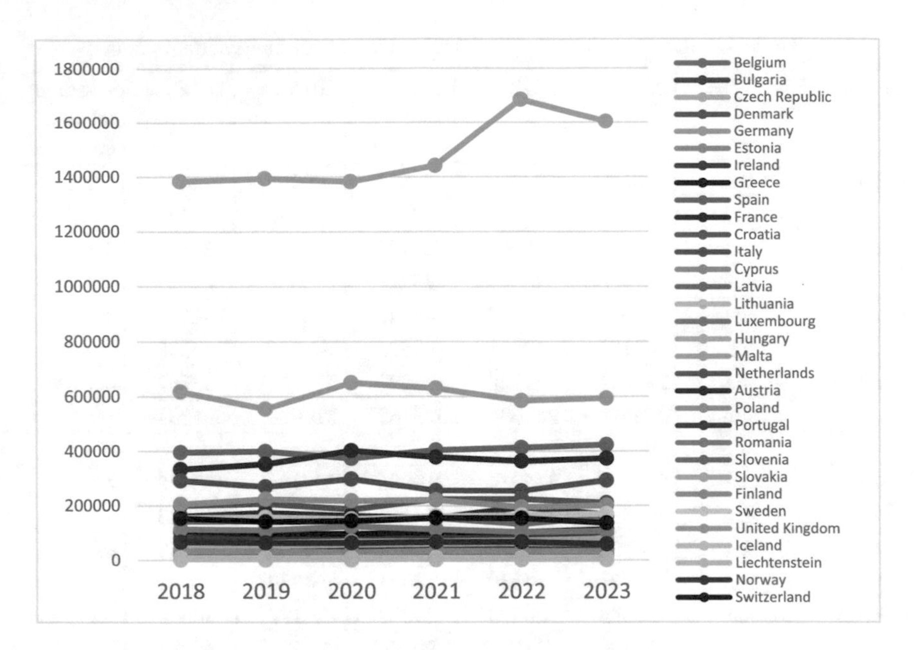

Fig. 3. Forecast of total immigration, person

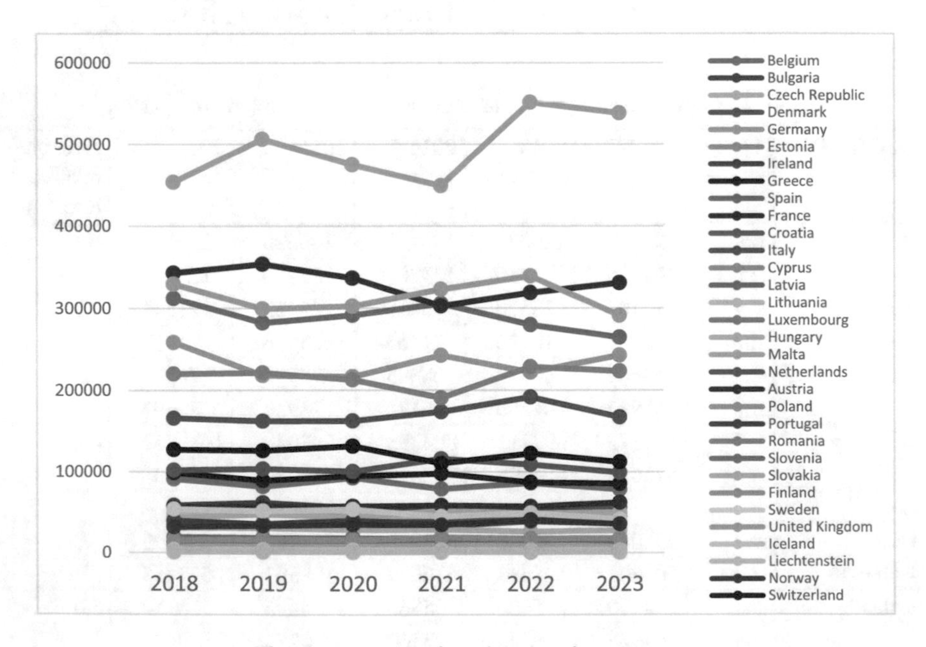

Fig. 4. Forecast of total emigration, person

6 Conclusion

The ongoing research is the development of the theory of decision support systems for the management of interregional migration. The development of this class of software is becoming ever more demand due to the increasing instability of the geo-political situation in the world.

The article gives attention on separate issues the development of a forecasting component of the decision-making system. The forecasting component is focused on the formation of different estimates of the situation in interregional migration depending on the scenario conditions on basis of simulation.

The authors propose the architecture of such a component and the principles of its work. The function of the predictive component is based on a combination of analytical and simulation modeling. Analytical entropy models offer to be configured using machine learning methods. The complex of simulation models is created on basis of formalized expert knowledge and statistics. According to the authors, the integration of such demo-economic entropy models and simulation models can improve the quality of forecasting results.

Acknowledgements. The study was financially supported by RFBR Project No. 16-29-12878 (ofi_m) "Development of methods for identification of dynamic models with random parameters and their application to forecasting migration in Eurasia".

References

1. Denisenko, M.B. Mathematical models of population migration. M.B. Denisenko, in the book Modern Demography. Under the Editorship of A.I. Kvasha, V.A. Iontsev. M.: MSU (1995)
2. Lukina, A.A.: Analysis and mathematical modeling of international labour migration. In: Lukina, A.A., Prasolov, A.V. (eds.) Management Consulting, vol. 10 no. 82, pp. 146–156 (2015)
3. Yudina, T.N.: Predictive models of migration to Russia: approaches and analysis. In: Yudina, T.N. (ed.) MIGRATION 6, Tbilisi, Ivane Javakhishvili Tbilisi State University Migration Research Center, pp. 25–43 (2013)
4. Schmertmann, C.P.: Stationary populations with below-replacement fertility. Demogr. Res. 26, 319–330, Article 14 (2012). https://doi.org/10.4054/demres.2012.26.14
5. Feichtinger, G., Prskawetz, A., Veliov, V.M.: Age-structured optimal control in population economics. Theor. Popul. Biol. **65**(4), 373–387 (2004)
6. Simon, C., Skritek, B., Veliov, V.M.: Optimal immigration age-patterns in populations of fixed size. J. Math. Anal. Appl. **405**, 71–89 (2013)
7. Popkov, Y.S.: Mathematical demoeconomic: macrosystem approach. In: Popkov, Y.S. (ed.) M.: LENAND (2013)
8. Database – Eurostat. http://ec.europa.eu/eurostat/data/database. Accessed 14 May 2018
9. Key Indicators for Asia and the Pacific 2017—Asian Development Bank. https://www.adb.org/publications/key-indicators-asia-and-pacific-2017. Accessed 12 Mar 2018

Strategic Decision for Investments in Real Estate Businesses

Fernando Freire Vasconcelos, George Henrique Tavares de Castro(✉), and Paulo Brasileiro Pires Freire

University of Fortaleza, Fortaleza, Brazil
nandofreireadm@yahoo.com.br, george.castro@jccbr.com, pfreire66@gmail.com

Abstract. The purpose of this article is to identify the criteria that allow a better decision making among managers of companies in construction segment related to investments in real estate business. In this scenario, we perceive the need to identify and classify the attributes for a strategic investment decision, through a multicriteria decision model that allows greater economic effectiveness, considering a limited rationality from managers and their capacity to observe the previous economic inefficiency and the dynamics of the real estate business, which is not obtained with the investment analysis techniques currently employed. This text presents some requirements for decision making through the construction of a cognitive map, with the purpose of subsidizing a later study aiming at the construction of a multicriteria investment analysis model.

Keywords: Real estate business · Investment decision
Strategic investment decision · Real estate investment decision

1 Introduction

As competition in the business environment grows, proportionately, pressure on companies increases to achieve effectiveness over their investments. This way, decision-making process is primordial in entrepreneurial action, requiring strategic positioning, seeking to address two elementary aspects of business activity: (1) Best capital return and (2) Lower risk associated to the business (Tobin 1958; Van Horne 1972).

According to de Souza (2015), the construction industry is strongly characterized by projects with long implementation terms, lack of flexibility, high exit barriers, high commitment of financial resources and exposure to inflationary pressures. Thus, the duration of the projects has a well-defined period and the survival of these organizations depends on the renewal of this cycle.

The construction sector has a very intense market dynamic, since the demand for its products is directly related to the country's macroeconomic conditions, as well as to the volume of public and private investments in this sector, where the managerial perspective of these organizations needs to shape the frequent changes, implementing actions that fit the new reality, for its own organizational survival.

This article has as main objective to identify the essential skills to the strategic decision making of investment in real estate in order to ensure satisfactory results in a

R. Silhavy et al. (Eds.): CoMeSySo 2018, AISC 859, pp. 72–80, 2019.
https://doi.org/10.1007/978-3-030-00211-4_8

sustainable way. The problematic in question is: What are the essential characteristics and abilities to support the decision of the builders in the process of selecting the best investments and how can this process influence the economic sustainability in this sector?

2 Strategic Investment Decision

This topic will explore basic concepts of **decision-making process,** as well as addressing specific studies in **strategic decision-making process.**

For Lindblon (1981), the decision-making process is a highly complex task with uncertain limits to understand how the characteristics of the various actors involved in the decision-making process, their roles, their authority, and the power of each agent deal with each other.

With the advance in making decision research, it moved to, more easily, identify the circumstances involved, which provided a further development of management techniques to better deal with risks and uncertainties. For this reason, the decision-making process began to be studied from the perspective of two theoretical fields: descriptive approach and normative approach, also called prescriptive.

In the descriptive approach, the analysis focuses on how the decision maker behaves in the actual decision process when assessing the difficulties and biases encountered. On the other hand, the prescriptive or normative theory addresses the structuring of the steps to be followed in the analysis and resolution of problems in order to find the correct way to decide (Ludkiewicz 2008).

The number of decision models is very large, all based on the developed theories of decision-making. However, in this article we highlight only three of them: rational, behavioral and political. The rational model is based on the prescriptive perspective and aims to provide the creation of technical instruments that promote the maximization of profit and results, without taking into account other qualitative factors such as time, uncertainty and direct influence of the decision maker (Raiffa 1977). In the behavior model, the decision maker seeks to make good enough choices, other than making optimal choices as in the previous model. Based on this, it is necessary to accept the interferences of the behavioral characteristics of the decision maker, understanding their motivations, emotions, perceptions and uncertainties. According to Yu (2011), the decision-making process is profoundly influenced by people or groups that have greater power within the organization to pursue their own interests. The author, according to Table 1, can define the structure and phases of the decision-making process.

This way, we realize that the steps are similar and involve the identification of the problem, categorization of alternatives, analysis of the evaluation criteria and the decision-making itself (Nonohay 2012).

According to Matheson and Matheson (1998) the decisions can be divided into operational or strategic, which makes important classification of the decision by its specific characteristics so that the methodology to be used in the decision process is defined. Then, Dacorso (2004) defines the characteristics of an operational decision,

Table 1. Examples of process stages by the author. Source: Nonohay (2012)

	Bazerman and Moore (2010)	Uris (1989 apud Gomes; Gomes and Almeida 2009)	Costa (1977 apud Gomes; Gomes and Almeida 2009)	Shablin and Stevens Jr. (1989 apud Gomes; Gomes and Almeida 2009)	Hogart (1980)	Courtney (2001)
1	Define the problem	Analysis and identification of the problem situation	Generation of the problem	Formulation of the problem	Structure the problem	Recognition of the problem
2	Identify the criteria	Development of alternatives	Problem formulation	Construction of a study model	Evaluate the consequences	Definition of the problem
3	Ponder the criteria	Comparison between alternatives	Identification of the problem solution	Suggestion of solution based on the study	Evaluate the uncertainties	Generation of alternatives
4	Generate alternatives	Classify the risks of each alternative	Implementation of the problem solution	Testing the model solution	Evaluate the alternatives	Development of the model
5	Rate each alternative according to each criterion	Choosing the best alternative		Establishing controls on the solution	Sensitivity analysis	Analysis of alternatives
6	Identify ideal solution	Implementation and evaluation		Solution implementation	Information grouping	Choice
7					Choice	Implementation

when the errors are not so expensive, because it involves few resources and has a quick response facilitating the learning, whose knowledge is referred in the own team. However, he conceptualizes the strategic decision because any errors are large, since it involves many resources and has a long-term response, usually guided by external or specialized personnel.

For the analysis of the strategic decision-making processes, we separated three more relevant studies regarding the revision of the state of the art in the case of the first two authors: Rajagopalan et al. (1993); Elbanna and Child (2007) and the Sousa metadata model (2006). The summary of the strategic decision-making research (SDM) by authors, in a more panoramic view of the process, is described in Table 2.

Regard to real estate investments, in recent years, the complexity of financial decisions has increased fast and the development and implementation of sophisticated and efficient quantitative analysis techniques to support and assist in strategic investment. One of the segments of operational research, based on multicriteria decision-making, provides resources to decision makers and analysts with a wide range of methodologies well served by the complexity of modem financial decision making.

Table 2. Summary of SDM - a process overview. Source: Sousa (2015)

SDM - a process overview		
Research	Ranking	Used items
Rajagopalan et al. (1993)	Organizational factors	Past strategies; organizational structure, power distribution; size of organization; organizational backlash; characteristics of high organization
	Characteristics of the decision-making process	Extension of rationality; degree of political activity; participation; duration/extension
	Decision - specific factors	Urgency; uncertainty and complexity
	Economic results	Return on investment; growth
Elbanna and Child (2007)	Dimensions of the strategic decision-making process	Intuition; rationality; political behavior
	Characteristics of the external environment	Environmental uncertainty; munificiency and hostility
	Internal company characteristics	Performance and company size
Sousa (2006)	Meta-decision of the process	Analytical structure; stage of the process; participatory dependency; influence
	Model "deciding how to decide"	Categorization of variables; methodology

3 Multicriterial Decision Analysis

A decision support system helps to obtain answers to questions presented to an intervener in a decision-making process, using a clear but not necessarily formalized model (Roy 1985). Multicriteria Decision Analysis (MDA) can be conceptualized as a set of tools, in order to analyze several alternatives, with several criteria and conflicting objectives.

Through this tool, it is possible to create a hierarchy of alternatives, according to the degree of sympathy of these to the decision maker (Gomes et al. 2002). The decision makers relate criteria and under each of them, the alternatives will be analyzed, not in a watertight perspective, but on the contrary, trying to measure all the complexity to be contemplated.

Therefore, this method seeks to combine both quantitative and qualitative criteria, often conflicting ones, for the purpose of analyzing complex problems (Alexandrini et al. 2009), being very useful within organizations, given the complexity already mentioned above. It is based on the assumption that there is no optimal decision criterion, but several conflicting and interdependent criteria and alternatives.

The multicriteria analysis lies between the purely exploratory and the few structured approaches, such as brainstorming and discussion groups, as well as rigidly structured quantitative models related to operational research (Ensslin et al. 2001).

In this way, the multicriteria analysis seeks a compromise solution, negotiated among several objectives that it intends to achieve, since it is not an optimal solution, but one that seeks a consensus among the adopted criteria (Gomes et al. 2004). A complexity that cannot be analyzed on just one optic.

Firstly, with a qualitative character, a clear definition of the problem to be solved is sought; identify valid alternatives to solve or answer the question; identifying decision makers; define with each decision maker the criteria or indicators and their respective weights to solve the problem and assign the value reached to each evaluation criterion for each identified alternative (Jannuzzi et al. 2009).

The choice of the problem will influence the structuring of the multicriteria model. The selection of an alternative or a set of alternatives among all the proposals is characteristic of a problem of choice. Within this context, the visual representation of knowledge from cognitive maps of analysis seeks to graphically represent the individual's perception of a problem situation (Alexandrini et al. 2009). Cognitive maps are representations or mental models built by people, resulting from their interactions and knowledge of a specific domain of the environment in which they are inserted, in the mission to make sense of reality and to deal with problems and challenges presented (Swan 1997).

Therefore, it is advantage to use a cognitive map to outline the solution of the multicriteria decision-making problem, starting from the definition of the problem, identifying the constraints, the criteria and the alternatives, even by crossing the criteria with the alternatives and the final objective with the criteria (Briozq and Musetti 2015).

The decision problem is divided into hierarchical levels, through the adoption of criteria and alternatives, in order to facilitate understanding and evaluation (dos Santos et al. 2011).

Therefore, the use of cognitive maps can greatly facilitate the resolution of decision problems within organizations, through the elaboration of a structured model for problem solving.

4 Method

The present study is based on a descriptive research, considering that it seeks to raise the opinions of a group delimited in relation to a theme (Gil 2002), namely, the decision analysis of investment in real estate business.

However, initially the research had an exploratory phase, because it sought to identify the main factors that influence the decision making of investment in real estate business, which will focus this article.

In the present research, the qualitative method was used, because in this approach the researcher seeks to interpret the phenomena in their natural environment, in the search to give meaning to the data collected through interviews, according to the meanings that people attribute to him (Denzin and Lincoln 1994, p. 22) (Fig. 1).

The population surveyed included managers of construction companies in the state of Ceará (Brazil). The sample was of non-probabilistic type, due to the difficulty of access to all population environments, due to the accessibility of respondents willing to answer the questionnaire.

The data collection instrument used an open questionnaire, elaborated based on the researched theoretical framework, seeking to raise the main factors that influence the decision making of investment in real estate business.

By analyzing the answers, we sought to create a cognitive map, structured based on the multicriteria analysis model, through the hierarchy of the main criteria obtained through the interviews and the alternatives based on the specialists' view. Cognitive maps are usually obtained through interviews to represent the subjective mental model of the interviewee (Eden 2004).

5 Multicriteria Model for Strategic Investment Decision in Real Business

In order to apply the different environmental impacts inherent in the construction segment, the result of this work, illustrated below in a cognitive map, drawn from the studied knowledge, the main issues to be addressed: To better segregate the direct impact on decision-making in real estate define the following variables:

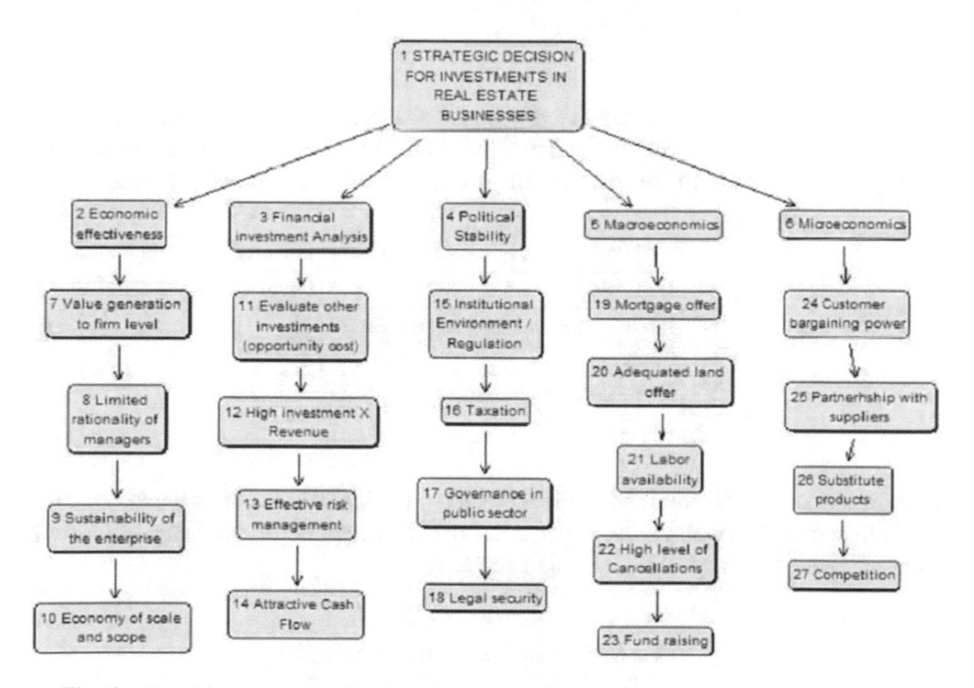

Fig. 1. Cognitive map of strategic real estate investment decision. Source: own author

Economic Effectiveness: Seeks to generate value of the enterprise for the sustainability of the business, justifying it for its implementation and permanence in the company's portfolio (Table 3);

Table 3. Descriptors for economic effectiveness

NI	Description	Order
N4	Value generation to firm level	1°
N3	Limited rationality of managers	2°
N2	Sustainability of the enterprise	3°
Nl	Economy of scale and scope	4°

Financial Investment Analysis: Use financial engineering to attest to the financial viability of the achievement, measured by financial performance indicators that can attest to the employability of the resource in the suggested product (Table 4).

Table 4. Descriptors for financial investment analysis

NI	Description	Order
N4	Evaluate other investiments (opportunity cost)	1°
N3	High investment X revenue	2°
N2	Effective risk management	3°
NI	Attractive cash flow	4°

Political Stability: Being a regulated product, the institutional environment, the tax regime to be applied, as well as the development of the master plan of the cities where these products operate make this variable a primordial for this process (Table 5);

Table 5. Descriptors for political stability

NI	Description	Order
N4	Institutional environment/regulation	1°
N3	Taxation	2°
N2	Governance in public sector	3°
NI	Legal security	4°

Macroeconomics: Unemployment levels, supply of skilled labor, land banks available for development, as well as the appetite of foreign investors to inject capital into this segment is another important variable (Table 6);

Table 6. Descriptors for macroeconomics

NI	Description	Order
N5	Mortgage offer	1°
N4	Adequated land offer	2°
N3	Labor availability	3°
N2	High level of cancellations	4°
NI	Fund raising	5°

Microeconomics: Finally, the microeconomic environment with the whole chain that participates directly in the development of these products, such as the predisposition of consumption, the relationship with suppliers and partners, as well as the effect of competition, close this first level of impacting variables in this process decision-making (Table 7).

Table 7. Descriptors for microeconomics

NI	Description	Order
N4	Customer bargaining power	1°
N3	Partnership with suppliers	2°
N2	Substitute products	3°
NI	Competition	4°

In order to "cascate" these impacts, it is evident in the map above the sub-variables that complement this decision-making process.

6 Conclusion

Focusing on the objective of the present study, the authors sought to develop a prototype multicriteria analysis model for decision-making, through the construction of a cognitive map, in order to guide a qualitative research with the managers responsible for making decision in real estate deals.

The application of the cognitive map for structuring a complex problem through the extraction of concepts is a widely used practice in the academic literature.

As can be seen, the concepts of economic effectiveness, financial analysis of investment, political stability and macroeconomic and microeconomic conditions were approached through the analysis of the cognitive map constructed by the members of the group.

The related variables were taken from the academic literature, as well as the experience of the study participants themselves.

It is important to emphasize that this is a previous study, for constructing a model that will be of great value to the civil construction managers, as well as to be able to effectively demonstrate how the cognitive model of decision making in the sector works. The elaboration of the cognitive map seeks to represent graphically how the decision making in the sector works.

References

Alexandrini, F., et al.: Multicritério de apoio a decisão e o aumento de equipe na vigilância sanitaria de agrolândia-sc. In: Simpósio de excelência em gestão e tecnologia, 2009, Resende. Anais. Seget, Resende (2009)

Briozq, R.A., Musetti, M.A.: Método multicritério de tomada de decisão: aplicação ao caso da localização espacial de uma Unidade de Pronto Atendimento - UPA 24 h. Gest. Prod. **22**(4), 805–819 (2015)

Eden, C.: Analyzing cognitive maps to help structure issues or problems. Eur. J. Oper. Res. **159** (3), 673–686 (2004)
Elbanna, S., Child, J.: Influences on strategic decision effectiveness: development and test of an integrative model. Strateg. Manag. J. **28**, 431–453 (2007)
Ensslin, L., Monttbeller Neto, G., Noronha, S.M.D.: Apoio à Decisão: metodologias para estruturação de problemas e avaliação multicritério de alternativas. Insular, Florianópolis (2001)
Gomes, E.G., Lins, M.P.E., de Mello, J.C.C.B.S.: Selecção do melhor município: integração SIG- Multicritério. Investigação Operacional **22**(1), 59–85 (2002)
Gomes, L.F.A.M., Araya, M.C.G., Carignano, C.: Tomada de decisão em cenários complexos: introdução aos métodos discretos do apoio multicritério à decisão. Pioneira, Thomson Learning, São Paulo (2004)
Jannuzzi, P.D.M., de Miranda, W.L., da Silva, D.S.G.: Análise multicritério e tomada de decisão em políticas públicas: aspectos metodológicos, aplicativo operacional e aplicações. Revista Informática Pública **11**(1), 69–87 (2009)
Roy, B.: Methodologie multicritère d'aide à la décision. Economica, Paris (1985)
dos Santos, P.R., Curo, R.S.G., Belderrain, M.C.N.: Aphcação do mapa cognitivo a um problema de decisão do setor aeroespacial de defesa do Brasil. J. Aerosp. Technol. Manag. **3**(2), 215–226 (2011)
Swan, J.: Using cognitive mapping in management research: decisions about technical innovation. Br. J. Manag. **8**, 183–198 (1997)
Lindblon, C.E.: O processo de decisão política. UNB, Brasilia, Ed. (1981)
Ludkiewicz, H.F.F.: Processo para a tomada de decisão estratégica: um estudo de caso na parceria banco e varejista. Dissertação (Mestrado em Administração). Universidade de São Paulo, São Paulo (2008)
Matheson, D., Matheson, J.: The Smart Organization: Creating Value Through Strategic R&D. HBS, Boston (1998)
Nonohay, R.G.: Tomada de decisão e os sistemas cerebrais: primeiros diálogos entre a administração, psicologia e neurofisiologia. Dissertação (Mestrado). Universidade Federal do Rio Grande do Sul, Porto Alegre (2012)
Raiffa, H.: Teoria da Decisão. Vozes, Petrópolis (1977)
Rajagopalan, N., Rasheed, A.M.A., Datta, D.K.: Strategic decision process: critical review and future directions. J. Manag. **19**(1), 349–384 (1993)
de Souza, A.S.: Processo de decisão de Novos Empreendimento Imobiliários para Pequenas e Médias Empresas. Tese de Mestrado em Engenharia Civil - Universidade Federal do Espirito Santo. Espirito Santo (2015)
Tobin, J.: Liquidity preference as behavior towards risk. Rev. Econ. Stud. **25**, 65–86 (1958)
Yu, A.S.O. (Coord.): Tomada de decisão nas organizações: uma visão multidisciplinar. Saraiva, São Paulo (2011)

Modeling Syntactic Structures of Vietnamese Complex Sentences

Co Ton Minh Dang(✉)

Saigon University, Ho Chi Minh City, Vietnam
dangsgu@gmail.com

Abstract. This paper is executed with the help of Vietnamese grammar works and Definite Clause Grammar to build an automatically information answer system. The author proposes a new experimental method for patternizing syntactic structures of Vietnamese complex sentences. Specifically, 6 general structure classes for Vietnamese complex sentences have been patternized. After building and testing, the system is shown to have an accuracy of 82%–100%.

Keywords: Vietnamese complex sentences
Information question-answer systems · Definite clause grammar
DCG · Swi-Prolog

1 Introduction

So far, information question-answer systems in Vietnamese have put attention on processing simple questions based on simple structures [6, 10–19]. It is obvious when one wants to upgrade the capacity of communication systems in natural languages, the syntactic analysis cannot just handle simple sentences, but it must involve complex ones. For other languages than Vietnamese, there exist different researches about syntactic analysis in Japanese, Croatian, English, … Therefore, this paper investigates patternising method for parsing syntactic structures of some complex sentences in order to apply on information command-answer systems in Vietnamese. However, it is possible neither to realise general syntactic structures parsing because of various modalities of Vietnamese complex sentences, nor to identify all syntactic structures for complex sentences as simple sentences [5, 6, 9–14, 16, 17] have been done. Only can we count on a Vietnamese grammar theory and use experimental approach to patternize syntactic structures of Vietnamese sentences in accordance to a grammatical pattern which can be processed on computer.

As the fore-mentioned problems, the paper has as goal developing a experimental approach for making use of grammar patterns DCG to patternize syntactic structures of some Vietnamese complex sentence classes founded on subject-predicate pattern as basic structure presented by Diep Quang Ban [1, 2]. With results of syntactic structure patternizing for Vietnamese complex sentences on DCG, this paper will build an automatically information answer system about some dairy products.

R. Silhavy et al. (Eds.): CoMeSySo 2018, AISC 859, pp. 81–91, 2019.
https://doi.org/10.1007/978-3-030-00211-4_9

New features of the paper:

- Proposition of an experiential approach for syntactic structure of Vietnamese complex sentences patternizing based on DCG and subject-predicate grammar pattern.
- Construction complex command processing mechanism for an information query system in conformity with investigated pattern in the paper.

We organize the rest of this paper as follows: in Sect. 2, The task of Parsing and patternizing syntactic structures of Vietnamese complex sentences is described; in Sect. 3 we report the program building; in Sect. 4 we present the experiment results, and finally, we conclude the paper and discuss possibilities for future work.

2 Parsing and Patternizing Syntactic Structures of Vietnamese Complex Sentences

2.1 Outlines of Vietnamese Grammar

In this paper, the author will choose the syntactic analysis pattern of Vietnamese sentences according to subject-predicate relationship (Diep Quang Ban [1, 2]).

- Simple sentence: is defined as a sentence with only one S – P (subject – predicate) combination considered as the core of syntactic structure.
  Ex 1: They are ex-students of this university.
- Compound sentence: is defined as a sentence with two S – P combinations or more, in which these combinations are independent of each other.
  Ex 2: "The tire was busted and the car stopped." (Source: Diep Quang Ban [1]).
- Complex sentence: is defined as a sentence with an S – P combination considered as the core and its other parts can also be S – P combinations which are incapsulated in each other.
  Ex 3: "The mice which were running broke the lamp" (Diep Quang Ban [1]).

2.2 Patternizing a Syntactic Structure of Vietnamese Complex Sentences

Symbols: S: complex sentence, CN: subject 1 (the principal subject of the sentence), VN: predicate 1 (the principal predicate of the sentence), S – SUB: C – V combination is the incapsulated sentence, CN – SUB: the second subject (subject of the incapsulated sentence), VN – SUB: the second predicate (predicate of the incapsulated sentence), VP: verbal phrase (verb phrase), AuV: verb complement phrase (complement of verb), NP: nominal phrase (noun phrase), AuN: noun complement phrase (complement of noun), AuS: adver-bial phrase (adverb complement of sentence).

From Diep Quang Ban's grammar works [1, 2] on Vietnamese complex sentence structures, we can see current complex sentences:

- Complex sentence whose CN is an C – V combination:
  Syntactic pattern 1: S(CN(S-SUB(CN-SUB, VN-SUB)), VN)
- Complex sentence whose VN is an C – V combination:

Syntactic pattern 2: S(CN, VN(S-SUB(CN-SUB, VN-SUB)))

- Complex sentence whose AuV is an C – V combination:
  Syntactic pattern 3: S(CN, VN(VP, AuV(S-SUB(CN-SUB, VN-SUB))))
- Complex sentence whose AuN is an C – V combination:
  Syntactic pattern 4.1: S(CN(NP, AuN(S-SUB(CN-SUB, VN-SUB))), VN)
  Syntactic pattern 4.2: S(CN, VN(VP, NP, AuN(S-SUB(CN-SUB, VN-SUB))))
- Complex sentence whose AuS is an C – V combination:
  Syntactic pattern 5: S(AuS(S-SUB(CN-SUB, VN-SUB)), <comma> , CN, VN)
- Complex sentence in passive voice: Complex sentence in passive voice is Vietnamese sentence whose subject is affected by an agent and satisfies the following requirements:
  - The verb "to be" ("bị" or "được") must occur.
  - "to be" is succeeded by an C – V combination whose subject is different from CN. In some cases, the subject concerned can be absent.

The verb of the incapsulated sentence is a transition verb exerting influence on the CN.

Ex 4:

- "The boat was they push faraway."
- "The boat was water pull very fast." (Source: Diep Quang Ban [1]).

Syntactic pattern 6: S(CN, VN(<passive_voice> , S-SUB(CN-SUB, VN-SUB))).
In this pattern, the verb in passive voice is "bị" or "được".
The syntactic pattern of Vietnamese complex sentences is presented in the Table 1

Table 1. The syntactic patterns of Vietnamese complex sentences.

ID	Syntactic Patterns
1	S(CN(S-SUB(CN-SUB, VN-SUB)), VN)
2	S(CN, VN(S-SUB(CN-SUB, VN-SUB)))
3	S(CN, VN(VP, AuV(S-SUB(CN-SUB, VN-SUB))))
4	S(CN(NP, AuN(S-SUB(CN-SUB, VN-SUB))), VN)
5	or S(CN, VN(VP, NP, AuN(S-SUB(CN-SUB, VN-SUB))))
6	S(AuS(S-SUB(CN-SUB, VN-SUB)), <comma> , CN, VN)

The complex sentence set is investigated by the author observing the following criteria:

- Only will the complex sentences with two C – V combinations in maximum be examined.
- The numbers of different modalities of complex sentences are different due to their frequencies of current usage in reality.

2.3 Syntactic Structure of the Parts of Speech in Vietnamese Complex Sentences

On construction of Vietnamese sentence structures, we should undertake it step by step: word → phrase → sentence.

Traditional grammar often classifies phrases as shown in Table 2 [3, 4, 7, 8].

Table 2. Current phrases in Vietnamese.

Phrases	Abbreviations	Features
Nominal phrase	NP	Noun considered the core
Verbal phrase	VP	Verb considered the core
Adjectival phrase	AP	Adjective considered the core
Prepositional phrase	PP	Preposition considered the core

Nominal Phrase (noun phrase - NP): According to Nguyen Tai Can [3], nominal phrase is a short phrase whose a noun is the core (the principal part is a noun) and other dependent parts (modifier elements) like numerals, adjectives,… Nominal phrase structure is composed of three parts: the principal part, the the preceding dependent part and the succeeding one. Each of them involves other more dependent components.

Verbal Phrase (verb phrase – VP): Verbal phrase is a short phrase whose a verb is the core (the principal part is a verb) and other dependent parts (complement elements) like nouns, adjectives, adverbs… (Nguyen Tai Can [3]).

Adjectival Phrase (adjective phrase – AP): Adjectival phrase is a short phrase whose an adjective is the core (the principal part is an adjective) and other dependent parts like nouns, adverbs… (Nguyen Tai Can [3]).

Prepositional Phrase (preposition phrase – PP): Prepositional phrase is a short phrase whose a preposition is the core (the principal part is a preposition) and other dependent parts like nouns, determiners… (Nguyen Tai Can [3]).

2.4 Building DCG Grammar for Syntactic Complex Sentence Patterns

The aim of building a syntactic rules set is to construct a sufficient grammar to process the analysis of complex sentences on computer. If grammar is too vast, it will not only be unnecessary, but complicate the problem. If it is insufficient, it will not englobe all the sentence structures needing to be analyzed, so the problem will not be solved fully.

Building a Vocabulary Dictionary

A vocabulary dictionary (vocabulary set) is the foundation for building syntactic rules set for components which form Vietnamese complex sentences patterns [20, 21]. The vocabulary set is made reference to http://tratu.soha.vn. In the vocabulary set, we should mention the parts of speech necessary for the system: noun, verb, adjective, preposition, pronoun, adverb, con-junction, numeral, quantifier, verb in passive voice, interrogative pronoun. Vietnamese grammar is far complicated, the classification of

Table 3. Table of numbers of parts of speech in vocabulary set.

ID	Part of speech	Quantity
1	Adjective	74
2	Conjunction	
3	Determiner	11
4	Quantifier	18
5	Noun	329
6	Measurement unit noun	12
7	Numeral	53
8	Pronoun	45
9	Preposition	15
10	Adverb	53
11	Verb	213
12	Interrogative pronoun	7

parts of speech is still controversial. In fact, this classification only bears a relative signification because sentences, cases, contexts can determine a word to be of a part of speech or of another one.

The numbers of parts of speech in the system are statistically put in Table 3.

Building Syntactic Rules for Adjectival Phrases

The syntactic rules for adjectival phrases complying with DCG grammar in Prolog language as below:

```
ap(ap(A)) --> a(A).
ap(ap(A1,A2)) --> a(A1), a(A2).
ap(ap(A,R)) --> a(A), r(R,suffix,adj).
ap(ap(A,NP)) --> a(A), np(NP).
ap(ap(R,A)) --> r(R,prefix,adj), a(A).
ap(ap(R,A,NP)) --> r(R,prefix,adj), a(A), np(NP).
ap(ap(R1,A,R2)) --> r(R1,prefix,adj), a(A),
r(R2,suffix,adj).
```

Building Syntactic Rules for Prepositional Phrases

The syntactic structures of prepositional phrases, we find out syntactic rules for prepositional phrases complying with DCG grammar in Prolog language as below:

```
pp(pp(PREP,NP)) --> prep(PREP), np(NP).
pp(pp(PREP,D)) --> prep(PREP), d(D).
pp(pp(PREP_cua,NP)) --> prep_cua(PREP_cua), np(NP).
```

Building Syntactic Rules for Nominal Phrases and Verbal Phrases

Same to below, we also build syntactic rules for nominal phrases and verbal phrases.

2.5 Processing Vietnamese Complex Sentences

The 1st Pattern Class

According to Sect. 2.2, the pattern for complex sentence whose CN is an C –V combination (CN complex sentence): S(CN(S-SUB(CN-SUB,VN-SUB)),VN).
DCG grammar interpretes it as follows:

```
s(complex1(CN1,VN1)) --> cn(CN1),vn(VN1).
cn(cn(S1)) --> s_sub(S1).
s_sub(s_sub(CN2,VN2)) --> cn_sub(CN2),vn_sub(VN2).
```

The 2nd Pattern Class

The pattern for complex sentence whose VN is an C –V combination (VN complex sentence): S(CN,VN(S-SUB(CN-SUB,VN-SUB))).

DCG grammar interpretes VN complex sentence as follows:

```
s(complex2(CN1,VN1)) --> cn(CN1), vn(VN1).
vn(vn(S1)) --> s_sub(S1).
s_sub(s_sub(CN2,VN2)) --> cn_sub(CN2), vn_sub(VN2).
```

The 3rd Pattern Class

S(CN,VN(VP,AuV(S-SUB(CN-SUB,VN-SUB)))).

DCG grammar interpretes this pattern as follows:

```
s(complex3(CN,VN)) --> cn(CN), vn(VN).
cn(cn(NP)) --> np(NP).
vn(vn(VP,S)) --> vp(VP), s_sub(S).
s_sub(s_sub(CN1,VN1)) --> cn_sub(CN1), vn_sub(VN1).
```

The 4th Pattern Class

41 pattern: S(CN(NP,AuN(S-SUB(CN-SUB,VN-SUB))),VN).

DCG grammar interpretes this pattern as follows:

```
s(s(CN,VN)) --> cn(CN), vn(VN).
cn(cn(NP,AuN)) --> np(NP), aun(AuN).
cn(cn(NP,C,AuN)) --> np(NP), c(C), aun(AuN).
aun(aun(AuN)) --> s_sub(AuN).
s_sub(s_sub(CN_SUB,VN_SUB) --> cn_sub(CN_SUB),
vn_sub(VN_SUB).
```

4.2 pattern: S(CN,VN(VP,NP,AuN(S-SUB(CN-SUB,VN-SUB)))).
DCG grammar interpretes this pattern as follows:

```
s(s(CN,VN)) --> cn(CN), vn(VN).
vn(vn(VP,NP,AuN)) --> vp(VP), np(NP), aun(AuN).
vn(vn(VP,NP,C,AuN))-->vp(VP), np(NP),c(C),aun(AuN).
aun(aun(AuN)) --> s_sub(AuN).
s_sub(s_sub(CN_SUB,VN_SUB)) --> cn_sub(CN_SUB),
vn_sub(VN_SUB).
```

The 5th Pattern Class
S(AuS(S-SUB(CN-SUB,VN-SUB)),[dấu phẩy],CN,VN)
DCG grammar interpretes this pattern as follows:

```
s(s(AuS,[dauphay],CN,VN))--> aus(AuS),[dauphay],
cn(CN),vn(VN).
aus(aus(AuS)) --> s_sub(AuS).
s_sub(s_sub(CN_SUB,VN_SUB))--> cn_sub(CN_SUB),
vn_sub(VN_SUB).
```

The 6th Pattern Class
S(CN,VN([passive_voice], S-SUB(CN-SUB,VN-SUB)))
DCG grammar interpretes this pattern as follows:

```
s(s(CN,V_pass,S_SUB)) --> cn(CN), v_pass(V_pass),
s_sub(S_SUB).
s_sub(s_sub(CN_SUB,VN_SUB)) --> cn_sub(CN_SUB),
vn_sub(VN_SUB).
```

3 Building the Program

The system architecture is built in the image of Fig. 1.

The system is designed to comprise two branches whose main functions are described as shown below:

The 1st branch:

- Syntactic structures of Vietnamese complex sentences are supposed as to be classified depending on their C – V grammatical structure.
- Classifying and parsing 300 complex sentences obeying syntactic structures supposed.
- Patternizing syntactic structures then classifying them into different classes.
- Building syntactic pattern classes complying with DCG grammar.

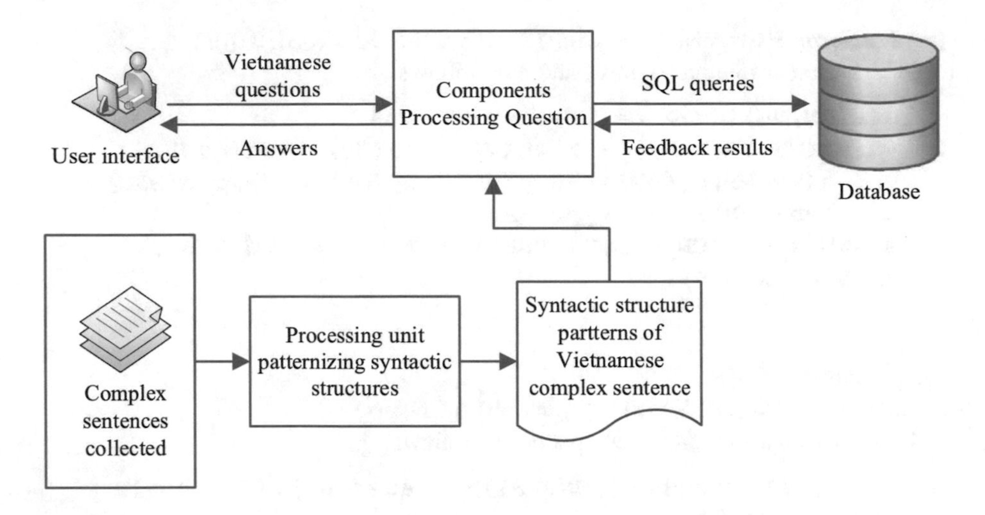

Fig. 1. System architecture pattern

The 2nd branch:

- Receiving Vietnamese questions about Nutifood dairy information through user interface.
- Based on syntactic structure patterns of Vietnamese complex sentences built in the 1st branch in order to classify complex questions into the above patterns.
- Processing question syntax to find out answers.
- Returning results to users.

4 Experiments

In this report, we try to adopt two machine learning techniques, the Support Vector Machine and Naïve Bayes, to classify the text commands, and then choose a more suitable method. With the classifier, the system can easily route a call to one of the appropriate agents. In this report, we present a case study with five agents as follows: Finance Department (FD); Aca-demic Registrar Office (ARO); Graduate Education Department (GED); Student Service Department (SSD) and Faculty of Information Technology (FIT).

4.1 Conditions for Experiments

Computer configuration	Intel ® Core™ i3 M380 @ 2.53 GHz, RAM 2.0 GB
Processing system	Windows 7 Ultimate 32-bit
Software tools	Visual Studio 2010 + SWI-Prolog

4.2 Data for Experiments

Data A	Affirmative complex sentence set in the paper used for parsing and building patterns	300 sentences
Data C	Number of accidentally affirmative complex sentences (whose words belong to vocabulary set and syntactic structures belong to 6 patterns)	73 sentences
Data B	Number of complex sentences about Nutifood dairy information	35 sentences

4.3 Results of Experiments

Rates of good results	Data A	300 good results returned (100%)
	Data C	60/73 good results returned (82.19%)
	Data B	Good results to 35 questions (100%)

4.4 Processing Time

Parsing complex sentence syntax	2.79 s/300 complex sentences
Responding queries	0.147 s/query

4.5 Evaluation

The system has checked correctly the syntactic structure of all 300 affirmative complex sentences and responded correctly 35 complex questions from the application to Nutifood dairy information queries as well as 73 accidentally complex sentences without making any mistake, therefore we can confirm its high reliability. In regard to affirmative complex sentences not belonging to Data A, the possibly of correct parsing is likely to occur provided that users input right syntactic structure patterns and vocabulary of sentences found in the vocabulary set of the system. Specifically, 73 accidentally complex sentences (data C) input are checked, parsed by the system and 64/73 syntactic sentences are returned (87.67%), out of which 60/73 (82.19%) syntactic sentences are estimated right, 4/73 (5.48%) are right but have different syntaxes and 9/73 (12.33%) are estimated to be false. This reveals that DCG syntactic rules built and the vocabulary dictionary have not embraced all possible cases, therefore some of which are handled ambiguously. If the vocabulary dictionary is completed and the DCG syntactic rules improved, the system's coverage and accuracy will reach higher degree.

5 Conclusion

The classification of Vietnamese sentences according to C – V structure is not new, but classificating, patternizing syntactic structures of Vietnamese complex sentences and building DCG grammar for patterns built present a new approach; this paper has obtained the results at the levels theoretical and experimental as mentioned below:

- Proposing a new experimental method for patternizing syntactic structures of Vietnamese complex sentences. Specifically, 6 general structure classes for Vietnamese complex sentences have been patternized.
- Applying DCG grammar for parsing syntactic structures of Vietnamese complex sentences based on C – V grammatical structure to classify them. Building DCG grammatical rules for the patterns built.
- Installing and experimenting successfully 6 pattern classes and 300 Vietnamese com-plex sentences.
- Building complex questions processing mechanism founded on 2 popular pattern clas-ses experimented for automatical answering to query information system.

Through experimenting and checking results, the paper has shown to have some restraints:

- The quantity of sentences experimented is not enough to evaluate the system's accuracy.
- The vocabulary set and DCG grammar built are not complete, general enough to em-brace all syntactic structures of the parts of speech in one Vietnamese complex sentence.
- The inability to process the cases in which complex sentences comprise more than 2 C – V structures and nominal, verbal, adjectival phrases of ordinal structure.

For future researches, we plan to:

- Improve vocabulary set and DCG grammar for patterns.
- Research and build more syntactic rules for processing parts of speech of ordinal structure.
- Keep on building query information system with complex questions for maintaining patterns.

Acknowledgment. This paper was supported by the research Project CS2017-60 funded by Saigon University.

References

1. Ban, D.Q.: Ngữ pháp Việt Nam "Vietnamese Grammar". Viet Nam Education Publishing House (2009)
2. Ban, D.Q.: Ngữ pháp tiếng Việt "Vietnamese Grammar". Viet Nam Education Publishing House (2011)
3. Can, N.T.: Ngữ pháp tiếng Việt "Vietnamese Grammar". VNU-Hanoi (1999)

4. Can, N.T.: Từ loại danh từ trong tiếng Việt hiện đại "Nouns in modern Vietnamese". Social Sciences Publishing House (1975)
5. Ha, L.T., Khoa, N.D.: Xây dựng parser ngữ nghĩa cho câu đơn tiếng Việt theo mô hình UBG "Constructing the semantic parser for the Vietnamese sentence based-on the Unification-Based Grammar model". Master thesis, VNU-HCM (2013)
6. Ha, L.Q.T.: Xây dựng công cụ tìm kiếm tài liệu học tập bằng các truy vấn ngôn ngữ tự nhiên trên kho học liệu mở tiếng Việt "Developing a search engine for learning materials using natural language queries in the Vietnamese Open Learning Materials". Master thesis, VNU-HCM (2009)
7. Hao, C.X.: Tiếng Việt mấy vấn đề ngữ âm, ngữ pháp, ngữ nghĩa "Vietnamese phonetics, grammar, semantics". Viet Nam Education Publishing House (2001)
8. Hao, C.X.: Tiếng Việt sơ thảo ngữ pháp chức năng "Vietnamese grammar". Viet Nam Education Publishing House (2004)
9. Kiet, N.V., Tín, P.T.: Ứng dụng mô hình PDCG vào việc xây dựng Treebank và parser Tiếng Việt "An application of Probabilistic Definite Clause Grammar". Thesis, VNU-HCM (2013)
10. Le, N.V.: Tìm kiếm thông tin theo ngữ nghĩa: ứng dụng trong lĩnh vực du lịch "Semantic search: an application in the field of tourism". Thesis, PTIT (2011)
11. Lien, N.D.: Phương pháp xử lý một số dạng câu hỏi tiếng Việt cho công cụ tìm kiếm thư viện điện tử "Method of processing some types of Vietnamese questions for electronic library search engine". Master thesis, Lac Hong Univ (2011)
12. Tai, P.T.: Hệ thống truy vấn tiếng Việt về các chương trình truyền hình "Vietnamese query system for television programs". Master thesis, PTIT (2012)
13. The, P.T.: Cơ chế xử lý câu hỏi tiếng Việt cho hệ thống truy vấn thông tin đào tạo hệ tín chỉ "Vietnamese Question Processing System for Credit Information Training System". Master thesis, PTIT (2012)
14. Thu, D.M., Thai, N.P.: Nòng cốt câu phức và ghép "Simple complex and compound sentences", (VLSP), pp. 82–85 (2009)
15. Trong, N.B.: Hệ thống hỏi đáp tiếng Việt cho công tác tuyển sinh đại học "Question and answer system for university entrance examination". Master thesis, Lac Hong Univ (2011)
16. Dang, N.T.: Công cụ trợ giúp tìm kiếm và sử dụng các tuyến xe bus nội thành TP. HCM dựa trên cơ chế xử lý câu truy vấn tiếng Việt "Tool to help find and use bus routes in the HCMC", Project No: B2010-26-06 (2010)
17. Dang, N.T.: Xây dựng công cụ tìm kiếm học liệu mở tiếng Việt dựa trên cơ chế xử lý các truy vấn ngôn ngữ tự nhiên "Develop a search engine Vietnamese open learning materials based on the mechanism of natural language processing query", Project No: B2010-26-01TĐ
18. Nguyen, D.Q., Nguyen, D.Q., Pham, S.B.: A Vietnamese question answering system. In: Proceedings of International Conference on Knowledge and Systems Engineering 2009 (KSE 2009), Hanoi, Vietnam, pp. 26–32 (2009)
19. Phan, T.T., Nguyen, T.C., Huynh, T.N.T.: "Question Semantic Analysis in Vietnamese QA System. In: The Advances in Intelligent Information and Database Systems, Series of Studies in Computational Intelligence, vol. 283, pp. 29–40. Springer (2010). ISSN 1860-949X
20. Tran, T.K., Phan, T.T.: A hybrid approach for building a Vietnamese sentiment dictionary. J. Intell. Fuzzy Syst. **35**(1), 967–978 (2018)
21. Tran, T.K., Phan, T.T.: Mining opinion targets and opinion words from online reviews. Int. J. Inf. Technol. **9**(3), 239–249 (2017)

Information Monitoring of Community's Territoriality Based on Online Social Network

Igor O. Datyev(✉) and Andrey M. Fedorov

Institute for Informatics and Mathematical Modeling-Subdivision of the Federal, Research Centre "Kola Science Centre of the Russian Academy of Sciences", 184209 Apatity, Russia
{datyev, fedorov}@iimm.ru

Abstract. The wide representation of real communities in the virtual space of online social networks opens new opportunities for sociological research. The focus is on possible methods for communities monitoring using the data of online social networks. The main research question is whether it is possible to investigate the real territorial characteristics of communities on the basis of their representations in a virtual environment? How accurately does the information received from the virtual space reflect the real territorial aspects of the chosen community? The article provides an overview of possible methods for assessing, analyzing, forecasting and monitoring the territorial characteristics of communities based on their representations in a virtual environment. The virtual environment in this study is the social networking service "VKontakte" popular in Russia. An example of using the developed software tool to analyze the communities' territoriality based on "VKontakte" is given.

Keywords: Virtual communities · Information monitoring · Territoriality Online social network "VKontakte"

1 Introduction

Analysis of online social networks can be viewed as a relatively new and effective tool for sociological research. Under the term "community" in this paper we mean a group of interacting participants sharing common interests, activities, intentions, beliefs, resources, preferences, needs, risks and other factors. Communities can exist both in the real world and have their own "reflections" in online social networks, i.e. in the virtual world. Real communities can be formed initially, and then can be formed corresponding to them virtual communities. Or maybe the opposite: initially, there are virtual communities, and then appear real ones. In any case, to date, virtual communities have a significant impact on real communities, and hence on the real world.

The study of real communities through their virtual "reflections" allows you to automate the big data' processing and to improve the speed of data acquisition.

Communities can be "good" or "bad". Examples of a "bad" community include extremist groups, gangs, and so on. An example of a "good" community are animal welfare groups, care for the elderly and other socially important projects. It is possible to introduce a definition of the community "goodness" based on fuzzy logic. This

R. Silhavy et al. (Eds.): CoMeSySo 2018, AISC 859, pp. 92–101, 2019.
https://doi.org/10.1007/978-3-030-00211-4_10

characteristic can be determined with respect to a certain subject - the state, society, police, a specific person, etc.

In this context, the task of information monitoring of communities is important for the following reasons. In case the community is "bad", we could monitor possible threats and, thus, increase security. Otherwise, if the community is "good", then we could promote community growth by conducting the agitation and propaganda, monitor the territory of distribution, etc.

The aim of this work is to identify methods suitable for the analysis of communities based on online social networking data, as well as the creation of basic framework software for studying the territoriality of communities.

2 Communities Monitoring and Analysis: Methods and Applications

One of the first works considering studying of computer-based social networks as peculiar social phenomena is "Computer Networks as Social Networks" by Wellman [1]. Author states that 'When computer networks link people as well as machines, they become social networks'. Such computer-supported social networks (CSSNs) which are currently referred to as social networking services (SNSs) have become important bases of new kinds of human activities - virtual communities, computer-supported cooperative work, and telework.

There are a lot of techniques and applications based on social network analysis: Recommendation Systems, Link Analysis, Expert Identification, Influence Propagation, Trust & Distrust Relationship Prediction, Opinion Mining, Mood Analysis and Community Detection [14–16].

There are many works devoted to monitoring and analysis of virtual communities. In these works, different research goals are pursued, and different methods are used to achieve them. In this section we will consider several methods and applications that are close in scope to our research.

2.1 Community Mining

A network community refers to a group of network nodes within which the links connecting them are dense but between which they are sparse. A network community mining problem (NCMP for short) can be stated as the problem of finding all such communities from a given network. In [12, 17], authors have introduced the network community mining problems, and discussed various approaches to tackling them. Most networks, such as social media, blogs, and co-authorship networks, are dynamic as they tend to evolve gradually, due to frequent changes in the activity and interaction of their individuals. In [13], authors present a framework encompasses community matching algorithm and an event detection model to capture all of the possible events that occur for communities. This includes tracing the formation, survival and dissolution of communities as well as identifying meta communities, series of similar communities at different snapshots, for any dynamic social network. Finding patterns of social interaction within a dynamic network has a wide-range of applications such as disease modeling and information transmission.

In [24] addresses the interactive collaboration in virtual communities for Open Source Software (OSS) development projects. In particular, Social Network Analysis (SNA) techniques will be used to identify those members playing a middle-man role among other community members.

Other authors [26] aims at understanding virtual communities of learning in terms of dynamics, types of knowledge shared by participants, and network characteristics such as size, relationships, density, and centrality of participants. They looks at the relationships between these aspects and the evolution of communities of learning.

With the increasing trend of online social networks in different domains, social network analysis has recently become the center of research. Online Social Networks (OSNs or SNS) have fetched the interest of researchers for their analysis of usage as well as detection of abnormal activities. Anomalous activities in social networks represent unusual and illegal activities exhibiting different behaviors than others present in the same structure. The paper [18] discusses different types of anomalies and their novel categorization based on various characteristics. A review of number of techniques for preventing and detecting anomalies along with underlying assumptions and reasons for the presence of such anomalies is covered in the paper [18].

SNS generate big data (profiles, links, and content are created). Analyzing these data, one can get a lot of useful information on different communities and discussions, and on each user individually [2, 3]. A great interest in SNSs is experienced by various commercial organizations that use them as an instrument of interaction with the audience. Using specialized services, companies analyze information about users, their activities and personalize offers for individual segments of their target audience, thereby increasing the conversion and reducing the cost of the advertising campaign [4].

2.2 Semantic Social Network Analysis

Social Network Analysis provides graph algorithms to characterize the structure of social networks, strategic positions in these networks, specific sub-networks and decompositions of people and activities [9]. Increasingly popular web 2.0 sites are forming huge social network. Classical methods from social network analysis have been applied to such online networks. Thus, in addition to the established term Social Network Analysis, today the direction of the Semantic Social Network Analysis [11], combining the technologies of the semantic web and the classical Social Network Analysis, is being developed to improve the analysis of the SNS.

The study of extremist groups and their interaction is a crucial task in order to maintain homeland security and peace. The work [23] addresses the topic-based community key members extraction problem, for which authors' method combines both text mining and social network analysis techniques.

As example of a major research project in the field of the semantic analysis of social networking data, it is necessary to note the monitoring system "Krybrum" [10] by Ashmanov and Partners, which provides a wide coverage of online resources: the system tracks over 800,000 Internet sources, about 150,000,000 individual authors, more than 20,000 online media, collects 100% of messages posted on the Twitter microblogging platform in Russian, as well as in VK.com and Instagram. Cribrum uses its own unique technologies for collecting, storing and analyzing users' big data, searching for the morphology of the Russian language, as well as automatically highlighting the emotional coloring of statements.

2.3 Spatial Social Network Analysis

To date, in social networking research, much attention has been paid to the geospatial data of virtual communities and its members. This is evidenced by the emergence of Location-based social networks, such as Brightkite, Foursquare, Gowalla. Besides this, some social media messages contain spatial information in the form of geotags, user profiles, or place names in the messages, which can be used to create spatial social networks [19]. Spatial social networks transform social networks into maps and information landscapes using real-world locations or abstract coordinate systems. After the transformation, scientists can conduct space–time analyses, understand the diffusion of information, and identify spatial cluster patterns of network elements. In [7] poses the important question "Can we develop a visualization environment to incorporate social network graphics with geo-visualization methods to not only reveal the social networks' spatio-temporal characteristics but also keep the features in traditional social node-link diagram?".

The concept of social position and the associated technique of structural equivalence in social network analysis are explored as a means to integrate two different kinds of embeddedness: relative location in geographic space and structural position in network space. The presence of crime is one of the major challenges for societies all over the World, especially in metropolitan areas. Using spatialized network data, in [20] the authors compare the geography of rivalry relations that connect territorially based criminal street gangs in a section of Los Angeles with a geography of the location of gang-related violence. In [22] authors find that Social Media works exceptionally well for description of certain crime types and thus is also likely to enhance the accuracy of delinquency prediction.

The potential of VKontakte (VK), the Russian equivalent of Facebook, as a data source is now acknowledged in educational research, but little is known about the reliability of data obtained from this social network and about its sampling bias. Authors [5] investigate the reliability of VK data, using the examples of a secondary school (766 students) and a university (15,757 students). They describe the procedure of matching VK profiles to real students. Authors found that the structure of "virtual" social relationships reproduced both the socio-demographic division of students into grades, years of study or majors, and the spatial division into different school buildings or university campuses. Their study contributes to the understanding of how reliable data from this SNS is, how its accuracy can be improved, and how it can be used in educational research.

Another work [6] focuses on the perspectives of the practical usage of the Russian Federation Virtual Population project data. The project represents a new database relative to Vkontakte social network. Data set relative to the network structure can be found on the website, as well as an interactive map builder that summarizes the data in a map and provides different geographical, temporal and demographical filters. The website disposes a large data set that allows studying Vkontakte audience's profile taking into account such parameters as sex, age, education and migratory behaviour. The website boasts its body of data on the users' virtual friendship that seems to be the first large-scale research on whether virtual friendship relations are territorially dependent. The database seems to be useful for studying not only virtual space but also offline social processes like youth migration in microlevel, serving in this case as an implicit data source.

2.4 Integration of Semantic and Spatial Social Network Analysis Methods

A number of works concern different approaches and applications of social media data analysis including ones combining geospatial and semantic analysis.

The work [8] has provided examples of typical geovisualization capabilities for social media data. It highlighted the tight coupling of spatial and social analysis that is needed to fully exploit the geosocial content of social media. Focusing only on the visualization of buildings and infrastructure no longer sufficient, as new social media sources require the visualization of abstract concepts like the flow of information in a society, contextual information associated with places, and the emergence of communities. The joint visualization of the spatial and thematic analysis (e.g., in terms of keywords, sentiment) is becoming a standard approach in today's social media geovisualization systems (e.g., SensePlace2, GeoSocial Gauge).

Social media analytics has become prominent in natural disaster management. In spite of a large variety of metadata fields in social media data, in [21] four dimensions (i.e. space, time, content and network) have been given particular attention for mining useful information to gain situational awareness and improve disaster response.

Mining the social networking service Twitter for geolocated messages provides a rich database of information on people's thoughts and sentiments about myriad topics, like public health. In [25] uses a data-mining framework with a use of social media data retrieval and sentiment analysis to understand how geolocated tweets can be used to explore the prevalence of healthy and unhealthy food across the contiguous United States.

3 Software for Community's Territoriality Analysis

VKontakte (VK) is a social networking service popular in Russia. It provides a multilanguage interface but its main auditory is Russian-speaking users. As of March 2018, more than 97 million people a month use Vkontakte. It has rating of the most popular website in Russia. Addition to Russia VK ranked as the most popular social networking website on postSoviet space. VK is actively used by people in Belarus, Kazakhstan, Estonia, Kyrgyzstan, Moldova, Ukraine, Latvia. Interface of the site is available in 83 languages.

As with most social networks VK allows users to message each other publicly or privately, to create groups, public pages and events, share and tag images, audio and video, and to play browser-based games. Most essential features of the SNS are:

Messaging. VK Private Messages can be exchanged between groups of 1 to 30 people. Each message may contain up to 10 attachments: photos, videos, audio files, maps (an embedded map with a manually placed marker), and documents.

- News. VK users can post on their profile' walls, each post may contain up to 10 attachments.
- Communities. VK features three types of communities: Groups, Public pages, Events. The most popular communities type is Groups. They are better suited for decentralized communities such as discussion-boards, wiki-style articles, editable by all members, etc.
- «Like» buttons. VK like buttons for posts, comments, media and external sites operate similar «like» in Facebook. Additional there is second button named as «share with friends». It is used to share an item (make a repost) on users walls or to send it as private message to friends.

Additionally, VK features an advanced search engine that allows complex queries for finding friends, as well as a real-time news search.

VK uses a rather extensive and sophisticated data model, the most important components of which in the context of our task are:

- «User profile» - contains user's information such as Name, date of birth, hometown, etc.
- «Friend list» - contains number of VK users, which agreed to be a friend to the user.
- «Community list» - contains number of VK communities (groups) which user selected to be a favorite.
- «Post» - it is a combination of text, images, audio, video, which share on the users wall.
- «Like» - is a sign, which user places on a particular wall items to mark own interest to them.
- «Comment» - is a special post linked to another post and being contained a user answer to it.
- «Favorite group» - is a community of users by some certain interests.

Ordinary VK users get a necessary functionality via web-interface (vk.com) or via mobile applications based on Android, IOS or Windows Mobile operating systems. For more peculiar tasks such as data collection and analysis an application programming interface (API) should be used. API provides access for all features of the social networking service and allows to organize high-speed automated processing of big data stored within the SNS.

One of the significant tasks of identity analysis based on SNA is an identification of home region (territory) for a given user. Further, possible approaches to solving that task are discussed on the basis of information available in the VK. Users of the VKontakte explicitly or implicitly evince their territoriality in different ways:

Explicit declarations in the user profile fields like 'Hometown', 'City', etc. However, some users do not fill these fields due to varying reasons.

'Friends List': Each friend has own territoriality based on his profile or activity analysis. We assume that territorialities of friends correlate with the territoriality of the user.

'Communities List': Each Community ('Group' in other terms) may have territoriality identifiers, which can be obtained from different sources: group description, posts on the group wall, as well as from group members' territorialities. As in the previous case, we assume that territoriality of the user correlates with territorialities of groups, which he belongs.

Content generated by the user («Post»/«Comment»/«Like»/Photo). Territoriality identifiers can be found both in the content of the post itself, as well as in the external context of the post. In the first case, some semantic analysis of the content should be being done; appropriate technique to be used depends on the type of the data of the post. For example, some photos have geotag that explicitly defines locality and can be used in the analysis. In the second case, the post's author territoriality may be obtained from its likers and reposters and other information associated with the post.

Particular technologies for identifying the user's home region, based on one or another source of territoriality identifiers, would differ in the speed and complexity of the algorithms, as well as in the reliability of the results obtained. The fastest approach, based on explicit territoriality identifier in the user profile, would be at the same time as the least reliable, since users, changing the home region do not always reflect these changes in their profile. On the other side, probably most reliable approach based on the content and context analysis of user's posts (analyzing a user's activity) would be most complicated and slow.

To illustrate this, we conducted brief experiment implementing 'direct' approach to identify territoriality of the user. In this work, we used a programming language Python 3.x with special library for VK. For the experiment, a community group «Murmansk is the Capital of the Arctic» (https://vk.com/club14273164) was selected. The group has more than 70 thousand subscribers. Obviously, it has explicit territoriality linked to Murmansk region (target region) but only 66.5% of members pointed Murmansk as a home region, while 23.4% pointed another region as a home one. Indirect approach based on the group memberships analysis of users of the considered community gave even worse results: only 30.2% of the group display Murmansk territoriality while 35.5% show other regions as a home one. It indicates that direct and not complicated indirect approaches to analyze real social entities on the basis of observations of 'virtual world' are not reliable enough and need to be carefully verified and justified with experiments.

In addition to complicated analysis of a content and a context of the user's posts, the promising way to improve accuracy of territoriality analysis is taking into account an actual activity of the SNS users. This will eliminate the 'information noise' associated with 'dead' users who have a profile but are weak in the network. Particularly, this will allow to introduce a measure of belonging of the user to the group, which will improve reliability of the indirect method of indicating a territoriality implemented in this work. Might be the best approach should combine different techniques and consider all available data: profile attributes, user activity indicators, semantic of the content and the context.

4 Conclusion

In this paper, we examined the possibilities of communities monitoring based on online social networks. It can be concluded that the most socially important sphere of virtual communities' monitoring and analysis application is to provide all types of security: environmental, industrial, national, economic, societal, food, military, information, emergency and disaster management, etc. For future practical use in the territoriality monitoring tasks, the corresponding framework is proposed.

Online social networks open up new opportunities for exploring the territorial aspects of different communities. To do this, they provide a suitable software interface for automate the acquisition of online social network data. Most modern communities have their own web pages, where they publish information about themselves, members (subscribers) and their activities. This information is open. Its publication and wide distribution among users of a social network is the goal of community activities in the virtual environment. The study of published information does not contradict either official laws or moral and ethical principles. It was this open information that was used in this study to analyze the territorial characteristics of community activities. It would be very interesting for the researcher to observe the dynamics of the "good" community characteristics projected on the map. This allows us to assess the scale of the positive impact of the studied community on the rest of society, to see the potential development of this influence, to predict it in space and time. Communities that we called "bad" can also sometimes publish information about themselves. Monitoring of such communities will increase the overall security of the society, predict and prevent the negative spread of this influence in space and time. On the other hand, the lack of open information about the community chosen for the research can only talk about its closeness to everyone around. The ability to attribute the community to "bad" or "good" in this study is missing. The results of this study are intended for use by the community of people to ensure their development in a safe environment, which at the moment is very much associated with the virtual space. However, the virtualization of any activity is only an intermediate stage. To get the result of the activity, it needs to connect with reality - with real people, with real material resources, with real geographic regions. In this study, a tentative step has been taken towards realizing this fact. For this purpose, the corresponding methodological and instrumental support has been established, on the basis of which further research on this topic will be developed.

References

1. Wellman, B., Salaff, J., Dimitrova, D., Garton, L., Gulia, M., Haythornthwaite, C: Computer networks as social networks: collaborative work, telework, and virtual community. Ann. Rev. Sociol. **22,** 213–238 (1996). https://doi.org/10.1146/annurev.soc.22.1.213
2. Tan, W., Blake, M.B., Saleh, I., Dustdar, S.: Social-network-sourced big data analytics. IEEE Internet Comput. **17**(5), 62–69 (2013)
3. Khotilin, M.I., Blagov, A.V.: Visualization and cluster analysis of social networks. In: CEUR Workshop Proceedings, vol. 1638, pp. 843–850 (2016)

4. Bakaev, V.A., Blessing, A.V.: Analysis of profiles in social networks. In: Information Technologies and Nanotechnologies, pp. 1860–1863. The Science of Data (2017)
5. Smirnov, I., Sivak, E., Kozmina, Y.: In search of lost profiles: the reliability of vkontakte data and its importance in educational research, (4), 106–122. Voprosy obrazovaniya/Educational Studies Moscow (2016)
6. Zamyatina, N.Y., Yashunsky, A.D.: Virtual geography of virtual population. Monit. Public Opin.: Econ. Soc. Chang. (1), 117–137 (2018). https://doi.org/10.14515/monitoring.2018.1.07
7. Ma, D.: Thesis2012: visualization of social media data: mapping changing social network. Enschede, The Netherlands (2012)
8. Croitoru, A., Crooks, A., Radzikowski, J., Stefanidis, A.: Geovisualization of social media. In: The International Encyclopedia of Geography, pp. 1–17 (2017). https://doi.org/10.1002/9781118786352.wbieg0605
9. Erétéo, G., Buffa, M., Gandon, F., Corby, O.: Analysis of a real online social network using semantic web frameworks. In: Bernstein, A., et al. (eds.) The Semantic Web - ISWC 2009. LNCS, vol. 5823. Springer, Berlin (2009). https://doi.org/10.1007/978–3-642-04930-9_12
10. Cribrum Monitoring System. https://www.ashmanov.com/tech-i-services/kribrum/. Accessed 21 Nov 2018
11. Nakatsuji, M., Zhang, Q., Lu, X., Makni, B., Hendler, J.A.: Semantic social network analysis by cross-domain tensor factorization. IEEE Trans. Comput. Soc. Syst. **4**(4), 207–217 (2017). https://doi.org/10.1109/TCSS.2017.2732685
12. Yang, B., Liu, D., Liu, J.: Discovering communities from social networks: methodologies and applications. In: Furht, B. (ed.) Handbook of Social Network Technologies and Applications, pp. 331–346. Springer, Boston (2010). https://doi.org/10.1007/978-1-4419-7142-5_16
13. Takaffoli, M., Sangi, F., Fagnan, J., Zäıane, O.R.: Community evolution mining in dynamic social networks. Procedia Soc. Behav. Sci. **22**, 49–58 (2011). https://doi.org/10.1016/j.sbspro.2011.07.055
14. Yuan, T., Cheng, J., Zhang, X., Liu, Q., Lu, H.: How friends affect user behaviors? An exploration of social relation analysis for recommendation. Knowl. Based Syst. **88**, 70–84 (2015)
15. Anjaria, M., Mohana, R., Guddeti, R.: A novel sentiment analysis of social networks using supervised learning. Soc. Netw. Anal. Min. **4**, 181 (2014)
16. Zhang, J., Fang, Z., Chen, W., Tang, J.: Diffusion of following links in microblogging network. IEEE Trans. Knowl. Data Eng. **27**(8), 2093–2106 (2015)
17. ChitraDevi, J., Poovammal, E.: An analysis of overlapping community detection algorithms in social networks. Procedia Comput. Sci. **89**, 349–358 (2016). https://doi.org/10.1016/j.procs.2016.06.082
18. Kaur, R., Singh, S.: A survey of data mining and social network analysis based anomaly detection techniques. Egypt. Inform. J. **17**, 199–216 (2016)
19. Tsou, M.-H., Yang, J.-A.: Spatial social networks. In: International Encyclopedia of Geography: People, the Earth, Environment and Technology, 06 Mar 2017. https://doi.org/10.1002/9781118786352.wbieg0904
20. Radil, S.M., Flint, C., Tita, G.E.: Spatializing social networks: using social network analysis to investigate geographies of gang rivalry, territoriality and violence in Los Angeles. Ann. Assoc. Am. Geogr. **100**(2), 307–326 (2010). https://doi.org/10.1080/00045600903550428
21. Zheye, W., Xinyue, Y.: Social media analytics for natural disaster management. Int. J. Geogr. Inf. Sci. (2017). https://doi.org/10.1080/13658816.2017.1367003
22. Bendler, J., Ratku, A., Neumann, D.: Crime mapping through geo-spatial social media activity. In: Thirty Fifth International Conference on Information Systems, Auckland (2014)

23. L'Huillier, G., Alvarez, H., Ríos, S.A., Aguilera, F.: Topic-based social network analysis for virtual communities of interests in the dark web. In: ACM SIGKDD Explorations Newsletter, vol. 12, no. 2 (2010). https://doi.org/10.1145/1964897.1964917
24. Toral, S.L., Martínez-Torres, M.R., Barrero, F.: Analysis of virtual communities supporting OSS projects using social network analysis. Inf. Softw. Technol. **52**(3), 296–303 (2010)
25. Widener, M.J., Li, W.: Using geolocated Twitter data to monitor the prevalence of healthy and unhealthy food references across the US. Appl. Geogr. **54**, 189–197 (2014). https://doi.org/10.1016/j.apgeog.2014.07.017
26. Fontainha, E., Martins, J.T., Vasconcelos, A.C.: Network analysis of a virtual community of learning of economics educators. Inf. Res. Int. Electron. J. **20**(1) (2015)

Automated Extraction of Deontological Statements Through a Multilevel Analysis of Legal Acts

V. V. Dikovitsky(✉) and M. G. Shishaev

Institute for Informatics and Mathematical Modeling - Subdivision of the Federal Research Centre "Kola Science Centre of the Russian Academy of Science", 24A, Fersman St., Apatity, Murmansk Region 184209, Russia
dikovitsky@gmail.com, shishaev@iimm.ru

Abstract. Currently, large amounts of data are available in text form. This makes the automatic semantic analysis of natural language texts a topical task. This very complex task can be rather effectively solved in case when consideration is limited by certain specific kind of texts. The paper describes the developed technology of extraction of deontological statements from legal acts through a multilevel analysis of text documents combining linguistic and semantic methods. The technology is tested on the example of Russian legislative documents and shows its potential effectiveness. Experiments have shown that certain types of Universal Dependencies relations with high accuracy identify the components of deontological statements.

Keywords: Content analysis · Deontological statements · TensorFlow SyntaxNet · Universal Dependencies

1 Introduction

With increasing number and volume of documents used to solve various kinds of applied problems, development of automated text analysis systems becomes actual. Approaches to automatic text analysis and constructing a formal representation of it have been undertaken for a long time. There are many works [1–4] are devoted to modeling of reasoning and formalization of statements. Analysis and formal representation of legal acts is considered in works [5–8]. Also nonclassical logics are used to implement logical inference on formalized knowledge, for example deontic logic, in which norm statements and deontic modality (such as "necessarily", "forbidden", "allowed", "indifferent") are studied [2, 9]. In work [2], the norm is understood as prescript addressed to social agents:

(a) Actions to be performed or required situations such that, according to the law (rule, ideal), must necessarily be achieved as a result of agents fulfilling the requirements addressed to them;
(b) Actions that must does not occur, also as a result of agents fulfilling the requirements addressed to them, or, conversely, abstaining from performance;

R. Silhavy et al. (Eds.): CoMeSySo 2018, AISC 859, pp. 102–110, 2019.
https://doi.org/10.1007/978-3-030-00211-4_11

(c) Actions that are permissible under certain conditions, or permissible results of agents' actions, as well as the empowerment of someone to carry out any actions.

In this work we made experimental evaluation of possibility to use modern means of semantic analysis of texts to extract knowledge from legislative documents. As an example, we used Russian migration legislation which processed with the multi-level analysis technique to automatically extract deontological statements. Representation of knowledge as a system of deontological statements is considered as a future basis for systems of artificial intelligence, automated text analysis, expert systems and other systems that use logical inference.

2 The Technology of Multi-level Text Analysis

The statement in mathematical logic is a sentence expressing a proposition. The normative expression goes beyond truth or falsity of statement. In normative documents, the concept of norm is prescriptive. The expression of the norm is prescriptive statements, which, unlike descriptive ones, do not describe actions or situations, but prescribe what actions are necessary, allowed or not allowed to perform, and which situation is mandatory or permissible, and which is forbidden.

The implicit nature of normative statements adds complexity of the automated analysis. The complexness is consists in references to other normative acts, which is not always available for text analysis systems. The explicit statement is completely described in a legislative act, understanding of which does not require additional knowledge. Therefore, the integrity of statement is important for the automatic analysis of normative statements.

Approaches to the formation of norm systems can be divided into 2 types. The first is use of a global formal system describing general rules, branching conditions of logical inference, hierarchy of constraints. This approach is accurate, but requires expert knowledge, both at the stage of the logical system formation, and at the stage of operation of such a system. An example of a campaign for the formation of an international group of experts in the field of migration legislation is presented by the organization International migration law, which is the international legal framework governing migration, is not covered by any one legal instrument or norm. Instead, International migration law is an umbrella term covering a variety of principles and rules that together regulate the international obligations of States with regard to migrants. Such broad range of principles and rules belong to numerous branches of international law such as human rights law, humanitarian law, labour law, refugee law, consular law and maritime law. International Organization for Migration works to increase knowledge and acceptance of the legal instruments that protect migrants' rights, as well as the ratification and implementation status of these instruments. It also assists States in developing migration policies and legislation that conform to International migration law in order to manage migration more effectively and in a manner consistent with international law. In order to more effectively carry out its work on International migration law, International Organization for Migration has established the International Migration Law Unit [10]. Second approach use only explicit statements and automatically integrate into a single formal system. The result is a system of objects and statements of the subject area, where each concept is compared with a constraints system expressed by logical statements.

A deontological statement contains 4 obligatory elements: the subject, the object, the content and the character. The content of a rule is a description of the action or situation governed by this rule. The character of a norm is a type of requirement expressed by it. The content and character of a norm is called the disposition of the norm. The object of the norm is a condition of its application. This condition is describe situation, people or objects in presence of which this norm is valid. The subject of norm is its addressee or agent. This is person or group of people. The rights of the object and subject of norm are combined in hypothesis of norm.

To determine all four elements of the deontological statement, a semantic text analyzer was used, which is based on SyntaxNet, a natural-language understanding library for TensorFlow [11]. SyntaxNet library allows to define dependencies among words within the sentence in Universal Dependencies [12] notation, that open an opportunity to analyze and integrate knowledge from texts in different languages.

The technology for text analysis includes several stages: graphematic, morphological, syntactic and semantic analysis. Results of each level are used by the next level of analysis as input. The purpose of the morphological analysis is to determine the morphological characteristics of the word and its initial form. The purpose of syntactic analysis is to determine the syntactic dependence of words in a sentence. In connection with the presence in the Russian language of a large number of syntactically homonymous constructions the procedure for automated text parsing based on syntax is insufficient to determine the dependencies between the concepts of a sentence. The complexity increases exponentially with an increase in the number of words in the sentence and the number of used rules. The semantic stage determines the formal representation of the meaning of words and constructions composing the input text. The formation of the semantic domain model (SDM) as a structure of weighted semantic relations on the basis of a collection of documents allows to account for and store the context of concepts, and take into account the various forms of syntax transmission, and solve the problem of the equivalence of words. SDM makes it possible to implement procedures for extracting and storing the contexts of words, partially solving the problem of compatibility of new information with knowledge already accumulated, and also to reveal contradictions in the semantic images of documents, in case new information contradicts the accumulated information. The procedure for formation of the SDM is presented below.

First the text is divided into sentences and subjected to a graphematical analysis and lemmatization. For the syntactic analysis and definition of the morphological characteristics of words, we use the grammatical dictionary of the Russian language [13], thesaurus WordNet [14], and also based on the tools for neural networks TensorFlow SyntaxNet library. A feature of the application of neural networks is the ability to analyze the morphology and syntax for words that are not presented in the thesaurus. TensorFlow is an open-source software library. The library functions at the level of specifying the architecture of the neural network and its parameters. The data in TensorFlow are represented as multidimensional data sets with variable size - tensors. The set of morphological features defined by SyntaxNet, grammatical categories, and types of dependencies are specified in the Universal Dependency notation. Universal Dependencies (UD) is a framework for cross-linguistically consistent grammatical annotation and an open community effort with over 200 contributors producing more than 100 treebanks in over 60 languages. Sentences are sent at the input of the neural network. Words of sentences are converted into a vector by the Word2Vec library. In the preliminary training phase Word2Vec [15] get the text corpus and provides vocabulary vectors as output. Vector word representations allow us to calculate the semantic closeness between words. Since Word2Vec algorithms are based on training a neural network, in order to achieve effective work, it is necessary to use large text corpus for training. Pre prepared vectors obtained on the Google News [16] dataset part are available. The model contains vectors for 3 million words and phrases. Phrases were obtained using the skip-gram approach described in [17]. Further sentences in vector form are submitted to the input layer of the neural network which implemented in TensorFlow and trained on the Universal Dependences corpus. The Russian text corpuses in the Universal Dependencies are represented by the converted SinTagRus [18] and Google Russian Treebank [19]. The result of SyntaxNet is a dependencies tree of sentence and morphological characteristics of words. The result of text analysis is a semantic image of document. The semantic image of the document is a weighted semantic net:

$$D = \{C^D, L^D\}, C^D \subset C, L^D \subset L \tag{1}$$

where C^D – set of concepts in document, L^D – set of relation.

SDM is formed as a result of the integration of the semantic images of documents. SDM formally is represented as an n-ary weighted semantic network:

$$KB = \{C, L, Tp\}, L = \{l\}, l = \langle c_i, c_j, tp, w \rangle,\ c_i, c_j \in C, tp \in Tp \tag{2}$$

where C – a concepts set, L – a relations set, w – an importance parameter of relation, Tp – a relations types set. The weight of relationship characterizes the significance of the semantic relation between concepts, determined on the basis of statistics of the occurrence of concepts. The structure of semantic analysis service is shown on the Fig. 1.

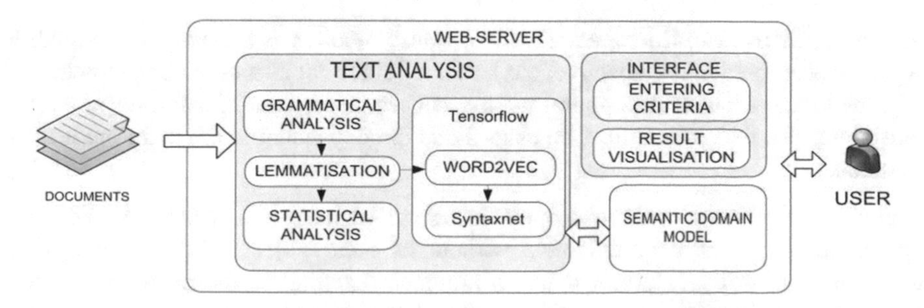

Fig. 1. The structure of the semantic analysis service

3 Experiment to Determine of Deontological Statements

We analyzed Federal laws regulating the residence and work of foreign nationals on the Russian Federation territory (Federal laws №. 109–115). The purpose of the experiment is to identify statements from the normative act text that contain modal statements that prohibit, permit or express the acceptability of an action.

To find the deontological statements, some types of syntactic dependencies and stable syntactic constructions which can be obtained by automatic parsing at the stage of multi-level text analysis, was expertly defined and used. The following types of Universal Dependencies [12] were used:

Negation Modifier. The negation modifier is the relation between a negation word and the word it modifies. It is used both for predicate negation (canonically, not) and nominal negation (canonically no). Dependents labeled "neg" in the current treebank are the following (in various lowercase/uppercase forms): n, n't, neither, never, no, non, not, nt, t. The negation modifier is the relation between the negation word "не" and the word it modifies. Negation in Russian is most of the time expressed using a bound morpheme (the prefix "не"). Occurrences of the morpheme as a separate word are rare in comparison to other languages, yet they exist.

Determiner. The relation determiner holds between a nominal head and its determiner. This relation is used for pronominal adjectival modifiers of noun phrases. Non-pronominal adjectives are tagged ADJ and the relation is labeled "amod".

Direct Object. The direct object of a verb phrase is the noun phrase which is the (accusative) object of the verb.

Nominal Subject. A nominal subject is a noun phrase which is the syntactic subject of a clause; in Russian, the phrase is in the Nominative Case. The governor of the nominal subject relation might not always be a verb: when the verb is a copular verb, the root of the clause is the complement of the copular verb, which can be an adjective or noun.

Passive Nominal Subject. A passive nominal subject is a noun phrase which is the syntactic subject of a passive clause.

Adverbial Modifier. An adverbial modifier of a word is a (non-clausal) adverb or adverbial phrase that serves to modify a predicate or a modifier word.

Adverbial Clause Modifier. An adverbial clause modifier is a clause which modifies a verb or other predicate (adjective, etc.), as a modifier not as a core complement. This includes things such as a temporal clause, consequence, conditional clause, purpose clause, etc. The dependent must be clausal and the dependent is the main predicate of the clause.

Open Clausal Complement. An open clausal complement of a verb or an adjective is a predicative or clausal complement without its own subject. The reference of the subject is necessarily determined by an argument external to the open clausal complement (normally by the object of the next higher clause, if there is one, or else by the subject of the next higher clause. These complements are always non-finite, and they are complements (arguments of the higher verb or adjective) rather than adjuncts/modifiers, such as a purpose clause.

Unspecified Dependency. Adependency can be labeled as unspecified dependency when it is impossible to determine a more precise relation. This may be because of a weird grammatical construction, or a limitation in conversion or parsing software.

Nominal Modifier. The nominal modifier relation is used for nominal modifiers of nouns or clausal predicates. Nominal modifier is a noun functioning as a non-core (oblique) argument or adjunct.

Fragments of the semantic network are shown on the figures. A fragment of the semantic network that visualizes deontological statements "A foreign citizen has the right(russ. «право») [to be on the territory(russ. «находится») …], [to carry out activities(russ. «осуществлять») …], [to receive(russ. «получение») …]" (Fig. 2).

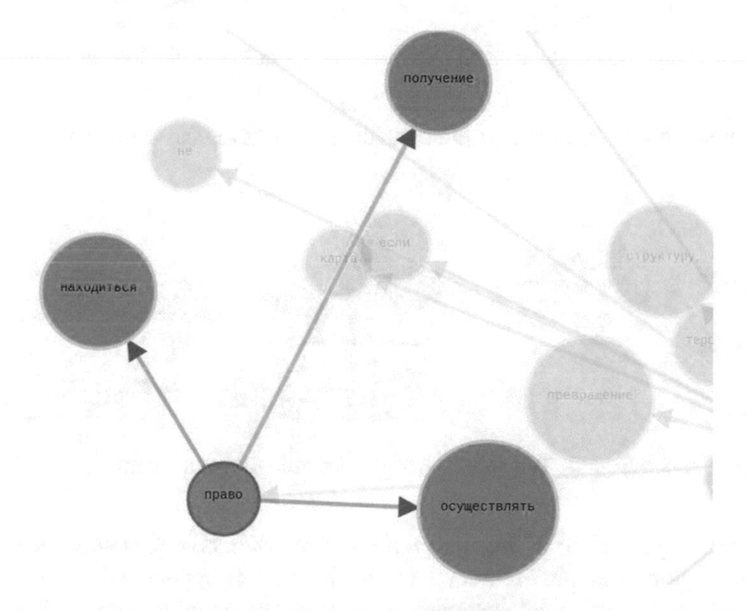

Fig. 2. A fragment of semantic network that visualizes deontological statements

Fragment of the semantic network that visualizes modal expressions "A foreign citizen must submit (russ. «предъявлять») [documents (russ. «документы»)], [card (russ. «карту») (-> migration (russ. «миграционную») (-> filled (russ. «заполненную»))]" (Fig. 3).

The difficulty of defining deontological expressions is an automatic detection of the difference between descriptive and prescriptive statements. This difference helps to distinguish the norm as a requirement or permission from the description of situation. As can be seen from the example, from the syntactic point of view, there is no difference between the descriptive and prescriptive expression of the properties of the "card" object - both properties are specified by the verbal adjectives "migration" and "filled".

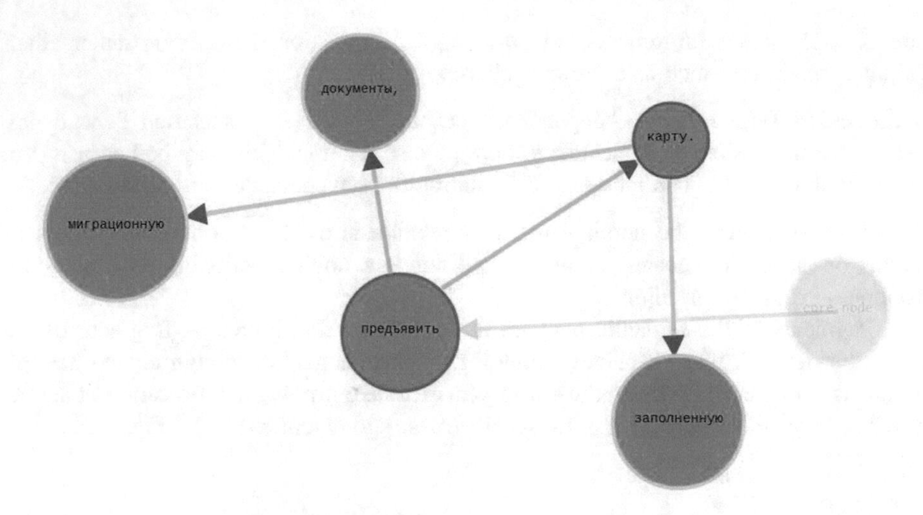

Fig. 3. Fragment of the semantic network that visualizes deontological statements

Common view of structure of the considered deontological statements is shown on Fig. 4.

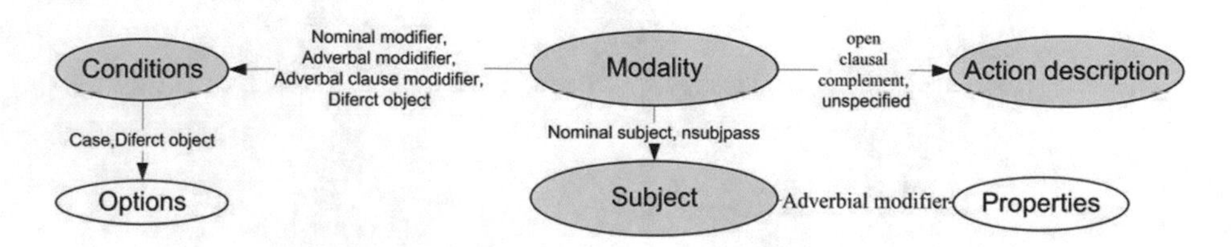

Fig. 4. The structure of the deontological statements

The resulting template of the analyzed statements can be represented as a sequence of lexemes, properties and relations in the Universal Dependencies notation. The character of the deontological statement is given by the defining verb, which is root of parse tree in each considered example. In the case of negation, the defining verb is related to the particle “not” by the Negation modifier. Examples of verbs and stable phrases that determine modality - (necessary, allowed, is required, not required, forbidden).

The description of the action is expressed by a clausal complement associated with the modality defining verb by the open clausal complement relation, or an additional syntactic construction associated with the determining verb with unspecified dependency.

The condition of norm application is expressed by a syntactic structure associated with adverbial modifier relation, or the direct object relation to the clausal complement of the defining verb.

The subject of the deontological statement is connected with the defining verb by Nominal subject relation in case when the subject is expressed by a noun. In case where the subject is represented by a verb, for example, in case of an action prohibition, the relation has the form Passive nominal subject. Also this type of relation is used to indicate the subject in sentences without a subject. The properties of the subject are defined by the relation Adverbial modifier.

The table shows statistics of syntactic role of the elements of deontological statements in sentences containing the norm, i.e. containing all four elements of statement (Table 1).

Table 1. Statistics of syntactic role of the elements of deontological statements in sentences.

Component of a deontological statement	Syntactic role in the sentence	Usage statistics, %
Subject of norm	Nominal subject	61
	Nominal subject passive	37
	Adverbial clause modifier	2
Norm action description	Open clausal complement	72
	Unspecified dependency	28
Clause of uses of norm	Nominal modifier	60
	Adverbal modifier	2
	Direct object	38

The result shows the possibility of automating search of deontological statements using the above technology. In addition, lexical complexity of the nature language is leveled by the obvious severity of normative acts language, for example, double negation does not take place in them.

In order to clarify, obtain and generalize other variants of representing syntactic constructions of deontological statements in further work, we plan to use a neural network classifier of statements, trained on an expertly prepared sample of classified deontological statements.

4 Conclusions

The resulting template of the modal expression can be used to automating search of deontological statements. The SyntaxNet library can be used to find object and subject of deontological statements. However, this is not sufficient for constructing a formal system and for carrying out a logical inference. It is necessary to develop a technique for identifying types of deontological statements, which is the subject of further researches.

As a result of the work done, we shown the possibility of automating search of deontological statements in legal acts using combination of linguistic and semantic analysis which include syntax analysis using SyntaxNet library. The subject of the future work is to provide automated classification of deontological statements types and to build a holistic formal system of limitations representing the sense of a certain set of normative acts.

Acknowledgements. The study was financially supported by RFBR project №. 16-29-12878 (ofi_m) "Development of methods for identification of dynamic models with random parameters and their application to forecasting migration in Eurasia".

References

1. Pospelov, D.A.: Modeling of reasoning. In: Experience from the Analysis of Cognitive Events (1989)
2. Alchourron, C.E., Bulygin, E.: Normative Systems. Springer, Vienna, New York (1971)
3. Martino, A.A., Socci Natali, F. (eds.): Automated Analysis of Legal Texts: Logic, Informatics, Law, North-Holland (1986)
4. Alchourron, C.E.: Philosophical foundations of deontic logic and the logic of defeasible conditionals. In: Meyer and Wieringa, pp. 43–84 (1993)
5. Allen, L.E., Saxon, C.S.: Analysis of the logical structure of legal rules by a modernized and formalized version of Hohfeld's fundamental legal conceptions (1985)
6. Mann, B.H.: The formalization of informal law: arbitration before the American Revolution. NYUL Rev. **59**, 443 (1984)
7. Francesconi, E., et al. (eds.): Semantic Processing of Legal Texts: Where the Language of Law Meets the Law of Language, vol. 6036. Springer, Cham (2010)
8. Brighi, R., Palmirani, M.: Legal text analysis of the modification provisions: a pattern oriented approach. In: Proceedings of the 12th International Conference on Artificial Intelligence and Law, pp. 238–239. ACM (2009)
9. Von Wright, G.H.: Explanation and Understanding. Cornell University Press, Ithaca (2004)
10. International Migration Law Unit. https://www.iom.int/migration-law/
11. TensorFlow. https://www.tensorflow.org/
12. Universal Dependencies. http://universaldependencies.org
13. Zaliznyak, A.A.: Grammatical dictionary of the Russian language. http://odict.ru/
14. Thesaurus of the Russian language WordNet. http://wordnet.ru/
15. Word2Vec. https://code.google.com/archive/p/word2vec/
16. Google News. https://drive.google.com/file/d/0B7XkCwpI5KDYNlNUTTlSS21pQmM/edit?usp=sharing
17. Mikolov, T., Sutskever, I., Chen, K., Corrado, G., Dean, J.: Distributed representations of words and phrases and their compositionality. In: Proceedings of NIPS (2013)
18. SinTagRus. http://www.ruscorpora.ru/search-syntax.html
19. Google Russian Treebank. https://old.datahub.io/dataset/universal-dependencies-treebank-russian

Using Decision Trees to Predict the Likelihood of High School Students Enrolling for University Studies

Zdena Dobesova(✉) and Jan Pinos

Department of Geoinformatics, Faculty of Science, Palacky University,
17. listopadu 50, 779 00 Olomouc, Czech Republic
{zdena.dobesova, jan.pinos01}@upol.cz

Abstract. This article presents the use of decision trees to identify the main factors which predict the likelihood of high school students matriculating at the Department of Geoinformatics, Palacky University in Olomouc (Czech Republic). The Department of Geoinformatics has been running a continuous and systematic information campaign about studying the fields of geoinformatics and geography within the department. In order to collect feedback about the information campaign, students who apply to study at the department are then given a questionnaire. Answers received from this questionnaire in two years, (2016 and 2017), were analyzed using decision trees that help us understand what specific type of information positively affects the likelihood of a student actually commencing studies at our department.

Keywords: Data mining · Decision tree · Questionnaire · Geoinformatics GIS · Advertisement · University applicant

1 Introduction

Universities in the Czech Republic offer numerous branches (fields) of study for students desiring a bachelor's degree. Every university has various way of advertising its study programs. One way is through face-to-face meetings engineered between a teacher/other university representative and secondary school students (and possibly their parents and friends). This method is still very important but has its limits when trying to reach higher numbers of possible applicants. Therefore, digital ways of advertising, specifically over the Internet, are used, in addition to classical advertising via printed materials. Some study fields are well known among the secondary school students (such as medicine, engineering, agriculture, law, etc.). On the other hand, newer fields such as geoinformatics have much less name recognition among secondary school students and therefore rely more on advertising in order to gain new students. Simply counting the total number of applicants to the program is not a fine-grained enough way to measure the impact of the department's measures. Therefore, at the department of Geoinformatics we decided to evaluate the information campaign in more detail.

R. Silhavy et al. (Eds.): CoMeSySo 2018, AISC 859, pp. 111–119, 2019.
https://doi.org/10.1007/978-3-030-00211-4_12

Valuable information was obtained by evaluating the information campaign using decision trees as one of the data mining techniques [1, 2]. The advertisement activities and collection of the data were described in a previous article [3]; this article also included information about secondary school students' most frequent sources of information about the department. Our current research extends the previous data set to add information about the commencement of studies, and from that extended data set, we construct a decision tree as a predictive model of whether students matriculate with us. The data set is described in Sect. 2. Section 3 is a short introduction to data mining by the decision tree. The statistical evaluation and decision tree are described in Sect. 4.

2 Public Presentations of the Study Fields and Consequent Data Collection

The delivery of information to secondary school students can be via multiple channels; Palacky University uses many of them. First, the University has extensive information about fields of study and department admission procedures on its web pages in digital form [4]. Additionally, information is presented by a representative of the Department of Geoinformatics at the special study fair Gaudeamus in Brno and Prague [5]. Third, the Open Day event at Palacky University could be considered as an opportunity to inform the students about the programs in person [4]. At this event, interested students can visit the university to speak with the teachers and collect information about the application procedure personally. At both events, the university provides printed leaflets and booklets with a list of fields of study and detailed descriptions.

Additionally, the Department of Geoinformatics at Palacky University organises several special presentations about bachelor's degrees in Geoinformatics and Geography. One presentation occurs as part of the worldwide GIS Day in November [6], then other ones are: GIS week in April, presentations at secondary school and excursions to the department to see laboratories and equipment. For all presentations, the printed booklets about Geoinformatics study are used.

Information about Geoinformatics is also spread indirectly by teachers, friends, schoolmates, siblings and parents. Other information channels like a radio broadcast, TV news, the Teachers' Newspaper (www.ucitelskenoviny.cz) could also be considered. A detailed description of the information channels, events and printed information materials is presented in the article "Discovering association rules of information dissemination about Geoinformatics university study" [3].

This research evaluates the influence of each information activity on the likelihood of the student actually starting the study of Geoinformatics. The research task is to measure the impact of each form of presentation: personal presentation, information on the internet, leaflets and other channels of advertisement, on an applicant's reaching the final decision to study Geoinformatics. Moreover, the event 'GIS weekend' is evaluated in this article. GIS weekend is solely organized for real applicants. The submission period for applications ends in February each year. Subsequently, all applicants are invited in April to visit the Department of Geoinformatics at the university. The GIS weekend event covers the detailed overview of study subjects, presentation of computer classrooms, and opportunities for studying abroad under the ERASMUS+ program.

Moreover, special laboratories are presented. There are eye-tracking laboratory (for cartographic testing), 3D modelling and printing laboratory (creation of earth surface models) and laboratory for the tangible creation of elevation models at the department. Students also create their first digital map on the computer. The students also practice taking an exam in mathematics and geography. Department members have assumed that the personal attendance of the department staff and the presentation of practical examples of Geoinformatics would be very influential for the final decision to actually begin studies at our department.

2.1 Source Data Matrix

The data were collected during admission periods in two years 2016 and 2017. Data from the questionnaire were gathered from applicants. The structure of the questionnaire contained 12 possible responses about the source of information [3]. Applicants check one or more sources of received information about the study:

1. Teacher of Geography at a secondary school *(TeacherG)*
2. Teacher of Information Science at a secondary school *(TeacherI)*
3. Lecture at a secondary school provided by the Department of Geoinformatics *(Lecture)*
4. Attending the GIS Day event at the Department of Geoinformatics *(GIS Day)*
5. The Open Day event at Palacky University (November and January) *(Open Day)*
6. Friends, schoolmates, siblings *(Friend)*
7. Parents and grandparents *(Parent)*
8. Gaudeamus Fair in Prague or Brno *(Gaudeamus)*
9. Leaflets about studying Geoinformatics *(Leaflets)*
10. TV and Newspapers *(TV News)*
11. Internet by your search *(Internet)*
12. Teacher Newspaper *(T News)*

Data were transformed to the matrix with binary values. Each applicant was assigned an identification number (ID) in the matrix. Information about attending the GIS weekend was manually added to the data matrix according to the attendance list.

Moreover, information about the start of the study was added into the matrix as the last column. This column is class (category or predicted) value in the prediction model. An example of input data for the construction of decision trees is shown in Table 1.

Table 1. Input data matrix (short example) for decision tree.

ID	TeacherG	Lecture	Open day	Internet	Friend	…	Gaudeamus	GIS weekend	**Start study**
1	1	1	1	0	0	…	0	1	1
2	0	0	0	1	0	…	1	1	0
3	0	0	1	0	1	…	0	0	1
4	0	0	0	0	0	…	1	1	0
5	0	0	0	1	0	…	0	0	0

3 Decision Tree Method and Prediction

The input data for classification task is a collection of records. Each record is characterised by tuples ($x_1, x_2, \ldots, x_n, \mathbf{y}$) where x_i is input data set, and ***y*** is a special attribute, designated as the predicted class (category or target attribute) [1, 2]. In case of the presented situation, the predicted class is 1/0 (yes/no) for the start of the study of Geoinformatics. The data could be used for the design of classification/prediction model that predicts the class for unknown records of data. A decision tree is a typical example of supervised learning [7, 8] where the aim is to predict resulting value.

The construction of optimal decision tree is computationally infeasible because of the exponential size of search space. Efficient algorithms have been developed to induce a reasonably accurate tree in a reasonable amount of time. The Hunt`s algorithm is a greedy strategy that grows a decision tree by making a series of locally optimal decisions of which attribute to use for partitioning data [1, 2]. Hunt's algorithm is a base foundation for algorithm ID3, C4.5 and CART. Decision trees, especially smaller/sized trees are relatively easy to interpret. The interpretation is by simple rule with a condition like "*If conditions Then.... Else*". The presence of redundant attributes does not adversely affect the accuracy of decision trees. The decision tree partitions the data set into the smaller group of records that are more homogenous than the group in a higher level of the tree.

The software WEKA v 3.8 was used for the construction of the decision tree. The algorithm J48 is accessible in WEKA. J48 is a Java implementation for generating a pruned or unpruned C4.5 decision tree. C4.5 was developed by Quinlan [9]. At each node of the tree, C4.5 chooses the attribute of the data that most effectively splits its set of samples into subsets enriched in one class or the other. The splitting criterion is the normalised information gain (difference in entropy). The attribute with the highest normalised information gain is chosen as a node in the tree to make the decision. The C4.5 algorithm then recurs on the smaller group of records [7, 8]. The outputs of the tree (leaves) are categorical values.

4 Statistical Evaluation and Data Mining

The survey collects data from the years 2016 and 2017. In 2016, the number of collected applicant questionnaires was 46 and 55 in the year 2017 (total 101). All applicants were admitted in both years. From 48 students who applied, only 22 matriculated in 2016 and 24 did not start the study (more than half of them). The proportion in 2017 was better, 35 students matriculated, and 20 students did not. The overview is displayed in Fig. 1.

Additionally, the influence of attending the GIS weekend event was statistically evaluated. Applicants can freely attend the GIS weekend in April (after submission of their application in February). The expectation is that students with a serious interest in the study of Geoinformatics are more likely to attend this excursion of the department and to prepare for the admission exams. Table 2 shows that only about half of the applicants that attended the event GIS weekend actually started the study. The

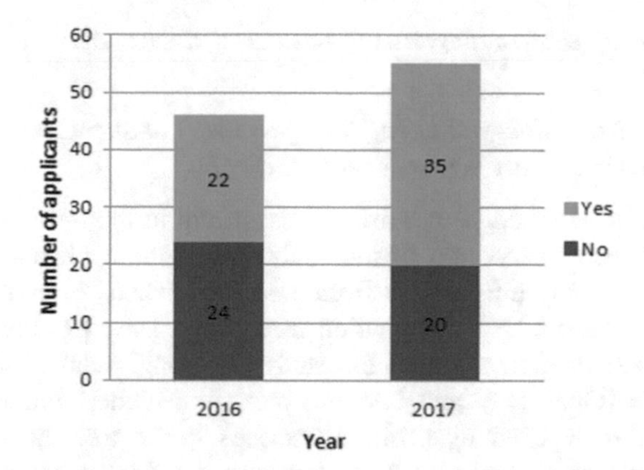

Fig. 1. Count of applicants and count of matriculants (start study Yes/No).

expectation about the positive influence of the GIS weekend event has not been verified. Also, many applicants who did not attend this event, did commence studies in our department (around half of them).

Table 2. Comparison of the number of applicants attending GIS weekend and number of students that start study

Year	Number attending GIS weekend	Number attending GIS weekend and start study	Number not attending GIS week and start study	Total number started study
2016	25	12	10	22
2017	29	17	18	35
Total	54	29	28	57

4.1 Decision Trees

The data matrix described in Sect. 2.1 was used as an input data for constructing a decision tree. Firstly, data for both years were taken as a training data set (totally 101 records). The pruned decision tree is shown in Fig. 2. The confidence factor has a default value 0.25. The confidence factor determines the pruning (smaller values incur more pruning, higher value less pruning). The tree predicts the commencement of studies in our department (rectangle leaf **Yes**) and not commencing studies in our department (rectangle leaf **No**). The total of correctly classified students is 69.3%, and the total of wrongly classified students is 30.6%.

The first node is a condition about attending the event **Open Day** (Yes or No). This condition divided the input data set, and it could be assumed as the most influential. The attribute at the first node produces the highest normalised information gain. When students attend the event Open Day, 38 of them matriculated (also 12 students are wrongly classified, they did not start to study at our department). The next node is a lecture of university staff or university students at a secondary school. The rule is interpreted as:

"When an applicant does not attend the ***Open Day*** *but attends the* ***Lecture****, then 17 applicants matriculate (5 are wrongly classified)."*

The next condition could be constructed according to the tree. The last node presents a **Friend**'s recommendation of the study. When the applicant has no previous information and has only information from a source labeled "Friend" then 9 of them start the study. In case of No information from Open Day, Lecture, GIS week and Friend, then 17 students do not start the study (7 wrongly classified). The first two nodes in the tree (Open Day and Lecture) mean the highest influence to start the successful study. It depicted by positive branches in the tree. In addition, positive branch at the fourth node Friend could be summarize in total number of matriculates: Yes = 38 (OpenDay) + 17 (Lecture) + 9 (Friend) = 64 (totally). The lower position of GIS week in the tree (Fig. 2) corresponds with the previous statistic finding that GIS week has only a partial influence on the motivation to start the study.

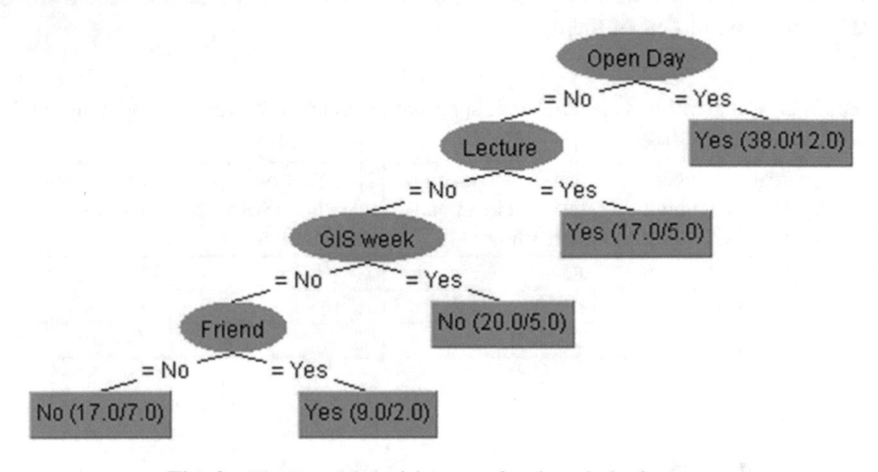

Fig. 2. The pruned decision tree for the whole data set.

Secondly, the construction of unpruned trees were generated for the year 2016 subset and 2017 subset separately. The confidence factor was set to the same value 0.25. The decision trees for these two subsets are showed in Figs. 3 and 4. The correctly classified instances were 35 (76%) and incorrectly classified instances 11 (24%) for the year 2016. The correctly classified instances were 41 (74.5%) and incorrectly classified instances 14 (25.5%) for the year 2017. In both cases, the correctness of decision trees is slightly higher than for the whole data set (above 74% in comparison with 69%). The first node for the whole data set and the data set for the year 2016 is the same; it is Open Day. For 2017 data set the highest node is Friend. The recommendation by Friend directly influences 16 applicants (right short branch in Fig. 4). The node Friend is also placed high in the tree structure (specifically at a second position) in the year 2016. In fact, the high position of nodes Open Day and

Friend in all resulting tree structures confirms the strong influence of a "personal way" of delivering information.

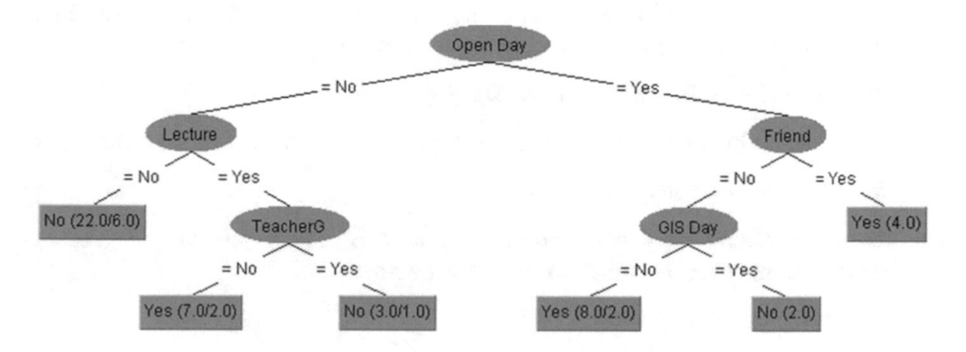

Fig. 3. The unpruned decision tree for data set in 2016.

Negative influence (left branch) can also be found in the decision tree generated for 2016 data set (Fig. 3). This influence is valid for the relatively high number of applicants (22 applicants). The interpretation of the rule is: *"When there is No information from Open Day and No information from Lecture, then 22 students will not start the study (6 are wrongly classified)."*

When all three decision trees are compared, there is low or no occurrence of the node **Internet**. In the 2016 data set and the whole data set, the node Internet is not present in the trees at all. In case of 2017 data set, this node is present but at a low position. The rule could be stated (left branch): "*When there is Friend = No and Lecture = No and Internet = Yes and Open Day = No then the 10 applicants will not start the study (4 are wrongly classified).*" It means that information from the internet does not assure the start of the study. However, the only positive influence is when we combine the Internet with the event Open Day which in this case produces 8 applicants that start the study.

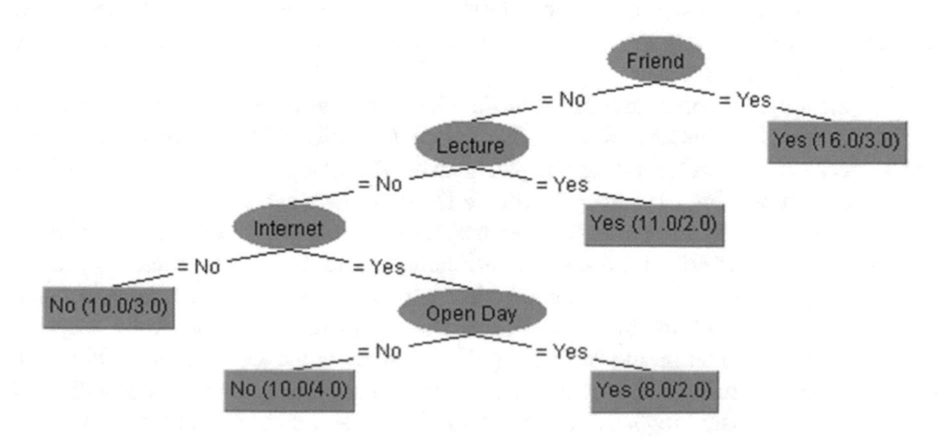

Fig. 4. The unpruned decision tree for data set in 2017.

Researched results could also be compared with the results which were presented in the previous article on this topic: about association rules [3] for the same input data set. Besides the rules with frequent combinations of several sources of information about the study, the rules with information about the start study were deduced in association with the source of information. Three rules were deduced:

- ***OpenDay = 1 and Internet = 1 => Study = 1***

If students attend Open Day and have information on the Internet, they start the study.

- ***Friend = 1 and Internet = 1*** $\geq$ ***Study = 1***

This rule is an exception. Students probably start study if they have the only recommendation of friends and information from the Internet.

- ***Internet = 1 => Study = 0***

This rule expresses that student probably does not start the study if he obtained the information solely from the Internet. The information about the study from Internet has a very low impact on deciding whether to start the study.

Concerning the rules about the positive start of the study corresponds with the information gained from the decision tree mentioned above.

5 Results

In this paper, we found that attendance at the Open Day event and recommendation of a friend (and schoolmates, siblings) have the greatest influence over whether students will commence studies at the Department of Geoinformatics. Both of these are considered a "personal way of delivering the information". The high position of nodes Open Day and Friend (followed by Lecture, GIS Day, GIS weekend) in the predictive decision trees means that the information gain divides the applicants into two groups. The first group consists of applicants that receive information by these personal ways, and they start the study of Geoinformatics. The second group consists of applicants that do not start the study. It has also been found that the influence of the Internet on the student's decision is very low. The Internet does not appear as a node in two of three presented decision trees.

Surprisingly, personal attendance at the GIS weekend event does not have a significant influence on the start of the study of Geoinformatics. The GIS weekend appears in the decision trees but in rather a negative way. The low position means that it does not bring minimisation of entropy to divide the input data set.

Comparison of decision trees with the previously presented data mining technique – association rules [3] brings the same main findings. The decision trees clarify a more detailed explanation of the combined elements of information dissemination and advertising. The resulting findings support the intensification of personal ways of delivering the information to potential applicants such as the Open Day, Gaudeamus, GIS Day and lectures at secondary schools in future. The personal presentation of information will assure higher numbers of applicants, and mainly, a higher number of students that actually decide to start the study of geoinformatics.

The construction of a predictive model based on questionnaire data brings a new approach. The use of the decision tree shows, in a straightforward manner, how to predict the likelihood of matriculation based on the way the student received information before enrolment if that data exists.

Acknowledgement. This article has been created with the support of the Operational Program Education for Competitiveness – European Social Fund (Project CZ.1.07/2.3.00/20.0170 Ministry of Education, Youth, and Sports of the Czech Republic) and with the support of the student Project IGA_PrF_2018_028 of the Palacky University Olomouc.

References

1. Tan, P.-N., Steinbach, M., Kumar, V.: Introduction to Data Mining, 1st edn. Addison-Wesley Longman Publishing Co., Inc., Boston (2005)
2. Šarmanová, J.: Methods of Data Analysis (Metody analýzy dat). Technical University of Ostrava, Faculty of Mining and Geology, Ostrava (2012). (In Czech)
3. Dobesova, Z.: Discovering association rules of information dissemination about Geoinformatics university study. In: Silhavy, R. (ed.) Artificial Intelligence and Algorithms in Intelligent Systems, vol. 764, pp. 1–10. Springer, Cham (2019)
4. Palacky University: Why study at Palacky University? In: Proč studovat na Univerzitě Palackého? www.studuj.upol.cz
5. MP-Soft, Gaudeamus Fair. https://gaudeamus.cz/
6. Esri, GIS day. http://www.gisday.com/
7. Witten, I.H.: Data Mining. Morgan Kaufmann, Burlington (2011)
8. Pavel, P.: Methods of data mining (Metody Data Miningu), part two. University of Pardubice, Faculty of Economics and Administration, Pardubice (2014). (In Czech)
9. Quinlan, J.R.: C4.5: Programs for Machine Learning. Morgan Kaufmann Publishers Inc., Burlington (1993)

Spatio-Temporal Analysis of Macroeconomic Processes with Application to EU's Unemployment Dynamics

Tomáš Formánek(✉) and Simona Macková

University of Economics, Prague, Czech Republic
{formanek,simona.mackova}@vse.cz
http://ekonometrie.vse.cz/english/

Abstract. Macroeconomic processes may not be properly evaluated and interpreted without accounting for their spatio-temporal nature. Interactions between neighbors, spill-over effects and the chronological autoregressive nature of observed variables need to be incorporated in macroeconomic modeling. This paper focuses on spatio-temporal aspects and dependencies in observed macroeconomic data. Corresponding quantitative analysis tools are outlined and an empirical application is presented: we provide a focused illustrative analysis of unemployment dynamics in EU-member states, along with insight into the robustness of such approach. We find significant spatio-temporal autocorrelation in the data. After accounting for such dependency, the econometric model shows significant robustness with respect to alternative neighborhood definitions used. Our results point at the necessity of cross-border cooperation and coordinated programs addressing unemployment in different countries.

Keywords: Spatio-temporal data · Spatial panel model
Unemployment dynamics · Spillover effects

1 Introduction

Spatial interactions play an increasingly important role in contemporary macroeconomic quantitative analysis. Spatial effects can be both positive (dissemination of high-tech advances) and negative (unemployment spillovers) and may change over time. Also, their dynamic properties depend on the scale of observation (small regions vs. state level data). For spatially augmented analysis, data are geo-coded by means of latitude/longitude coordinates system that is used to determine distances (or common borders), identify neighboring units and to estimate spatial dependencies. Spatial and spatio-temporal models have many diverse applications: social analyses, biological & ecological studies, applications in epidemiology and other non-economic fields.

R. Silhavy et al. (Eds.): CoMeSySo 2018, AISC 859, pp. 120–131, 2019.
https://doi.org/10.1007/978-3-030-00211-4_13

This contribution focuses on the autoregressive nature of observed macroeconomic processes; both spatial and time dimensions are considered. Also, appropriate statistical and econometric framework is outlined and discussed. We use unemployment rates in EU countries (2002 – 2016 annual data) to provide an illustrative empirical example. Our analysis is augmented by accounting for spatial and temporal variance in the observed data. Model evaluation and stability analysis is also provided. Figure 1 provides a simple insight into the data used: the color-coded unemployment rates show that spatially closer regions exhibit similarities in observed unemployment rates, while increasing distance is associated with more variance in the data. Similarly, it may be shown that data variance increases with the time lag (due to space limitations, only the first and the last time periods of our panel are shown in Fig. 1).

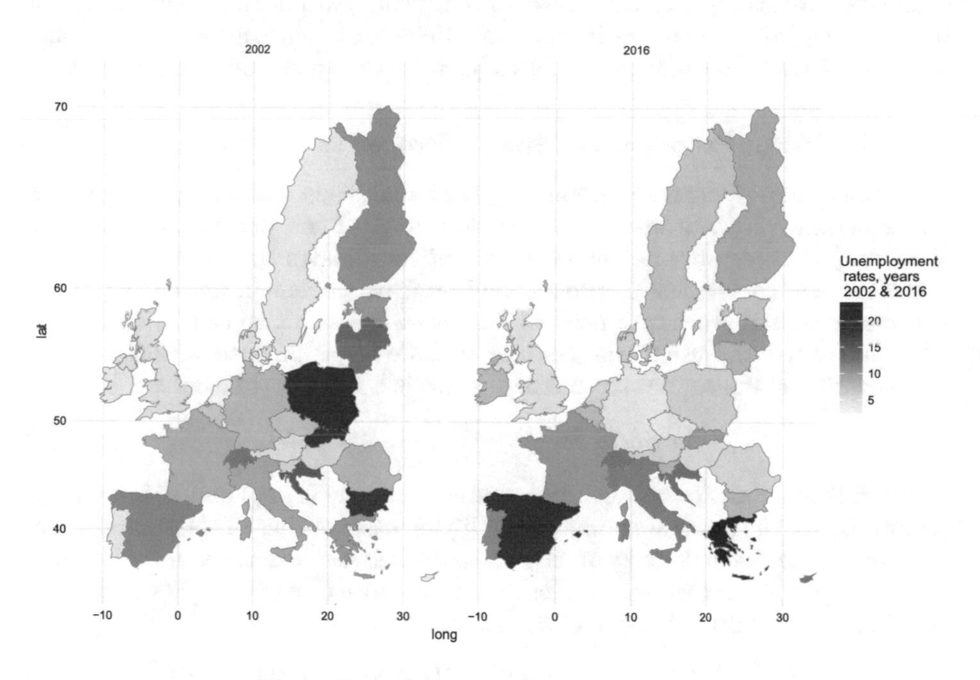

Fig. 1. Unemployment rates: 2002 & 2016

The remainder of this paper is structured as follows: section two covers selected key topics in the field of spatio-temporal data analysis (we use a geo-coded panel dataset): both statistical and econometric frameworks are described, along with references to fundamental literature. Section three features an empirical quantitative analysis, focused on unemployment dynamics. The presented model is evaluated in terms of efficiency and stability (robustness). Section four and the list of references conclude our paper.

2 Quantitative Analysis of Spatio-Temporal Data

Spatial stochastic process (random field) is an extension to the usual stochastic process, its realizations are indexed by sets of the underlying space where it is defined. Spatial stochastic process is often denoted as

$$Z(\boldsymbol{s}) : \boldsymbol{s} \in D \tag{1}$$

where D is a domain in $\mathbb{R}^n$. Typically, $n = 2$ for most economic and econometric applications, $n = 3$ is often used in fields such as geology or astronomy. Usually, D is fixed (e.g. EU) and $\boldsymbol{s}$ are the spatial locations $s_1, \ldots, s_n$ in D. Individual s_i units are points in space (say, with GPS-based latitude and longitude coordinates). Sometimes, such points can be associated with non-zero surface area elements - basically, they can serve as representative locations (centroids) for diverse types of observed variables: crop yields from agricultural fields, house prices in districts, unemployment rates in counties, regions and states.

2.1 Statistical Properties of Spatial Processes

A common simplifying assumption in spatial analysis is that the spatial process under scrutiny repeats itself over the domain D. Such process is said to be stationary - if we translate the entire set of coordinates by a specific distance and direction, the stochastic process and its features remain unchanged. For a stationary spatial stochastic process, its mean and variance can be calculated and interpreted the usual way (say, as in time series analysis). Similarly, the autocovariance of a spatial stochastic process is a function of distance:

$$cov[Z(s_i), Z(s_j)] = C(s_i - s_j) = C(h), \tag{2}$$

where h is the distance between two spatial locations (s_i, s_j) and C is a covariogram. Under the above mentioned simplifying assumptions of stationarity, spatial covariance depends only on the distance between locations s_i and s_j and not on the locations themselves. Hence, we can write $C(h) = cov[Z(s+h), Z(s)]$ and the spatial autocovariance $C(h)$ is defined as

$$C(h) = E\left\{[Z(s) - E(Z(s))] \cdot [Z(s+h) - E(Z(s+h))]\right\}. \tag{3}$$

Although expressions (2) and (3) carry useful information, most statisticians tend to favor the variogram and *semivariogram* over the covariogram. The reasons are historical and – more importantly – the empirical semivariogram is an unbiased estimator of the true semivariogram, while the covariogram estimates are biased. Variogram is defined as the expected squared difference between two observed realizations of a spatial stochastic process:

$$2\gamma(s_i, s_j) = E\left\{([Z(s_i) - E(Z(s_i))] - [Z(s_j) - E(Z(s_j))])^2\right\}, \tag{4}$$

where $2\gamma(s_i, s_j) \geq 0$ is the variogram. Generally, the value of (4) increases with growing distance between the two locations. The semivariogram is denoted as

$\gamma(h)$ and it equals to half the variogram (i.e. expected squared difference between two spatial observations). Since (4) is the expectation of a square, $\gamma(h) \geq 0$. Also, at $h = 0$, $\gamma(0) = 0$.

To accurately discuss variogram and semivariogram properties, we need to formalize the concept of stationarity in stochastic spatial processes. Second order stationarity assumes that the first two moments of the joint density exist, are invariant, finite and covariance only depends on distance:

$$\begin{aligned} E[Z(s)] &= \mu\,, \\ var[Z(s)] &= \sigma^2\,, \\ cov[Z(s+h), Z(s)] &= C(h), \end{aligned} \tag{5}$$

where $C(h)$ only depends on the distance h. If a stationary stochastic process has no spatial dependence at all (i.e. $C(h) = 0$ for $h \neq 0$), the semivariogram is constant: $\gamma(h) = var[Z(s)]$ everywhere, except for $h = 0$.

2.2 Spatio-Temporal Data

Socio-economical processes often exhibit both spatial and temporal dependency in their distributions. Given the frequency and density limitations of empirical sampling (measurements of variables that are continuous in space and time), we often regard our observations as realizations of a spatio-temporal stochastic function

$$Z(\boldsymbol{s}, t), \qquad (\boldsymbol{s}, t) \in \mathbb{R}^2 \times \mathbb{R} \tag{6}$$

where the 2D spatial domain is described by index vector $\boldsymbol{s} \in \mathbb{R}^2$ and time is identified by the index $t \in \mathbb{R}$. The separation between spatial and time dimensions is substantial, which is reflected in the notation used in (6). Assuming that second moments of $Z(\boldsymbol{s}, t)$ exist and are finite, we may use simple OLS for kriging (interpolating predictions). For other applications (presumably more sophisticated and useful), we often rely on simplifying assumptions – usually not very restrictive, yet leading to convenient properties of the processes studied:

Covariance stationarity: There are two aspects to be considered for covariance stationarity in spatio-temporal processes: First, Z has a spatially stationary covariance if $cov[Z(\boldsymbol{s}_1, t_1), Z(\boldsymbol{s}_2, t_2)]$ depends only on the distance between any arbitrarily chosen spatial points: $h_s = ||\boldsymbol{s}_1 - \boldsymbol{s}_2||$. Second, Z has a temporally stationary covariance if $cov[Z(\boldsymbol{s}_1, t_1), Z(\boldsymbol{s}_2, t_2)]$ only depends on the temporal lag: $h_t = t_1 - t_2$. If both spatial and temporal stationarity conditions hold, then Z has a stationary covariance and

$$cov[Z(\boldsymbol{s}_1, t_1), Z(\boldsymbol{s}_2, t_2)] = C(\boldsymbol{s}_1 - \boldsymbol{s}_2,\ t_1 - t_2) \tag{7}$$

holds for any pair of spatio-temporal coordinates $(\boldsymbol{s}_1, t_1)$ and $(\boldsymbol{s}_2, t_2)$ having the same corresponding h_s and h_t distances defined in $\mathbb{R}^2 \times \mathbb{R}$.

Intrinsic stationarity: This type of stationarity is based on the traditional approach of differencing observed data in order to achieve a stationary process.

The random field Z as in (6) is intrinsically stationary in space and time (has stationary increments in space and time) if the increment

$$Z(\boldsymbol{s}+\boldsymbol{s}_0;\ t+t_0) - Z(\boldsymbol{s};t), \qquad (\boldsymbol{s},t) \in \mathbb{R}^2 \times \mathbb{R} \tag{8}$$

is a spatio-temporal process, stationary in space and time for any fixed origin $(\boldsymbol{s}_0, t_0)$. For an intrinsically stationary process Z, covariance might not be well defined, but the *spatio-temporal semivariogram* is:

$$\gamma(\boldsymbol{s};t) = \frac{1}{2} var\left[Z(\boldsymbol{s}_0+\boldsymbol{s};\ t_0+t) - Z(\boldsymbol{s}_0;t_0)\right], \qquad (\boldsymbol{s},t) \in \mathbb{R}^2 \times \mathbb{R}. \tag{9}$$

For intrinsically stationary random fields Z, the semivariogram (9) does not depend on the selection of the $(\boldsymbol{s}_0, t_0)$ starting point, $\gamma(\boldsymbol{s};t)$ is nonnegative and $\gamma(\boldsymbol{0};0) = 0$. Additional in-depth aspects of spatio-temporal data analysis are covered e.g. in [4] or [5] and the estimation toolbox for R software is provided in [8].

2.3 Spatial Econometrics and Panel Data Models

As the observed geo-coded macroeconomic variables are often repeated in time, spatial panel models can be used to depict interactions among variables across spatial units as well as over time. This section provides a brief spatial panel model description. The general form of a static panel model that includes a spatial effects (spatial lag) for the dependent variable may be outlined as

$$\begin{aligned} \boldsymbol{y} &= \lambda\left(\boldsymbol{I}_T \otimes \boldsymbol{W}_N\right)\boldsymbol{y} + \boldsymbol{X}\boldsymbol{\beta} + \boldsymbol{u}, \\ \boldsymbol{u} &= \left(\boldsymbol{\iota}_T \otimes \boldsymbol{I}_N\right)\boldsymbol{\mu} + \boldsymbol{v}, \end{aligned} \tag{10}$$

where $\boldsymbol{y}$ is a $(NT \times 1)$ vector of observations of the dependent variable y_{it} ($i = 1, 2, \ldots, N$ denotes cross-sectional units and $t = 1, 2, \ldots, T$ relates to the time dimension). $\boldsymbol{X}$ is a $(NT \times k)$ matrix of k exogenous regressors, it has full column rank and its elements are uniformly bounded in their absolute values. $\boldsymbol{I}_T$ and $\boldsymbol{I}_N$ are identity matrices (with dimensions given by their subscripts) and $\boldsymbol{\iota}_T$ is a "unit" vector $(T \times 1)$ where all elements equal one. We follow the common practice of denoting spatial weight matrices as $\boldsymbol{W}_N$. Those matrices are used for modeling spatial interactions among neighbors (for technical discussion, see e.g. [3]). Spatial panel model (10) features the $\otimes$ Kronecker product operator. λ as well as the elements of vector $\boldsymbol{\beta}$ are model parameters (estimated by maximum likelihood methods). The disturbance vector $\boldsymbol{u}$ $(NT \times 1)$ is a sum of two terms: the unobserved individual effects vector $\boldsymbol{\mu}$ holds the time-invariant and spatially uncorrelated individual effects and it is composed by stacking N individual-specific $\boldsymbol{\mu}_i$ vectors $(T \times 1)$. $\boldsymbol{v}' = (\boldsymbol{v}_1', \ldots, \boldsymbol{v}_T')$ is a vector of spatially independent innovations that vary both over cross-sectional units and across time with $v_{it} \sim IID\left(0, \sigma_v^2\right)$. Model (10) can be easily generalized to encompass spatially autocorrelated regressors $\boldsymbol{X}$, individual effects $\boldsymbol{\mu}_i$ and/or random elements. For detailed technical discussion, see e.g. [7].

Following the standard approach to panel data analysis as in [1], individual effects μ_i are treated as either fixed or random. With the "random effects" (RE) model, we assume that unobserved individual effects are not correlated with other regressors of the model. However, this rather strong assumption may be relaxed using the Mundlak-Chamberlain approach (see [9] for technical discussion).

For spatial lag models such as (10), the estimated coefficents $\hat{\beta}_j$ do not constitute marginal effects. This can be shown by taking partial derivatives of the function (10) along any chosen j-th regressor $\boldsymbol{x}_j$: $\partial \boldsymbol{y}/\partial \boldsymbol{x}_j$ is a complex, non-symmetric matrix of "expected effects", differing for each y_{it} unit. In order to interpret the ceteris paribus effects of changes in exogenous regressors, we use *impacts*: either direct (the expected effect taking place within a spatial unit where regressor value is changed) or indirect (average spillover effects, affecting neighboring units). For derivation and description of the impact effects in spatial and spatio-temporal models, see e.g. [2] or [3].

3 Empirical Analysis

This section provides an empirical analysis of unemployment in EU-member countries (countries being EU members as of 2016). We take advantage of the repeated geo-coded observations (i.e. spatio-temporal data) and our estimates are augmented by the corresponding tools of spatio-temporal analysis. The estimated model is discussed and robustness of our results is analyzed.

3.1 Data Description

To ensure consistency in observed macroeconomic variables, we use data provided by The World Bank on-line database (https://data.worldbank.org). Our dependent variable – the unemployment rate – is the share of non-working labor force willing to find employment. For the sake of reproducibility, we provide identification code for the dataset used: SL.UEM.TOTL.NE.ZS. Gross domestic product (GDP) at purchaser's price is the sum of gross value added by all resident producers with added taxes and diminished by subsidies (NY.GDP.MKTP.CD). Foreign direct investment (FDI) refers to direct investments net inflow in business from other countries into the referred country (BX.KLT.DINV.CD.WD). Both FDI and GDP are recorded in current prices (USD). To provide scale-compatibility and interpretability to the estimated model (11), GDP is log transformed and rendered in terms of year-on-year percentage changes. Also, FDI is expressed in relative terms with respect to GDP, i.e. as FDI/GDP.

A basic unemployment data visualization is provided in Fig. 1. Please note that non-EU countries such as Norway, Serbia, Switzerland, etc. are excluded. While the basic similarities and dissimilarities among unemployment rates over spatial and time dimensions might be intuitively inferred from Fig. 1, we provide an empirical spatio-temporal semivariogram (STSV) for a more accurate

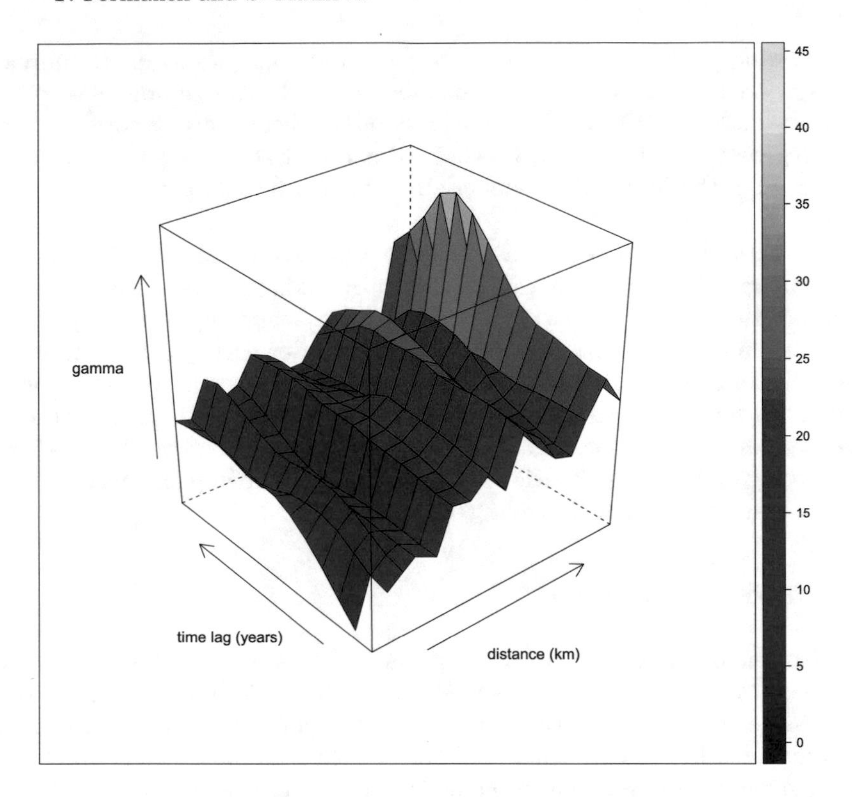

Fig. 2. Spatio-temporal semivariogram of EU's unemployment, years 2002–2016

description. Figure 2 is an empirical visualization of the $\gamma(\boldsymbol{s};t)$ statistic from Eq. (9).

While the surface of the empirical STSV is irregular (reflecting the stochastic and discrete nature of sampling from a continuous spatio-temporal process), we can clearly identify increasing $\gamma(\boldsymbol{s};t)$ values along with increasing time and spatial distances among observations. Figure 2 may be intuitively interpreted as follows: "variance" in observed data (measured by $\gamma(\boldsymbol{s};t)$) increases with growing time and space distances among observations. At the same time, our empirical STSV reflects the fact that observations located close together (in space and/or in time) are more alike than those far apart. Overall, the STSV as in Fig. 2 implicates the relevance and necessity of using repeated spatial observations (spatio-temporal data) for assessing macroeconomic dependencies in the observed unemployment variable.

3.2 Empirical Model and Its Estimation

Our unemployment model is derived along the concept of regional competitiveness, which is usually perceived as some combination of productivity, employment (unemployment) rate, living standards, FDI attractiveness and other related factors. We follow the approach to competitiveness as summarized in [6]. Therefore, unemployment is modeled as a function of GDP dynamics and FDI indicators. Using the general spatial panel model specification (10), our empirical model can be outlined as follows:

$$\begin{aligned} UN_{it} &= \lambda\left(\boldsymbol{w}_i \boldsymbol{UN}_t\right) + \beta_0 + \beta_1 \log GDP_{it} + \beta_2 \log \overline{GDP}_i \\ &\quad + \beta_3 FDI_{it} + \beta_4 \overline{FDI}_i + u_{it}, \\ u_{it} &= \mu_i + v_{it}, \end{aligned} \tag{11}$$

where UN_{it} is the unemployment level observed in country i at time t, $\boldsymbol{w}_i$ is the i-th row of the spatial weights matrix $\boldsymbol{W}_N$ and it describes the spatial dependency of UN_{it} on the observed unemployment levels in other countries at time t, as recorded in the $\boldsymbol{UN}_t$ vector (please note that only neighboring countries have a non-zero value assigned in the vector of weights $\boldsymbol{w}_i$). *GDP* and *FDI* have been introduced in Sect. 3.1 and $\overline{GDP}_i$ is country's individual average GDP value (usually referred to as the "within mean" of a variable). This term is included in our equation for technical reasons and it follows from the the above mentioned Mundlak-Chamberlain approach to panel data estimation. The construction and inclusion of the term $\overline{FDI}_i$ into Eq. (11) is analogous to $\overline{GDP}_i$. The remaining elements of Eq. (11) follow directly from the general form (10).

Our spatial panel data model (11) was estimated using the maximum likelihood method described in [7]. During the estimation and statistical inference process, the term FDI_{it} was dropped from Eq. (11) because of a lack of statistical significance under all reasonable neighborhood definitions. However, the within mean of FDI ($\overline{FDI}_i$) is statistically significant at $\alpha = 0.1$ and it was kept in the amended regression model (11) as a theoretically well based contrasting factor (see e.g. [6] for discussion).

Due to space limitations – and given the lack of interpretability of the estimated β_j coefficients – raw estimation results are omitted here. All relevant estimated models, tests and figures mentioned in this contribution are available from the authors upon request. Table 1 shows the estimation output: direct effect

Table 1. Estimated direct, indirect & total impacts and λ

Effects	log *GDP*	log $\overline{GDP}$	$\overline{FDI}$	λ
Direct	−2.6280	2.2022	−5.4486	
z-score	−3.8058	2.7907	−1.8803	–
p-value	0.0001	0.0053	0.0601	
Indirect	−4.6897	3.9299	−9.7232	
z-score	−2.4622	2.0698	−1.5741	–
p-value	0.0138	0.0385	0.1155	
Total	−7.3177	6.1321	−15.1717	0.6631
z-score	−2.9958	2.3880	−1.7195	10.3032
p-value	0.0027	0.0169	0.0855	0.0000

and spillovers (indirect effects) are shown, along with the overall (total) effect and corresponding statistical significance levels (z-scores and p-values).

The estimated model (11) as presented in Table 1 is statistically significant at the 5% significance level, residual elements $\hat{v}_{it}$ do not exhibit spatial nor temporal autocorrelation (at $\alpha = 0.05$). The estimated impacts along with their significance levels are based on a $\boldsymbol{W}_N$ matrix defined for maximum neighbor distance threshold of 1,440 km – such specification leads to the highest log-likelihood of the estimated regression function (see Fig. 3).

Please note that distances between states are measured using representative coordinates (centroids) and the actual labor market interactions – commuting, temporary or permanent relocation of workers, production outsourcing & cooperation, etc. – may take place at various distances, up to the reported maximum distance threshold.

Finally, the value of λ shown in Table 1 is the actual estimated coefficient from Eq. (11) and it describes the extent of spatial dependency ("spatial lag") in the spatio-temporal observations of unemployment. Hence, λ is not an impact and its placement into Table 1 is motivated by overall space limitations to the contribution.

3.3 Results Evaluation and Stability Analysis

The estimated spatial dependency coefficient λ as in Table 1 is rather strong and highly statistically significant. This important result has a twofold interpretation: First, it supports the overall validity of the methodology (based on spatio-temporal data) chosen for analyzing unemployment dynamics. Second, our results emphasize the prominent role of cross-border cooperation in macroeconomic policies aimed at unemployment & related negative socio-economic consequences. Please note that this paper provides a technical analysis of unemployment dynamics; actual EU's policies and their effectivity are not addressed here.

The impacts as in Table 1 can be summarized as follows: given a one percent increase in GDP, a small but statistically significant decrease in the rate of unemployment can be expected. The expected spillover effect of GDP is about twice as strong as the direct effect. Again, this underlines the importance of international cooperation in the field of macroeconomic policy making. Provided GDP_{it} observations are explicitly present in Eq. (11), the impact of $\overline{GDP}_i$ has a very limited & technical interpretation: $\overline{GDP}_i$ is used to control for correlation between GDP_{it} and the individual unobserved effects μ_i that would otherwise render the estimation of model (11) inconsistent (see [9] for technical discussion of the Mundlak-Chamberlain method). The same rather technical interpretation applies to $\overline{FDI}_i$ impacts. The individual observations FDI_{it} were excluded from model (11) on grounds of being statistically insignificant at any reasonable significance level, while the impacts of $\overline{FDI}_i$ are significant at $\alpha = 0.1$. Broadly speaking, the negative sign of FDI impacts is in line with underlying macroeconomic theory (while keeping in mind the technical nature of $\overline{FDI}_i$ in our model) and it corresponds to the competitiveness paradigms outlined in [6].

Besides direct impacts and spillovers, total impacts are included in Table 1 as well. A total impact is just a sum of the direct and indirect impacts, its z score and statistical significance level is also calculated by aggregating the underlying variabilities in the estimated direct/indirect impacts. However, in many empirical applications of spatio-temporal analysis, the direct and indirect effects may come with opposite signs. Therefore, at some higher level of spatial aggregation, direct impacts and spillovers could cancel out. For example, positive direct effects may come at the "price" of equally prominent negative spillovers. Therefore, total impacts are usually reported along their direct/indirect constituents – even if there are no contradicting signs of direct/indirect impacts in the results.

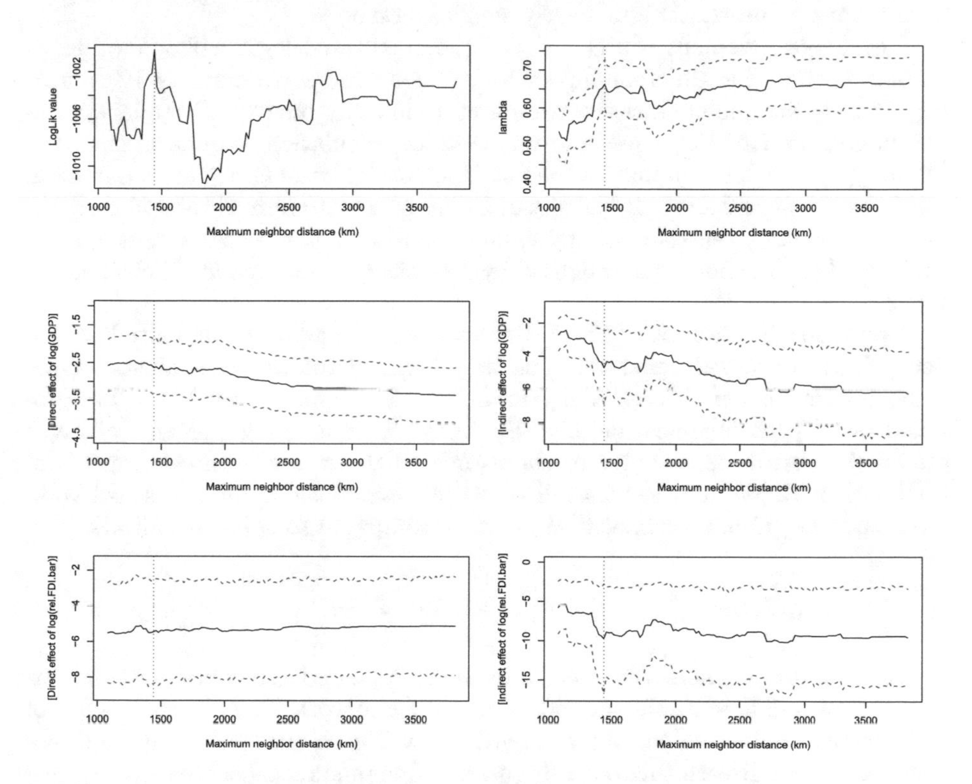

Fig. 3. Evaluation of impact stability with varying threshold distance

Neighborhood definitions and the corresponding spatial weight matrices $\boldsymbol{W}_N$ cannot be estimated along with the model. Rather, $\boldsymbol{W}_N$ as in model (10) has to be pre-specified. Importantly, the estimated β_j coefficients and impacts may be strongly influenced by the choice of a particular neighborhood definition ($\boldsymbol{W}_N$). Next, we discuss and apply a simple yet effective algorithm for evaluating model robustness against varying neighborhood definition and for choosing an optimal $\boldsymbol{W}_N$ setting from a group of alternatives.

Visually, model (11) robustness evaluation can be summarized in Fig. 3: we start with a relatively sparse $\boldsymbol{W}_N$ matrix, where each country has at least one neighbor. "Island" states, i.e. states without neighbors (where the $\boldsymbol{w}_i$ vector only contains zero elements) are not compatible with the maximum likelihood estimation of spatial models such as (11). Next, we iteratively increase maximum neighbor distance thresholds by 20-km increments (a total range of 1,020 km to 3,800 km threshold distances is used), estimating model (11) and calculating impacts at each step. The higher distance thresholds shown in Fig. 3 might not be realistic in the sense that they are not consistent with the commonly assumed regional unemployment interactions and spillovers. However, they provide illustrative information regarding model stability under varying neighborhood specifications. At the same time, our interpretations and conclusions are based on interactions bounded at significantly smaller distances.

The top-left element of Fig. 3 shows the maximized log likelihood values of models estimated at different threshold iterations and corresponding $\boldsymbol{W}_N$ matrices. The log likelihood statistics were used to identify our "best" model specification and the 1,440 km threshold was used for calculation of impacts shown in Table 1 (again, we re-emphasize the fact that spatial interactions are assumed to take place at significantly lower distances, up to a maximum threshold distance measured among centroids of individual countries). For reader's convenience, the 1,440 km threshold is highlighted by a vertical dotted line in all elements of Fig. 3.

The top-right element of Fig. 3 describes λ: the magnitude of spatial dependency in the observed panel data and its prominent robustness against changes in neighborhood definitions. Also, simulated ± 1 standard error bands are provided (see [7] for technical details). Similarly, the remaining elements of Fig. 3 are used to evaluate stability in the estimated direct and indirect impacts of GDP and FDI. We may see that direct effects are considerably less sensitive to changes in neighborhood specification when compared to spillovers effects.

4 Conclusions

Spatio-temporal econometric models based on repeated geo-coded observations of macroeconomic variables provide a useful framework for numerous types of quantitative analyses. Spatial panel data and well specified spatio-temporal models allow us to discern theoretically justified dynamic relationships among variables from spatial and spatio-temporal autoregressive dependencies.

This contribution describes selected aspects of the methodological framework for performing quantitative analyses using spatio-temporal data. Also, an empirical illustrative analysis of unemployment is provided, supporting the importance of cross-border cooperation in macroeconomic policies addressing unemployment. Stability analysis of our results suggest a robust nature of the macroeconomic model used.

Acknowledgement. Supported by the grant No. IGA F4/60/2018, Faculty of Informatics and Statistics, University of Economics, Prague. Geo-data source: GISCO - Eurostat (European Commission), Administrative boundaries: © EuroGeographics.

References

1. Baltagi, B.H.: Econometric Analysis of Panel Data. Wiley, New York (2005)
2. Elhorst, J.P.: Spatial Econometrics: From Cross-Sectional Data to Spatial Panels. Springer, Berlin (2014). https://doi.org/10.1007/978-3-642-40340-8
3. Formánek, T.: Spatially augmented analysis of macroeconomic convergence with application to the Czech Republic and its neighbors. In: Silhavy, R., Silhavy, P., Prokopova, Z. (eds.) Applied Computational Intelligence and Mathematical Methods, CoMeSySo 2017. Advances in Intelligent Systems and Computing, vol. 662, pp. 1–12. Springer, Cham (2018). https://doi.org/10.1007/978-3-319-67621-0_1
4. Gräler, B., Pebesma, E., Heuvelink, G.: Spatio-temporal interpolation using gstat. RFID J. **8**(1), 204–218 (2016)
5. Journel, A.G., Huijbregts, C.J.: Mining Geostatistics. Academic Press, London (1978)
6. Martin, R.: A study on the factors of regional competitiveness – a draft final report for the European Commission Directorate-General Regional Policy. Cambridge University Press, Cambridge (2004)
7. Millo, G., Piras, G.: splm: spatial panel data models in R. J. Stat. Softw. **47**(1), 1–38 (2012). https://doi.org/10.18637/jss.v047.i01
8. Pebesma, E.: spacetime: spatio-temporal data in R. J. Stat. Softw. **51**(7), 1–30 (2012). https://doi.org/10.18637/jss.v051.i07
9. Wooldridge, J.M.: Econometric Analysis of Cross Section and Panel Data. MIT Press, Cambridge (2010)

Using Blockchain Technology to Improve N-Version Software Dependability

Denis V. Gruzenkin(✉), Anton S. Mikhalev, Galina V. Grishina, Roman Yu. Tsarev, and Vladislav N. Rutskiy

Siberian Federal University, Krasnoyarsk, Russia
gruzenkin.denis@good-look.su, asmikhalev@yandex.ru, ggv-09@inbox.ru, tsarev.sfu@mail.ru

Abstract. Being a technique ensuring the dependability and fault tolerance of software, the N-version programming has proven its effectiveness. A formal definition and some practical experience support the idea that redundancy and diversity are the key points of the N-version software dependability. The implementation of N functionally equivalent versions allows to resist different types of faults, including residual ones. However, due to some peculiarities of N-version software design interversion and intermodule dependences can arrive. It results in the dependency of potential faults in versions or modules of the N-version software. The recently appeared blockchain technology can be applied to increase the dependability of N-version software. In the paper the authors suggest an approach to log N-version software faults by the means of the blockchain technology. As a result, the blockchain technology provides complete data on operation of the N-version software that is used to improve the N-version software dependability. An example illustrating the proposed approach is provided.

Keywords: N-version software · Software dependability · Blockchain Software reliability · Logging

1 Introduction

The information technology industry is characterized with the disappearance of out-of-date methods and the arrival of up-to-date ones. The blockchain technologies are considered one of the latest achievements in the IT [1, 2]. Nowadays they attract great attention because they have revolutionized the fields of information security and distributed data processing [3, 4].

Blockchain is a multifunctional and multilevel information technology assigned for the reliability of both tangible and intangible assets registration [5, 6]. All information and assets operations are coded. Then they are converted into the so-called block. Any asset corresponds to both a private key and a public key. The public key is essential for the asset operations and the private key is necessary for the asset operations validity checking [7, 8]. An example of such operations is financial transactions.

The transaction is approved in case of private and public keys matching. The keys themselves are individual for any asset. They are generated according to an algorithm

R. Silhavy et al. (Eds.): CoMeSySo 2018, AISC 859, pp. 132–137, 2019.
https://doi.org/10.1007/978-3-030-00211-4_14

and look like a set of symbols. To complete the block (fix an event) a user fits the key which is connected to the previous transaction only. It results in a chain of blocks or a blockchain system. Such method eliminates the substitution of former data. This peculiarity is an issue of interest for many researches.

The illustration of blockchain functioning is the cryptocurrency Bitcoin [1, 9, 10]. A user gets a certain amount of bitcoins for a block completion. This amount is reduced with each further completion but, at the same time, a fee for a transaction increases. The users do not complete the blocks manually, but they use computer technologies instead. In other words, people direct computer power to solve difficult mathematical tasks.

However, blockchain technologies are not restricted by the financial sphere only. They are widely used in other fields of human life. There is a hypothesis about the application of blockchain technology for increasing the N-version software reliability, which is a part of dependability. The problem of software dependability is crucial in such fields as nuclear power, finance, space exploration, etc.

2 Research Hypothesis

2.1 Hypothesis Description

The concept of N-version software implies parallel or sequential execution of diversified program components (multiversions) [11]. These components should be functionally equivalent, and they should belong to the same module of the common environment [12]. The results of program components calculations are assessed, and their correctness is defined by a voting method [13]. The correct results are considered as the results of the whole module operation and the executed multiversions belonging to that module. The voting results are the more correct the more versions are implemented by the module. As multiversions are diversified, they have considerable differences in their implementation [14]. It results in the independence of their potential failures. If one of the versions gives an error, the others will return correct results. It guarantees fault tolerance of the whole software system.

Failure exposure and elimination are issues of the day in the N-version programming. The blockchain technology can be used to solve these problems. Besides, it can identify the dependence between failures of different versions and modules. We suggest applying the blockchain technology as a logging tool.

A log is a record of the system information about the operation of the N-version software system. This record is implemented in the form of files or the chains of blocks. During the N-version software run-time, all executing operations are recorded. It allows the software programmers to identify the exact place of failure and to eliminate it by the means of proper tools.

Logs registration is not safe. The data can be changed or even removed by a programmer or a hacker. Moreover, the operations with data have risks of data loss and data distortion. Due to blockchain it is possible to log transactions without falsification and data loss. It increases the amount of logs and decreases the amount of residual errors.

Still there is a possibility for hackers to connect their equipment to the information system of an enterprise. The power of the equipment should be at least 1% higher than the power of the enterprise as a whole. In this case they can eliminate or distort logs. Nevertheless, this article is concerned with another way of information security. In this context, the possibility of correct and full logging of all transactions is equal to 100%.

2.2 The Theoretical Ground for the Hypothesis

Errors can be both residual and obvious. They can appear at any stage of the software lifecycle. Therefore, the task is to define the log error types which are useful for finding and eliminating the errors. Special attention should be paid to residual errors as they appear only during program execution. Its appearance can be unnoticed. It means that there are no messages about errors, but some logs can indicate the correlation of multiversions or even modules operation faults.

Log-messages can be classified in different ways. It depends on the character of the task. This study is concerned with the following log classification:

1. Standard error messages provided by a multiversions design engineer - H_1;
2. Unexpected system failures messages - H_2 (generated by the programming environment);
3. Messages indicating a joint application of shared hardware by some versions implicitly - H_3;
4. Messages indicating a joint application of shared hardware by some modules implicitly - H_4;
5. Messages indicating the lack of correct data exchange between some versions and execution environment implicitly - H_5;
6. Other messages - H_6.

The sum of error *detection probabilities* of every type is equal to the following equation:

$$\sum_{i=1}^{m} P(H_i) = 1$$

where m is a number of message classes (in our case $m = 6$).

A is taken for the detection event of a residual error in logs. The probability of this event appearance *P(A)* is the indicator that defines the probability of increase in multiversion software system reliability. The error elimination depends on the probability of error detection during software testing and operation. This increases N-version software dependability.

It is possible to calculate the error detection as follows:

$$P(A) = 1 - \prod\nolimits_{i=1}^{m} (1 - P(H_i)) \qquad (1)$$

where event A may take place during the execution of one of the H_i-events, i.e. while the detection of one of the classified messages according to the results of a logs set analysis.

In common case the probability of errors appearance for each message class is calculated using the equation:

$$P(H_i) = 1 - (1 - p_i)^n$$

where n is the amount of messages in a log-file or a logs chain, p is the probability of error messages appearance in logs. As far as the N-version software system is a modular-based one, it is possible for versions and modules to operate simultaneously. It means that the events of logs records are independent.

3 Experiment Results

The effectiveness of blockchain technology application for the residual error detection has been assessed. The value of residual error detection probability is used as an effectiveness criterion. Residual errors can be diagnosed according to the number of messages in logs. Logs analysis of the operation results of a series of multiversion software allowed to distinguish some types of independent messages. They can indicate some faults in operation (abnormal behavior):

1. the allocation of a correct total operation result of one of multiversions without intermediate data output;
2. the exceeding of a permissible limit of some multiversions operation time;
3. the increase of a random access memory capacity consumed by the module;
4. the impossibility for the N-version software environment to address to the memory containing a module operation result;
5. calculating multiversions errors on some sets of processing data.

The abundance of faults appearance in software operation is set by the probabilities:

$p_1 = 0.00004$;
$p_2 = 0.00002$;
$p_3 = 0.00003$;
$p_4 = 0.000008$;
$p_5 = 0.00006$.

The probability of a residual error detection $P(A)$ can be calculated with the Eq. (1). Logs analysis allows to discover one of possible abnormal behavior messages in software operation. The probabilities values are shown in Table 1.

Some logs chains contain more messages about multiversion software. This allows to detect and eliminate a residual error with higher probability. It is vital at an early testing stage.

Table 1. The experiment results.

Amount of messages (n)	$P(H_1)$	$P(H_2)$	$P(H_3)$	$P(H_4)$	$P(H_5)$	$P(A)$
100	0.0039920	0.0019980	0.0029956	0.0007997	0.0059822	0.016
250	0.0099504	0.0049876	0.0074721	0.0019980	0.0148885	0.039
500	0.0198017	0.0099503	0.0148883	0.0039920	0.0295553	0.076
1000	0.0392113	0.0198015	0.0295549	0.0079681	0.0582372	0.146
5000	0.1812725	0.0951635	0.1392940	0.0392107	0.2591885	0.546

4 Conclusion

The logging of information on multiversions execution along with extra multiversions intended to improve the software dependability increase a time period for software design and development. Besides, there is a negative effect of the blockchain technology, such as a low speed of transactions.

However, the blockchain technology application allows both to detect errors and to identify interversion and intermodule dependences at the stages of software testing and operation. The more multiversions provide a correct result the more dependable the N-version software is. Finding an exact problem source during the multiversions execution allows not only to debug and improve the program code, but also to prevent a program component from failure when the whole multiversions set is implemented.

References

1. Beck, R.: Beyond bitcoin: the rise of blockchain world. Computer **51**(2), 54–58 (2018)
2. Gatteschi, V., Lamberti, F., Demartini, C., Pranteda, C., Santamaria, V.: To blockchain or not to blockchain: that is the question. IT Prof. **20**(2), 62–74 (2018)
3. Hawlitschek, F., Notheisen, B., Teubner, T.: The limits of trust-free systems: a literature review on blockchain technology and trust in the sharing economy. Electron. Commer. Res. Appl. **29**, 50–63 (2018)
4. Khan, M.A., Salah, K.: IoT security: review, blockchain solutions, and open challenges. Future Gener. Comput. Syst. **82**, 395–411 (2018)
5. Hughes, T.M.: The global financial services industry and the blockchain. J. Struct. Finance **23**(4), 36–40 (2018)
6. Xu, C., Wang, K., Guo, M.: Intelligent resource management in blockchain-based cloud datacenters. IEEE Cloud Comput. **4**(6), 50–59 (2018)
7. Chen, Z., Zhu, Y.: Personal archive service system using blockchain technology: case study, promising and challenging. In: Proceedings - 2017 IEEE 6th International Conference on AI and Mobile Services, AIMS 2017, pp. 93–99. IEEE (2017)
8. Drescher, D.: Blockchain basics: a non-technical introduction in 25 steps. Apress, New York (2017)
9. Hong, K.H.: Bitcoin as an alternative investment vehicle. Inf. Technol. Manag. **18**(4), 265–275 (2017)

10. Kaushal, P.K., Bagga, A., Sobti, R.: Evolution of bitcoin and security risk in bitcoin wallets. In: 2017 International Conference on Computer, Communications and Electronics, COMPTELIX 2017, pp. 172–177. IEEE (2017)
11. Avizienis, A., Chen, L.: On the implementation of N-version programming for software fault-tolerance during program execution. In: Proceedings of IEEE Computer Society International conference on Computers, Software and Applications, COMPSAC, pp. 149–155 (1977)
12. Gruzenkin, D.V., Chernigovskiy, A.S., Tsarev, R.Y.: N-version software module requirements to grant the software execution fault-tolerance. In: Advances in Intelligent Systems and Computing, vol. 661, pp. 293–303. Springer, Heidelberg (2017)
13. Durmuş, M.S., Eriş, O., Yildirim, U., Söylemez, M.T.: A new bitwise voting strategy for safety-critical systems with binary decisions. Turk. J. Electr. Eng. Comput. Sci. **23**(5), 1507–1521 (2015)
14. Baudry, B., Monperrus, M.: The multiple facets of software diversity: recent developments in year 2000 and beyond. ACM Comput. Surv. (CSUR) **48**(1), 16 (2015)

Enhancing Feature Selection with Density Cluster for Better Clustering

Yang Chen[1,2], Hui Li[1,2(✉)], Mei Chen[1,2], Zhenyu Dai[1,2], Huanjun Li[3], and Ming Zhu[4]

[1] Department of Science and Technology, Guizhou University, Guiyang, China
gm.chenyang@gmail.com, {cse.HuiLi,gychm}@gzu.edu.cn
[2] Guizhou Engineering Lab of ACMIS, Guizhou University, Guiyang, China
[3] Guizhou Academy of Aerospace Technology, Guiyang, China
[4] National Astronomical Observatories, Chinese Academy of Sciences, Beijing 100016, China

Abstract. Feature selection is an important data analysis technique that used to reduce the redundancy of features and exploit hidden information in high-dimensional data. In this paper we propose a similarity metric based feature selection method named *Fesim*. We use the Euclidean distance to measure the similarity among all features, and then apply the density based DBSCAN algorithm to clustering features which to be relevant. Moreover, we present a strategy which choose representative features of each cluster accurately. We conducted comprehensive experiments to evaluate the proposed approach, and the results on different datasets are demonstrated its superiority.

Keywords: Feature selection · Density clustering · Euclidean distance Similarity

1 Introduction

Feature learning is one of the critical area in machine learning. Generally, it's very hard to achieve satisfactory efficiency and effectiveness goals in high dimensions data clustering or classification. Furthermore, with the increase of data dimensionality, it would result in heavy performance degradation. In order to solve this problem, feature selection techniques often used before train the machine learning model, and the effectiveness of feature selection becomes critical to data analysis.

At present, many researchers have studied feature selection methods which can be divided into four categories. The first category of approaches is the filter based methods, such as Relief [1], mutual information and maximum information coefficient [2] based techniques. This type of methods assigns weight [3] to each feature. Its main features are uncomplicated and ease of operation. However, it is unsatisfactory for calculating continuous variables. And the results are sensitive to discretization methods and generally lower in classification accuracy than the other three categories. The second category is wrapper methods, such as recursive feature elimination, Las Vegas Wrapper [4]. This kind of methods treat the selection of subsets as a search optimization problem, then generates different combinations which will be evaluated, and

R. Silhavy et al. (Eds.): CoMeSySo 2018, AISC 859, pp. 138–150, 2019.
https://doi.org/10.1007/978-3-030-00211-4_15

compares them with other combinations finally. Therefore, the obvious failing of the wrapper methods is the risk of overfitting and computationally expensive. The third category is the embedded methods, such as regularization, random forests. They try to reduce the computational time required to reclassify different subsets in the wrapper methods. It is in the process of determining the model to select those properties that are important to the training of the model. The last one, hybrid method, is the combination of filter and wrapper methods and attempts to take advantage of both the filter and wrapper methods to achieve optimal performance with specific algorithms.

There have been several clustering methods used for feature selection. They divide all features into a number of different subsets or clusters and the features in same cluster are highly related to each other while the features in different clusters are less relevant. DBSCAN [5, 6], as we all know, is a density-based spatial data clustering method and has been widely used in data mining. When feature selection and clustering are both applied to gather similar features into a cluster, DBACAN can meaningfully find irrelated features which has a great influence on the accuracy of clustering result.

In this paper, we use DBSCAN algorithm to cluster features. And we propose a similarity based feature selection strategy for data clustering. First of all, we normalize the data to make all values in a specified interval. Secondly, considering Euclidean distance with transitivity is one way of estimating how close between two features, we use it to measure how similar between features. The feature of transitivity guarantee that non-adjacent features may also be similar. As a result, the features in same cluster are more unrelated to other clusters'. Finally, we choose a superior strategy to select the most representative features.

The rest of the paper is organized as follows. Section 2 reviews the related work of feature selection. Section 3 presents our proposed algorithm. Section 4 describes the experiments and discusses the results. Section 5 concludes the paper.

2 Related Work

In the past several years, there have been various feature selection algorithms. Only the most typical methods are briefly discussed here. Many of these feature selection algorithms use metrics such as correlation coefficients and mutual information to evaluate whether features are relevant. These univariate feature selection methods' principle is to separately calculate values of each feature with one of statistical indicators above and only retain the top k features in rank. For the regression problem, the correlation coefficient indicator is commonly used. For the classification problem, the chi-square test, variance analysis, and mutual information can be used as the metric. The main advantages of these methods are simplicity and low computational costs. Tahir et al. [7] proposed Trees Classifier which can use the decision tree to calculate the importance of features and discard less important features. Given an external estimator that assigns weights to features (e.g. the coefficients of a linear model), the goal of recursive feature elimination (RFE) is to select features by recursively considering smaller and smaller sets of features. Firstly, it establishes a model on the original data and assign each feature a weight, then eliminate the feature with the smallest absolute weight, and perform this process recursively until the desired number of features is reached.

There are not many studies that use clustering for feature selection so far. Most of them are used to cluster data instances. When combined feature selection and clustering, distance metrics is used to measure the similarity between features. Au et al. [8] proposed a method called ACA in genetic engineering, which groups the features that are dependent on each other into clusters by optimizing a criterion function derived from an information measure that reflects the interdependence between attributes. Liu [9] propose a new method to select important features using clustering with information metric. They considered each feature as a data point cluster with between-cluster and within-cluster distances. Maji [10] proposed a new mutual information based quantitive measure, to compute the similarity among features. The proposed measure incorporates the information of sample categories while measuring the similarity between features. Eshaghi [11] proposed FFS algorithm which used mutual information to measure similarities and clustering features. And the strategies of selecting feature is considering the target class for the impact on attributes for classification.

As shown in recent studies, it is effective when using a feature clustering based method to solve the general feature selection problem. We will cluster features and choose some of them as the best representation of full feature which can make the clustering results more accurate.

3 The Fesim Algorithm

The process of feature selection approach works like data clustering, where each cluster contains features that are more relevant and not related to other clusters' features. Therefore, we consider using clustering for feature selection. We apply the DBSCAN clustering algorithm and divide the features into different clusters based on the similarity. An important point is that one of the advantage of it is the ability to identify noise points which are extended as the atypical features that would be preserved because they have less neighbor features and is far away from other clusters, which has an important influence on clustering results. Moreover, DBSCAN performs better under massive data. Therefore, we consider using DBSCAN to implement the large amount of features clustering and selection.

The proposed *Fesim* method uses distance metric and DBSCAN clustering algorithm to choose a subset of relevant features. Given a dataset D which have M instances and N features, $F = \{f_1, f_2, \cdots, f_N\}$. The feature selection algorithm identifies a subset of features F' with dimension n where $n \leq N$ and $F' \subseteq F$, that minimizes the redundancy between features. The *Fesim* has three phases: (1) constructing similarity matrix, (2) clustering features, (3) selecting representative features. The following part of this section will introduce the details of these three phases in turn.

3.1 Similarity Matrix

Prior to constructing similarity matrix, it is essential to implement data normalization as data preprocessing step. The advantage of normalization is to improve the accuracy. This is significant when it comes to distance calculation based algorithms. In our study, the algorithm needs to calculate the Euclidean distance as similarity degree to have same effect on the result.

Fesim apply the MinMaxScaler to scaler features. It needs clear given maximum and minimum values. The transformation is given by:

$$x_{std} = \frac{x - x_{min}}{x_{max} - x_{min}} \tag{1}$$

$$x_{scaled} = x_{std} \times (max - min) + min \tag{2}$$

where *max* and *min* are the feature range, we choose 1 and 0 as the maximum and minimum in our algorithm.

When processed all features of the dataset D, we will get a new dataset D' in which all value is at least 0 and at most 1. Each feature of dataset D' can be measured in Euclidean distance with other features. The formula measuring the distance of $f_i(x_1, x_2, \cdots, x_M)$ and $f_j(y_1, y_2, \cdots, y_M)$ is defined as:

$$dist(f_i, f_j) = \sqrt{\sum_{k=1}^{M} (x_k - y_k)^2} \tag{3}$$

The result of calculating all features can be represented by a matrix called similarity matrix. The form of the matrix is as follows:

$$matrix_s = \begin{bmatrix} 0 & & & & \\ E_{21} & 0 & & & \\ E_{31} & E_{32} & 0 & & \\ \vdots & \vdots & \vdots & \ddots & \\ E_{N1} & E_{N2} & E_{N3} & \cdots & 0 \end{bmatrix}_{(N \times N)} \tag{4}$$

where the $matrix_s$ is a square matrix, and the values of the matrix represent the Euclidean distance of the $f_{row^{th}}$ and $f_{column^{th}}$. Obviously, the similarity matrix is a symmetric matrix since the distance between f_i and f_j are the same as which between f_j and f_i. The algorithm of this section is as follows:

Algorithm 1. FESIM_Preprocess

Function FESIM_Preprocess (Dataset D)
1: $D' = MinMaxScaler(D)$
2: **For** each $f_i \in D'$ **do**
3: **For** each $f_j \in D'$ **do**
4: $matrix_s[i][j] = matrix_s[j][i] = distance(f_i, f_j)$
5: return $matrix_s$

3.2 Feature Clustering

In order to grouping features $f_1, f_2, \cdots, f_N$ into different clusters, we convert the concept of points in DBSCAN to the features. Since *Fesim* uses the DBSCAN for feature clustering, most definitions in this section are the extension of concepts in DBSCAN. Therefore, its two hyper parameters *minPts* and *Eps* applied to the feature representation becoming *minFts* and *Eps*, which respectively denoted as radius for the neighborhood and the minimum number of features in the neighborhood. And we divide all features into three categories namely: core feature, border feature, and atypical feature.

Definition 1. The **neighborhood**, denoted as $Nei(f_N)$, means that the Euclidean distance of two features f_i and f_j is less than *Eps*, and defined by:

$$Nei(f_i) = \{f_j | dist(f_i, f_j) \leq eps, f_j \in F\} \tag{5}$$

Definition 2. Afeature is a **core feature** if its neighborhood number was not less than *minFts*.

Definition 3. **Aborder feature** has fewer neighborhood number than *minFts*, but is the neighborhood of a core feature.

Definition 4. An **atypical feature**, which extended from noise point, is any feature that is neither a core feature nor a border feature.

Definition 5. Afeature f_j is **directly density-reachable** from a feature f_i if f_i is a core feature and $f_j \in Nei(f_i)$.

Definition 6. Afeature f_i is **density-reachable** from a feature f_j if there is a chain of features $f_1, \cdots, f_N$, with $f_1 = f_j$, $f_N = f_i$ such that f_i is directly density-reachable from f_{i+1}.

The details of the *FESIM Clustering* algorithm in the section are shown in Algorithm 2.

Algorithm 2. FESIM_Clustering

Input: *minFts*, *Eps*
Output: clusters, atypical_set
1: matrix$_s$ = *FESIM_Preprocess()*
2: **For** each $f_i \in F$ **do**
3:   **If** f_i have not been classified **then**
4:     **If** Count($Nei(f_i)$) >= *minFts* **then**   // f_i is a core feature
5:       search all features density-reachable from f_i
6:       assign new clusters
7:   **Else**
8:     add all feature which not classified to atypical_set

3.3 Representative Features Selection

In this section, we propose a strategy to choose features that can represent clusters. To obtain the optimal feature selection scheme, we propose to choose suitable hyper parameters in the step of features clustering so that the cluster size is as small as possible, and at least two or more atypical features which not belong to any cluster are marked. Because in real-world datasets, there is rarely a case where a large number of features are linearly related.

There are three cases shown in Algorithm 3 in terms of the number of clusters, atypical features and the number of features to select. The best case when the *Fesim* is much more accurate is that the number of clusters plus the number of atypical features is exactly equal to the n features to select (line 3–8). In this case, there are exactly n clusters as the atypical features are seen in a single feature cluster. The subset of features F' consist of one feature from each cluster. In this case, the feature that is farthest from all atypical features in each cluster can be selected and added to the atypical feature set so that the later selected features can be guaranteed to be the least related with other features in the same cluster.

Since most of the scenes do not make that the number of clusters plus the number of atypical features is exactly equal to the number of required features. If the feature number required to select is smaller than the number of remaining clusters (line 9–16), then the features with the most distant features in each cluster are calculated, and these features are sorted to select in descending order.

If the feature number is greater than the number of remaining clusters (line 17–22), the selected feature number is allocated according to the feature number of each cluster, and the corresponding number of border features are selected in each cluster. Because the features in the same cluster are already relevant, the simpler the selection strategy is, the more efficient the algorithm can be.

This strategy can be verified by testing the accuracy of clustering results in the experimental section. The details of the *FESIM Selection* algorithm are shown in Algorithm 3.

Algorithm 3. FESIM_Selection

```
Input: the number of features to select
Output: F′, an optimal subset of features
1:  clusters, atypical_set = FESIM_Clustering( )
2:  append all atypical features to F′
3:  If Count(clusters) + Count(atypical_set) == n_features then
4:    For each cluster ∈ {clusters} do
5:      For each atypical_feature ∈ {atypical_set} do
6:        For each feature ∈ cluster do
7:          select the feature with max distance between atypical_feature
8:      assign the feature to F′
9:  Else if Count(clusters) + Count(atypical_set) > n_features then
10:   For each cluster ∈ {clusters} do
11:     For each atypical_feature ∈ {atypical_set} do
12:       For each feature ∈ cluster do
13:         select the feature with max distance between atypical_feature
14:         append the feature to set<f>
15:   sort set<f> and choose top t maximum needed features
16:   assign t features to F′
17: Else
18:   calculate the number of features which need to be select
19:   For each cluster ∈ {clusters} do
20:     calculate the proportion of features to select for each cluster
21:     while proportion > 0 do
22:       F′ ← Search(border feature ∈ cluster)
23: return F′
```

3.4 Clustering Methods

To evaluate the clustering performance of feature subsets, we used k-Means algorithms, the most widely used and successful clustering method, to verify the results.

The k-Means algorithm is a partition-based method and its idea is very simple. But it is necessary to determine the number of clusters which finally grouped into several categories, then pick several points as the initial center point, and then iteratively reset the data points according to predefined heuristic algorithms until reaching the final goal, "the points within the class are close enough and the points between the classes are far enough". Since the similarity between all samples and each centroid is calculated every time, the k-Means algorithm converges slowly on large-scale datasets.

In order to have reference to the clustering results, we use the datasets are all labeled. So, the k value of k-means is the number of class labels for classification.

4 Experiments and Results

The following experiments are run on a personal computer with 3.6 GHz Intel 2 Core CPU, 4 GB memory, and 64-bit Centos7 operating system. All datasets of this section are from UCI dataset repository [12]. Detailed parameters are shown in Table 1.

Table 1. Experimental datasets

Dataset	Instances	Attributes	Classes	Attribute types
Mice Protein Expression [13, 14]	1080	77	8	Real
SCADI [15]	70	205	7	Integer
Epileptic Seizure Recognition [16]	11500	178	5	Integer
Gene expression cancer RNA-Seq [17]	801	20531	5	Real

5 Experiment Methods

We use scikit-learn, a free software machine learning library for Python, to compare the *Fesim* algorithm with several feature selection methods, namely: *SelectKBest* with analysis of variance, *Extra Trees Classifier* and *RFE*. Among them, the univariate feature selection, *SelectKBest* in our experiment, cannot analyze all types of values, so we take the *f_classif* metric in scikit-learn to measure discrete data and the *f_regression* metric to continuous data.

It should be noted that the following *Fesim* results are the outcome of repeated experiments to obtain optimal *minFts* and *Eps* values. And we don't specifically explain what the values of two hyper parameters are.

5.1 Clustering Results

To evaluate the accuracy of our algorithm, we use the well know clustering methods k-Means. We compare the clustering results measured with four metrics namely: Rand index [18], V-measures [19], Silhouette Coefficient [20] and Calinski-Harabaz score [21]. The larger the values of the four metrics are, the better the clustering effect is.

Above all, the TreeClassifier can automatically select the number of features, so the lines of the TreeClassifier in figures below only show its results with floating number of features, which aim to be compared with other methods clearly. The reason why the TreeClassifier get an unfixed number of features is that its main idea is to train a series of different decision tree models and use a random subset of the feature set in each tree.

Figures 1 and 2 shows the comparison the accuracy of different feature selection methods on different number of selected features in the two small datasets, Mice Protein Expression and SCADI. There are 1080 samples, 77 feature attributes and 8 labels in Mice Protein Expression dataset and 70 samples, 205 feature attributes and 7 labels in SCADI.

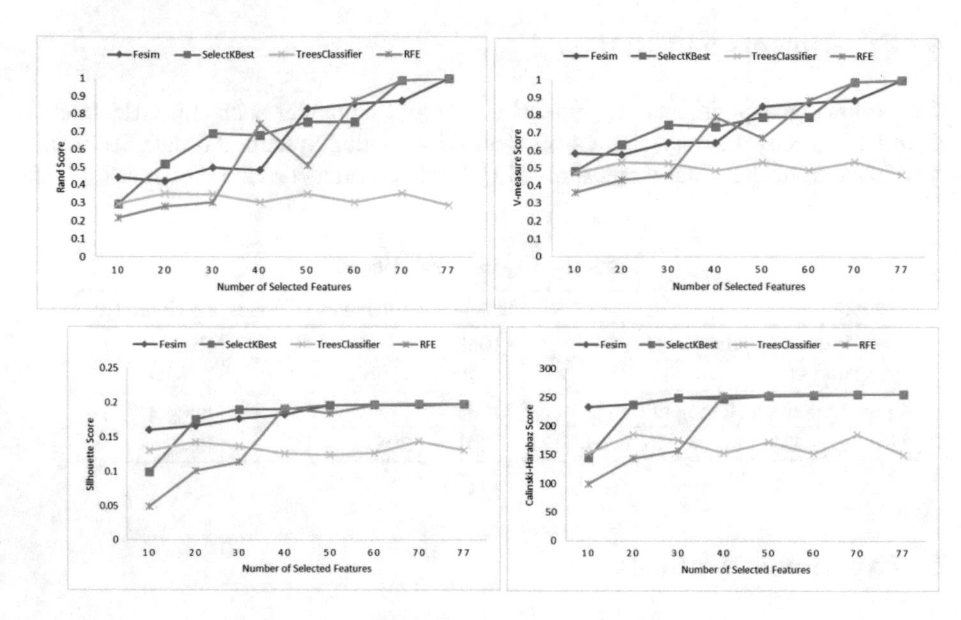

Fig. 1. Comparison of *Fesim* and other methods in Mice Protein Expression dataset

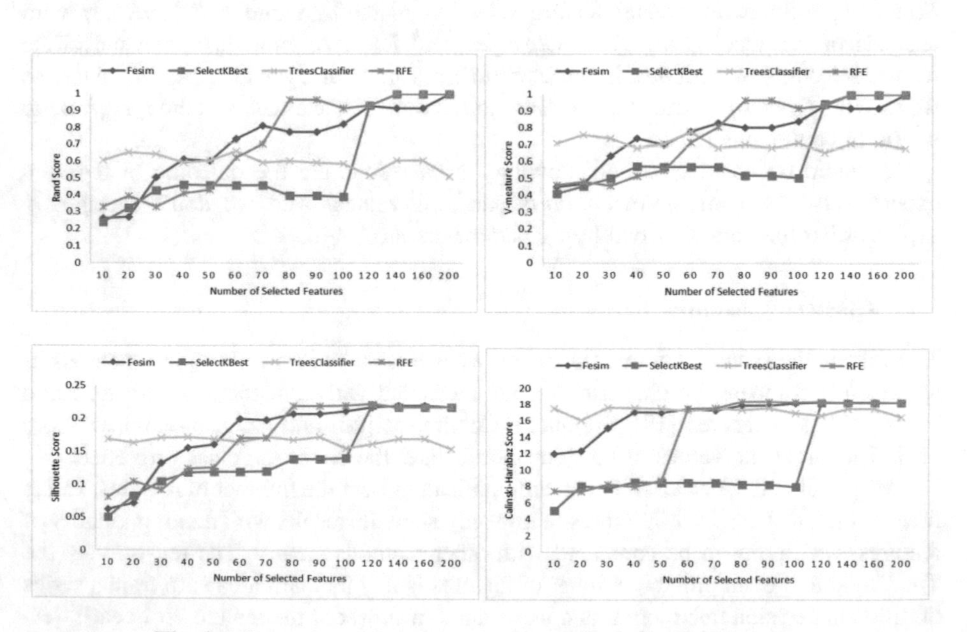

Fig. 2. Comparison of *Fesim* and other methods in SCADI dataset

From Fig. 1, we observe that all four metrics in TreeClassifier are significantly lower than *Fesim*. SelectKBest and RFE perform poorly when selecting fewer features. Moreover, *Fesim* can obtain more accurate results than other three methods when scored in Silhouette Coefficient and Calinski-Harabaz score.

From the Fig. 2, it is known that the accuracy of SelectKBest is higher than the accuracy of the first data set in Calinski-Harabaz metric. *Fesim* has a high degree of accuracy in most cases and the precision of it will be improved with the number of selected features increasing. The TreeClassifier selects about 50 features on the dataset, but it has worse accuracy in greater 50 feature numbers than other methods. SelectKBest has poor results until 120 features are selected to achieve a good accuracy.

The comparison of the accuracy of results when selecting different feature numbers in two larger datasets, Epileptic Seizure Recognition and gene expression cancer RNA-Seq dataset, where Epileptic Seizure Recognition dataset contains a large number of instances while the gene expression cancer dataset has many attributes, is shown in Figs. 3 and 4.

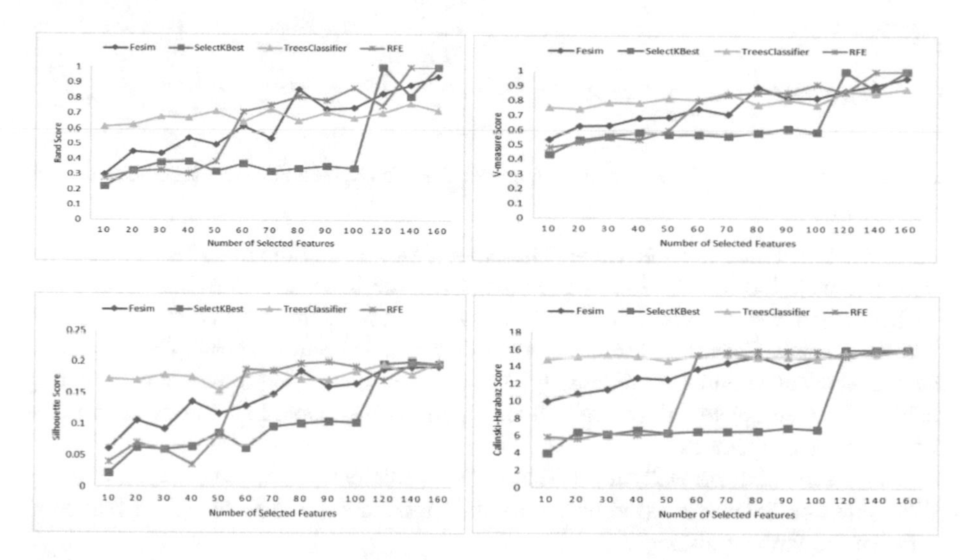

Fig. 3. Comparison of *Fesim* and other methods in Epileptic Seizure Recognition dataset

As can be seen from Fig. 3, TreeClassifier will automatically select more than 70 features to get the result, but it has lower scores in this situation. Because the TreeClassifier model cannot set the required number of features, it will lose the advantage of high accuracy when the number of feature selections is greater than the average feature number in TreeClassifier.

Figure 4 illustrates the impact of dimensions of the dataset on four methods. *Fesim* is less accurate than other algorithms when selecting a small number of features. When the number of features is over 600, all the measurements of four methods are close to the maximum.

All the above experiments illustrate that the clustering evaluation indexes show upward trend as the number of features increases. One reason is that the more the dimension features are, the more accurate the probability estimation is and the scores

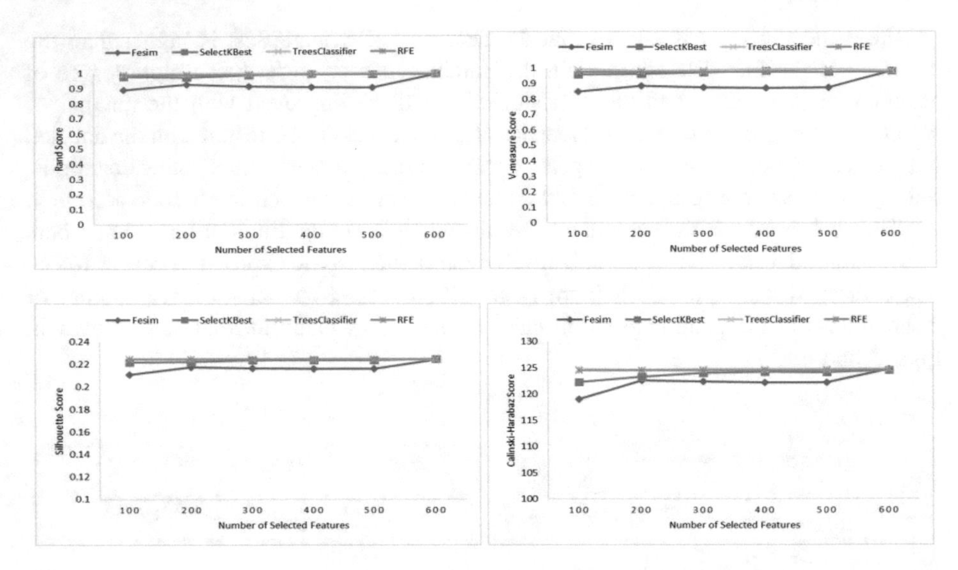

Fig. 4. Comparison of *Fesim* and other methods in gene expression cancer RNA-Seq dataset

are easier to obtain the larger value. The other reason is that the more features are, the more complete the information of the data is and the better the clustering results are.

Moreover, we can see that the accuracy of *Fesim* is poor while the number of features is small. Because when the number of selected features is small, *Fesim* tends to choose atypical features preferentially, which often cannot reproduce the clustering results of the complete dataset well. But its performance is still similar or even better than most other methods.

And *Fesim* perform well on the Silhouette Coefficients, since the calculation of the Silhouette Coefficients is based on the Euclidean distance, and *Fesim* select feature in terms of the farthest distance.

On the other hand, *Fesim* spends more time on calculating the similarity matrix in high-dimensional data, which takes longer than SelectKBest and TreeClassifier. But the average execute time of our method is only one tenth of RFE.

6 Conclusion

In this paper, we proposed a novel density clustering based feature selection algorithm named *Fesim* to solve the common curse of dimensionality problem in feature selection of high dimensional data and achieve better clustering. During the evaluation, several UCI datasets are used to validate the effectiveness of the *Fesim* in both the data size and data dimensions aspects. The experiment results indicate that *Fesim* improved the clustering accuracy on the UCI data. However, although *Fesim* have excellent performance in feature selection, it need to find the appropriate values of two specific parameters, thus we plan to improve the adaptivity of parameters in our future work, which will make *Fesim* can be easier employed in real scenarios.

Acknowledgements. This research is supported by the National Natural Science Foundation of China (Grant No. 61462012, No. 61562010, No. U1531246), the Innovation Team of the Data Analysis and Cloud Service of Guizhou Province (Grant No. [2015]53), Science and Technology Project of the Department of Science and Technology in Guizhou Province (Grant No. LH [2016] 7427).

References

1. Kira, K., Rendell, L.A.: The feature selection problem: traditional methods and a new algorithm. In: Proceedings of the 10th National Conference on Artificial Intelligence, pp. 129–134 (1992)
2. Guyon, I., Elisseeff, A.: An introduction to variable and feature selection. J. Mach. Learn. Res. 1157–1182 (2003)
3. George, F.: An extensive empirical study of feature selection metrics for text classification. J. Mach. Learn. Res. 1289–1305 (2003)
4. Brassard, G., Bratley, P.: Fundamentals of Algorithmics, 1st edn. Pearson, London (1995)
5. Ester, M., Kriegel, H.P., Xu, X.: A database interface for clustering in large spatial databases. In: KDD, pp. 94–99 (1995)
6. Ester, M., Kriegel, H.P., Xu, X.: A density-based algorithm for discovering clusters a density-based algorithm for discovering clusters in large spatial databases with noise. In: International Conference on Knowledge Discovery and Data Mining, pp. 226–231. AAAI Press, Palo Alto (1996)
7. Tahir, N.M., Hussain, A., Samad, S.A.: Feature Selection for Classification Using Decision Tree. Research and Development, Malaysia (2006)
8. Au, W.-H.: Attribute clustering for grouping, selection, and classification of gene expression data. IEEE Trans. Comput. Biol. Bioinform. 83–101 (2005)
9. Liu, H.: A new feature selection method based on clustering. In: 2011 Eighth International Conference on Fuzzy Systems and Knowledge Discovery, vol. 2. IEEE (2011)
10. Maji, P.: Mutual information-based supervised attribute clustering for microarray sample classification. IEEE Trans. Knowl. Data Eng. **24**(1), 127–140 (2012)
11. Eshaghi, N., Aghagolzadeh, A.: FFS: an F-DBSCAN clustering- based feature selection for classification data. J. Adv. Comput. Res. Sari Branch, Islamic Azad University, Sari, I. R. Iran, pp. 43–54 (2017)
12. Asuncion, A., Newman, D.J.: UCI Machine Learning Repository, University of California, Department of Information and Computer Science, Irvine (2007). http://www.ics.uci.edu/~mlearn/MLRepository.html
13. Higuera, C., Gardiner, K.J., Cios, K.J.: Self-organizing feature maps identify proteins critical to learning in a mouse model of down syndrome. PLoS One **10**(6), e0129126 (2015)
14. Ahmed, M.M., Dhanasekaran, A.R., Block, A., Tong, S., Costa, A.C.S., Stasko, M., et al.: Protein dynamics associated with failed and rescued learning in the Ts65Dn mouse model of down syndrome. PLoS One **10**(3), e0119491 (2015)
15. Zarchi, M.S., SMM Fatemi Bushehri, Dehghanizadeh, M.:. SCADI: a standard dataset for self-care problems classification of children with physical and motor disability. Int. J. Med. Inf. (2018)
16. Andrzejak, R.G., Lehnertz, K., Rieke, C., Mormann, F., David, P., Elger, C.E.: Indications of nonlinear deterministic and finite dimensional structures in time series of brain electrical activity: dependence on recording region and brain state. Phys. Rev. E **64**, 061907 (2001)

17. Weinstein, J.N., et al.: The cancer genome atlas pan-cancer analysis project. Nat. Genet. **45** (10), 1113–1120 (2013)
18. Hubert, L., Arabie, P.: Comparing partitions. J. Classif. (1985)
19. Rosenberg, A., Hirschberg, J.: V-Measure: a conditional entropy-based external cluster evaluation measure (2007)
20. Rousseeuw, Peter J.: Silhouettes: a graphical aid to the interpretation and validation of cluster analysis. Comput. Appl. Math. **20**, 53–65 (1987)
21. Calinski, T., Harabasz, J.: A dendrite method for cluster analysis. Commun. Stat. (1974)

Segmentation Algorithm for the Evolutionary Biological Objects Images on a Complex Background

Yury Ipatov[1(✉)], Alexand Krevetsky[1], Yury Andrianov[1], and Boris Sokolov[2]

[1] Volga State University of Technology, Yoshkar-Ola, Russia
ipatovya@volgatech.net
[2] St. Petersburg Institute for Informatics and Automation of the Russian Academy of Sciences (SPIIRAS), Saint Petersburg, Russia

Abstract. In the paper, characteristic features of color images for evolving biological objects are investigated. To create an effective and fast algorithm, the choice of the color space for subsequent processing is justified. A mathematical model of images for localized objects is proposed. Estimates of the accuracy for image segmentation are received.

Keywords: Evolving objects · Image segmentation · Non-uniform background Dynamic scenes · Convex hull · Color space · Precision segmentation Color image · Clustering · Calorimetry · Model approximating an ellipsoid

1 Introduction

An important step in image recognition systems, machine vision is a spatial localization of objects in inhomogeneous background. To solve this problem, there are many approaches, the choice of which is conditioned by the context of the problem being solved and the limitations associated with the baseline data [1].

With the growth of computer technology becoming increasingly important question of image recognition show changes in the properties of objects in dynamic. Objectives of the dynamic images analysis are diverse and their content is essentially determined by the specific application. Dynamics of objects, target tracking, detection of changes, reconstruction of 3D models here are some examples of tasks that use fundamentally different mathematical approaches [2–4].

Among the wide range of dynamic objects can allocate a special type of evolved objects, examples of which are organisms (Fig. 1) or the crystal structure as they grow, other self-organizing systems, time changes in the landscape and etc. Image analysis tasks such objects a priori complicated by the uncertainty of template images classes due to variability in time the shape and size of objects, and sometimes their texture.

To automate the analysis and prediction of plants growth authors created software and hardware registration of such image classes (see Fig. 2). Scenes with evolving objects obtained with this complex served as the basis for the development of basic measurement algorithm evolving objects - segmentation of images, which is explored in this article.

R. Silhavy et al. (Eds.): CoMeSySo 2018, AISC 859, pp. 151–171, 2019.
https://doi.org/10.1007/978-3-030-00211-4_16

Fig. 1. An example of a dynamic scene with evolving object

Existing segmentation techniques can be divided into two areas: "supervised learning" and "unsupervised learning" [5, 6]. Thus, in the first case, the user manually selects areas in the image with the desired subject and the background for which the calculated value of the discriminatory features (such as color, shape, texture). On the set of values derived parameters are calculated decision rule. In the second case these parameters are calculated automatically, such as sample distribution histogram characteristics [1, 7].

Thresholding [7] a simple method of image segmentation on the primary characteristics (color or brightness of pixels), some homogeneous areas which differ in average brightness. It has a low computational complexity. There are approaches to global and local (adaptive) thresholds. The main drawback of these approaches is associated with the inability of distribution laws separation implementations color and background, dependent brightness or color when the form of these laws in colorimetric space is not known or is not symmetrical.

The method of the watershed [8] is a method of mathematical morphology, where the image is considered as a geological landscape, and the line of the watershed—a border separating objects. The levels of each brightness pixel serve as the height of this point. The transformation of the watershed catchment areas and calculates the line of ridges, and the relevant catchment areas of the image, and the lines are the boundaries of these areas. Application of this method to the considered class of images does not provide an adequate level of segmentation accuracy for the same reasons as the threshold processing [9].

Methods based on contour extraction [10] allow us to limit the scope of similar items. Using the methods of fill areas [11], it is possible to segment the scene investigation. However, frequent breaks contours in noisy areas are not allowed to directly use this approach.

Graph approaches are actively developing trend in image segmentation [12]. They consist in the representation of the scene as a weighted graph with vertices of the image. Weight edges of the graph reflects the "similarity" of points (determined by the metric, e.g., Euclidean). Dividing a scene into segments is calculated sections of the graph. The results depend on the initial segmentation algorithm parameters, and the question of how to optimize remains open.

Also, there are approaches based on cluster analysis of selective colors distribution in a given feature space [13]. Examples are the nearest neighborhood method, the method of potential functions [6]. This is calculated Mahalanobis distance, which is a difficult task in terms of computing. These approaches have a high computational complexity. The applicability of these approaches are also limited by the need to know the probability distribution of color is not only an object, but the background that cannot be observed in practice.

Fig. 2. Hardware-software complex of registration dynamically changing scenes

Direct application of the known approaches according to preliminary estimates ineffective in the case of plants evolving images. Therefore, the proposed solution of this problem is urgent.

2 Problem Definition

The object of research are dynamic characteristics of the changes in the specific area of the observed objects. The subject will be the discriminatory features of the objects images, as well as algorithms for processing and analysis.

As a criterion for assessing the quality of the scenes segmentation take the probability of erroneous assignment of pixels to the object or the background. Disincentives serve all non-target objects (background), with similar brightness and color tone, and noise of the image sensor and digitization.

The proposed segmentation algorithm in evolutionary biological objects based on the selection of colorimetric space, providing the smallest intersection of the color distribution laws for objects and background. At the same time to reduce the complexity of determining the belonging to the object or the background region, the formation of the separating surface by an approximation of the equation order and constraint surface. A property of evolution is taken into account through the use of objects at different time intervals agreed separating surfaces.

3 Statistical Characteristics of the Evolving Objects Images

For visual methods of biological research today use digital images. Modern digital cameras have high resolution in brightness and color tone. Noise video channels and digitizing in this case can be approximated by an independent centered normal noise for each color component [1]. In addition to the noise of various biological objects have their own color blur, increasing by glare, different angles of illumination and other conditions of observation.

Figure 3 shows the histogram distribution of brightness of the object and background images Fig. 1, obtained via hardware and software (Fig. 2) with 2 Megapixel image sensor. It can be seen that the luminance component of the background has considerable statistical scatter values. It should be added that the identified spatial heterogeneity of the brightness statistical characteristics in the field of sensor view. Brightness value will not be the basic selective feature for objects of this class images.

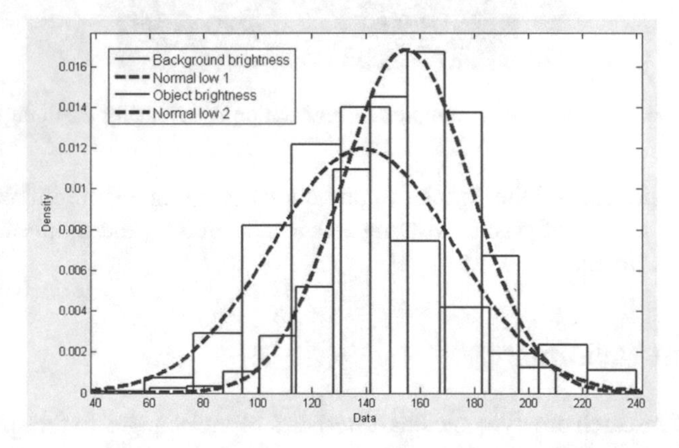

Fig. 3. Theoretical and empirical distributions brightness of the object and the background

Basic statistical characteristics of the background and object on the dynamically changing scenes are shown in Table 1. The sample size is $0.75 * 10^6$ points.

Table 1. Statistical characteristics of the background and object.

	Average brightness	Median brightness	Mode brightness	Minimal brightness
Background	138	135	94	48
Objects	155	157	166	64
	Maximum brightness	Variance	Skewness	Kurtosis
Background	233	1278	0.62	3.31
Objects	252	1120	−0.51	3.17

An analysis of the statistical data for observed images leads to the following conclusions. The average brightness of objects background ranges from 48 to 233, value $\sigma = 33.5$ is not contrary to the rule of 3σ. Investigated object plants have a pronounced elliptical shape, and border transition "object/background" stand high contrast. The objects that make up a heterogeneous background does not have a distinct shape, color tone is typical for large statistical variations in the brightness. Some background objects has hue similar to the object under study. Figure 4 shows the cumulative distribution function for the empirical values of elective areas background and objects, also shows the theoretical law under the same values of the expectation and variance. The theoretical curves in Figs. 3 and 4 are indicated by dotted lines.

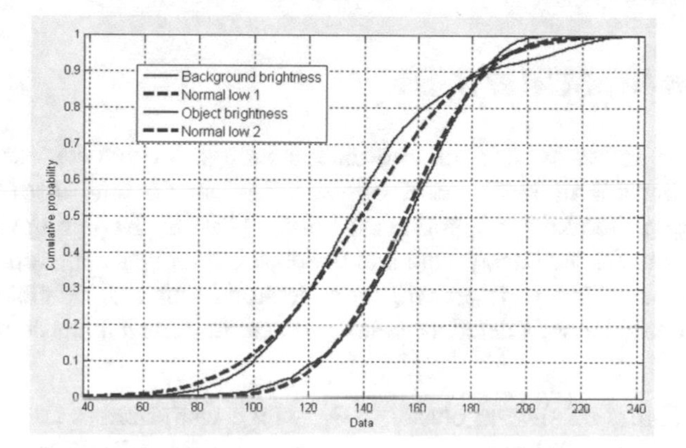

Fig. 4. Cumulative distribution function for the empirical and theoretical distribution of the background and object brightness

Check the normal random distribution laws luminance background of Kolmogorov-Smirnov test [14]:

$$\lambda = \sqrt{n}\left(\max_{x} |F_x(x) - \tilde{F}_x(x)\right) \tag{1}$$

where $\tilde{F}_x(x)$ – empirical distribution function, $F_x(x)$ – theoretical distribution function, n – sample size. For a level of significance $\alpha = 0.01$ threshold criterion $\lambda_\alpha = 1,628$ [14], and the value of statistics λ, calculated by the formula (1), is $\lambda = 4,389$ (at n = 0.75 * 10^6 points).

Therefore, since $\lambda > \lambda_\alpha$, the hypothesis of normal distribution for background is rejected in this class of images. A similar result for the distribution of objects.

The asymmetry coefficient Pearson

$$As = \frac{\sum_{i=1}^{m} (x_i - \overline{x})^3 * n_i}{\sigma^3 \sum_{i=1}^{m} n_i} \tag{2}$$

where m – number of bands, $\overline{x}$ – the average value and n_i – rate shows slight left-sided asymmetry of objects and right asymmetry for background. Indicator of kurtosis

$$Ex = \frac{\sum_{i=1}^{m} (x_i - \overline{x})^4 * n_i}{\sigma^4 \sum_{i=1}^{m} n_i} - 3 \tag{3}$$

It indicates slight peakedness distributions in both cases.

In general we can conclude that the brightness of the background points is not typical, well-defined structure and is insufficiently informative.

4 The Choice of Color Space

Since brightness information is not informative enough for this class segmentation of images, we consider in more detail of the color components distribution for the observed image in the known colorimetric spaces [15–18]. Exploring the sample distribution of colors for the background and the object color space appropriate to use the notion of a cluster. Under cluster will be understood union of several homogeneous elements that could be considered as an independent unit, which has definite properties [19].

RGB Space. Figure 5a shown: cluster – H_1, which characterizes the objects of flora and cluster statistically inhomogeneous background – H_2 in RGB space.

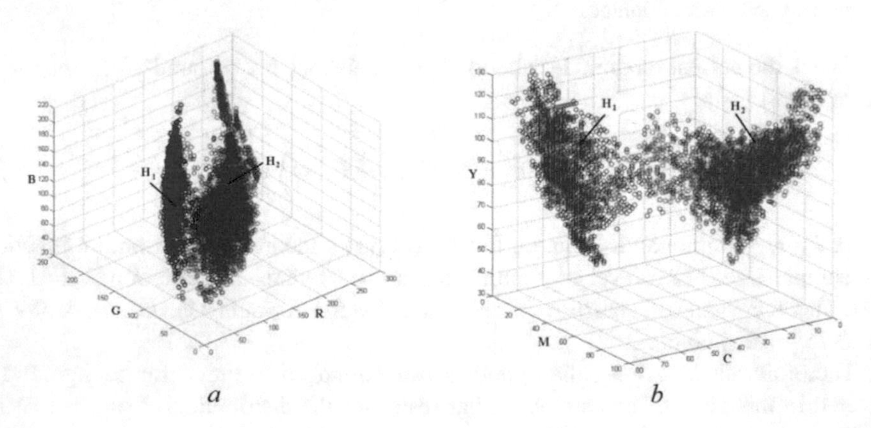

Fig. 5. Cluster distribution of the background and the object in the color space: *a* – RGB, *b* – CMY

MYK Space. Hardware-dependent standard CMY formed on the principles of subtractive synthesis and methods for non-emitting display [20]. It is characterized by simplicity and obtaining computer clusters without additional calculations to the appropriate plane (Fig. 5b).

XYZ Space. To obtain vector XYZ conversion matrix, there are several systems from RGB [21, 22]. The choice of the transformation matrix depends on the use of primary colors and the selected standard light source, that is, the percentage composition of the primary colors needed to produce white.

To select a three-part look at transformation matrix vector, where each element is described by its coordinates in three-dimensional space of the base. The measure of vectors similarity may serve by their scalar product.

The choice of the transformation matrix was carried out by the criterion of the lowest similarity to the basic standard CIE RGB 1931. This condition is in good agreement with the model of *Wide Gamut RGB* and *ProPhoto RGB*. Chart in Fig. 6 shows that they have a minimum scalar product (0.91–0.92). In addition, the models provide maximum coverage of the chromaticity diagram [20]. Figure 6 vertical postponed the value of the scalar product, and the horizontal the name of the model.

Figure 7a shows the clusters in the XYZ space for RGB ProPhoto model.

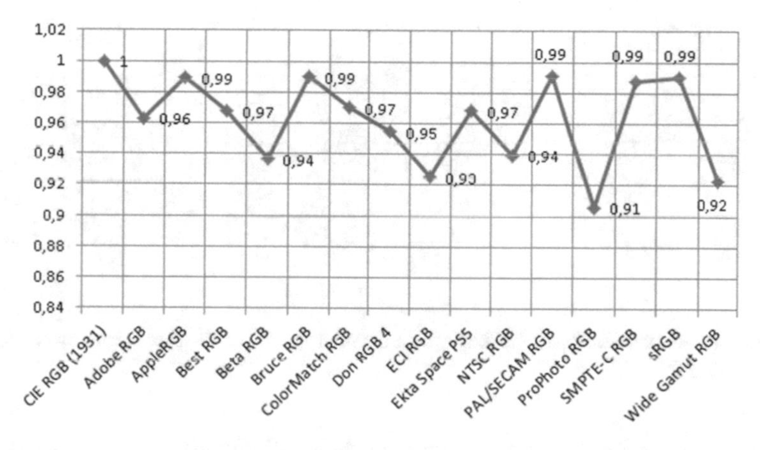

Fig. 6. Chart similarity of different XYZ spaces with CIE RGB 1931

UVW Space. Standard equal-contrast UVW colorimetric system, characterized in that all areas on the chromaticity diagram by any pair of colors at the same distance corresponding to the same color contrast [23]. The result of building clusters in this color space is represented in Fig. 7b.

HSV space (hue (color), saturation, value) [5] was created as a natural and closer to human color perception. There are versions analogue: HSI, HSB HSL [24, 25]. This group color space has some unique performance [26], in which the tone is invariant under certain types of lighting, shading and shadows, allowing segmentation threshold procedures in one coordinate colors. Figure 8 shows the clusters in the space of hue, saturation and value. Presentation of clusters in the data model has a number of distinctive features.

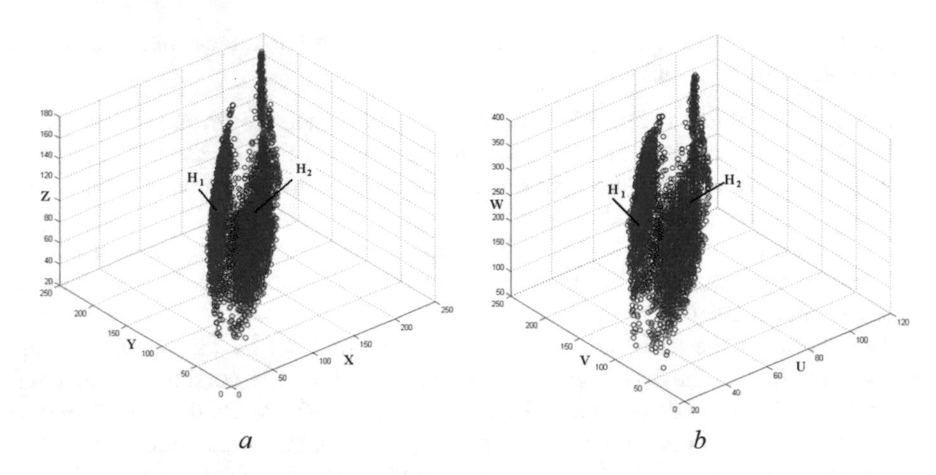

Fig. 7. Cluster distribution of the background and the object in the color space: *a* – XYZ, *b* – UVW

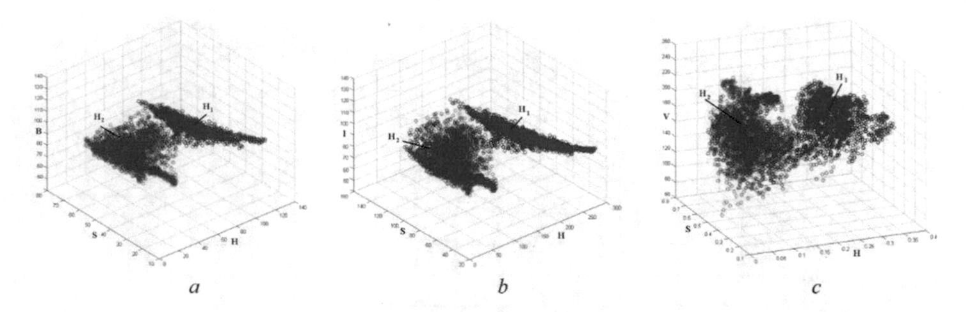

Fig. 8. Cluster distribution of the background and the object in the color space: *a* – *HSB, b – HSI, c – HSV*

L** space. The next class of device-independent and equally contrasting spaces Lab, Luv and LHC is based on brightness and two chromatic components [27, 28]. Since the components are ab linear transformation (X,Y) CIE, and uv associated with the (X,Y) – linear transformation. LHC – cylindrical coordinate system. It is consistent with both the empirical Munsell, and are consistent with the physiological model of color vision.

Figure 9 in the calculations for the illuminant D65 was taken the point of equal energy (0.9642, 0.8249, 1). It can be seen that the light clusters and the background object is insignificantly overlap.

*Y** space* are perceptual, device-dependent came from TV [29, 30], where luminance component and two color difference components. Figure 10 shows the clusters in spaces YUV, YIQ and YCrCb.

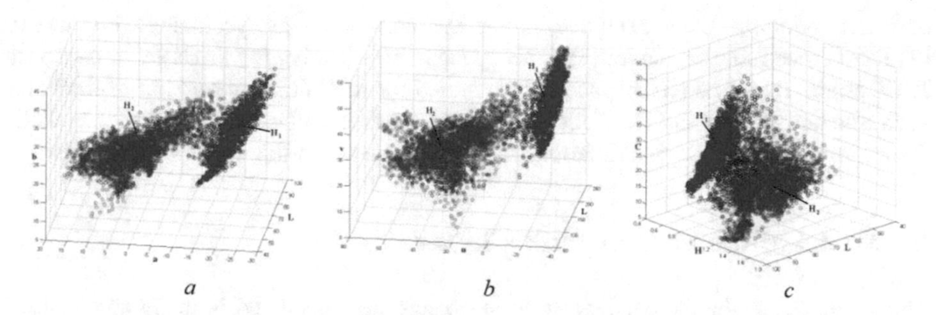

Fig. 9. Cluster distribution of the background and the object in the color space: *a* – *Lab*, *b* – *Luv*, *c* – *LHC*

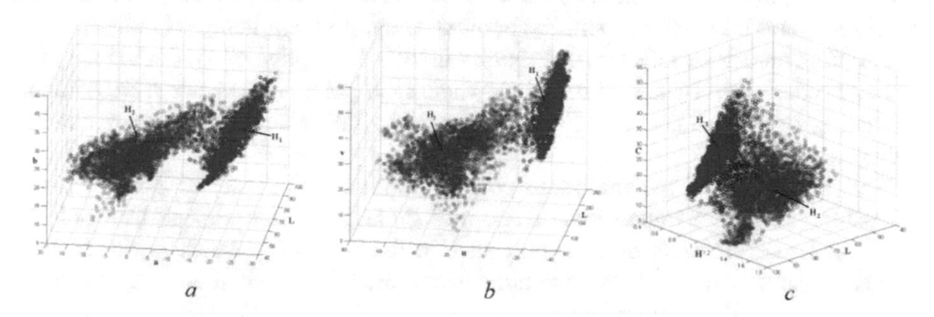

Fig. 10. Cluster distribution of the background and the object in the color space: *a* – *YCrCb*, *b* – *YIQ*, *c* – *YUV*

To analyze the shape and degree of the relative position of the cluster intersection for the background and object in each color space, according to the algorithm QuickHull [31] built a minimum convex hulls and analytic approximation of individual clusters. To evaluate the sensitivity for variations in the color space of the object and the background calculated the volume of shells inside the clusters (see Table 2). In order to assess the selectivity of specific volumes for the intersection region of space bounded by membranes.

Comparison of cluster shells subject and background leads to the following conclusions. (1) variations in the color of the object and the background only in the RGB and XYZ form elongated along its main axis almost axially symmetric shapes with

Table 2. The volume of the clusters in a single color cube.

RGB	HSB	HSI	HSV	Lab	LHC
0.044	0.019	0.017	0.043	0.003	0.007
Luv	YCrCb	YIQ	YUV	XYZ	UVW
0.033	0.007	0.011	0.011	0.017	0.026

collinear principal axes extended along the corresponding vectors of brightness; (2) Data space more selective; (3) RGB space more sensitive to variations in the space XYZ color. Thus, the RGB space is best suited for the initial submission of the color segmentation algorithm. The following describes the mathematical model for shells formation and cluster approximations, based on that make the above conclusions.

5 Model of the Convex Hull Clusters

The models describe the clusters of point objects are considered in many applications [32, 33]. Many of them are based on binding of specific point objects in the group, for example by a graph (a complete graph, a minimum spanning tree [34], a wire model [35]) associated continuous manner [36], and others. All these models are used in the specific recognition tasks where you need to take into account the features of the scene description, and take into account the context of the problem being solved.

Building models of selective distribution in their description of a point to imply that the objects described by the set of points where each point has a vector representation. Depending on the choice of colorimetric model, the value of the vector quantity change.

The calculation of the minimum convex hull on a finite set of points in two-dimensional and three-dimensional case, well studied and investigated in detail [31, 37]. Under the minimum convex hull shell will understand, which is the smallest convex polygon perimeter in the two-dimensional case, and a minimum size of external faces of the three-dimensional case.

Today there are many methods to build a minimum convex hull: Graham [38], Jarvis [39] Chen [40] and others. So if the dimension n – defines a set of points, and the value h – the number of convex hull vertices, the computational complexity of the algorithms have to be Graham $\Theta(n \cdot \log n)$, Jarvis – $\Theta(n^2)$ and Chen – $\Theta(n \cdot \log h)$. It should be noted that the Chen algorithm is unique synthesis of Jarvis and Graham, allowing give better computational performance.

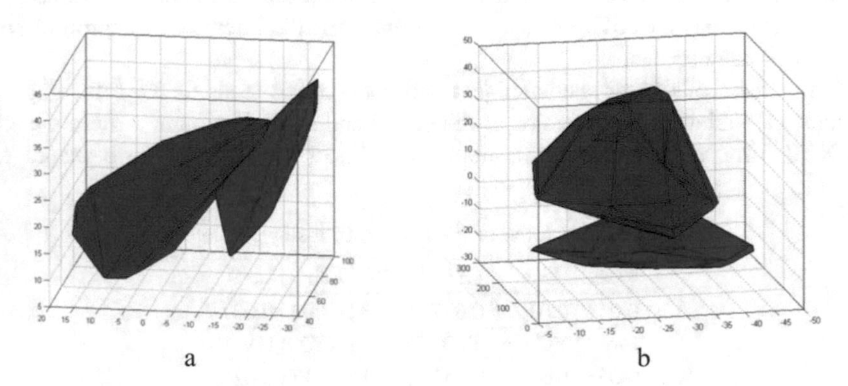

Fig. 11. Models of the convex hulls for sampling distributions: a – model Lab (Fig. 9a), b – model YCrCb (Fig. 10a).

To construct a minimum convex hull by the representation of points in the colorimetric spaces were experimentally selected four the most diverse forms of selected distribution laws. Among these color spaces: RGB, Lab, HSV and YCrCb (see Figs. 11 and 12).

Visualization of the results shows the most characteristic features of selective distribution in these color spaces. So the area of overlap joint spatial clusters of

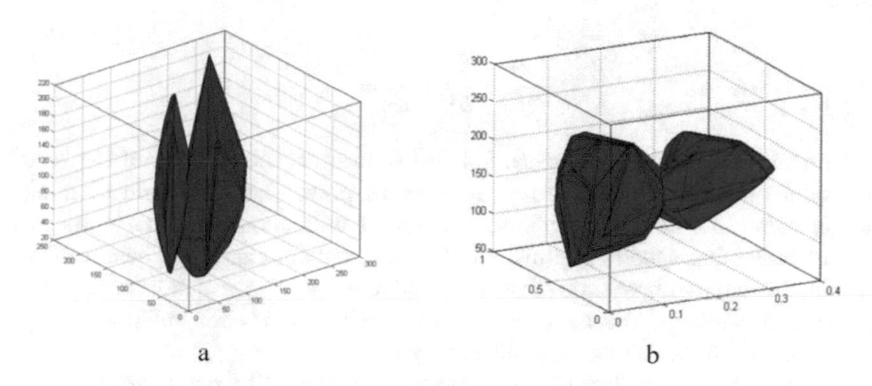

Fig. 12. Models of the convex hulls for sampling distributions: a – model RGB (Fig. 5a), b – model HSV (Fig. 8c).

complex spatial form, which requires the establishment in each case, their approach to the division of the observed sets. Creation of models based on the convex hull, requires the solution of problems belonging to the set, to solve a computationally intensive task. Therefore, it is advisable to reduce the dimension of the object described by approximation, any geometric primitives, which has an analytic representation. At the same time, it has the form of convex hulls clusters may be judged on their concentration and elongation along its major axis. Let's try to introduce analytical approximation that preserves these features clusters.

6 Model of the Convex Hull Clusters

Issues of approximation cluster polyhedra and smooth surfaces are considered in the works [41, 42]. The known approaches to the approximation can be divided into geometric and algebraic [43–46]. An algebraic approach is applied to the square of the distance between points and admits an analytic solution of the optimization problem with minimal computational overhead, so it stopped.

For the problem to be solved are restricted to surfaces of the second-order approximation problem considered set. The surfaces of the second order in a general form given by equations of the form

$$A \cdot x^2 + B \cdot y^2 + C \cdot z^2 + 2 \cdot D \cdot x \cdot y + 2 \cdot E \cdot x \cdot z \\ + 2 \cdot F \cdot y \cdot z + G \cdot x + H \cdot y + I \cdot z + J = 0. \tag{4}$$

where x, y, z – coordinates of points in space.

Equation (4) provides a six types of surfaces second order, but the form of sample sets give reason to stop the election on the ellipsoid. Canonical equations for this figure is given by

$$\frac{x^2}{a^2} + \frac{y^2}{b^2} + \frac{z^2}{c^2} = 1 \tag{5}$$

A review of existing approaches [44–46] shows that the creation of the "best" in terms of the of approximation to a given set of points there ellipsoid of minimum volume, ellipsoid principal component, ellipsoid with minimal trace of its matrix, ellipsoid containing specified terms, an ellipsoid contained in a polyhedron, Deakins ellipsoid, the maximum volume of the ellipsoid. Choice of approach may be due to the a priori uncertainty of the data, or computational methods for solving timing and accuracy of decision-making or problem context.

To reduce the computational cost decrease the power of points set for approximating clusters, leaving only the point of the convex hull shell. The result of this operation is shown in Figs. 13a, 14, 15 and 16a.

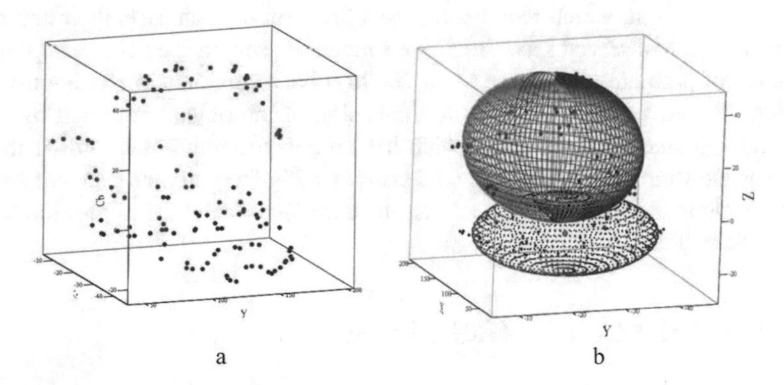

Fig. 13. Construction of the model for YCrCb color space: a – points of external shell (Fig. 11b), b – approximation of ellipsoids.

Equation of ellipsoid in the form with a given center:

$$\frac{(x - x_0)^2}{a^2} + \frac{(y - y_0)^2}{b^2} + \frac{(z - z_0)^2}{c^2} = 1 \tag{6}$$

where x_0, y_0, z_0 – center coordinates of the second order curve, and reduce it to a canonical form:

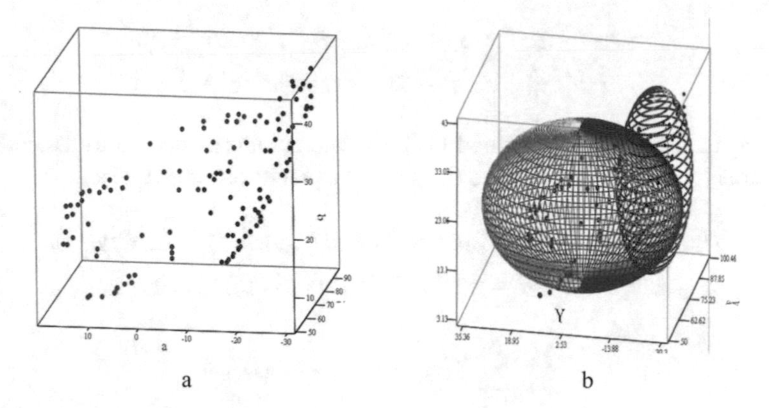

Fig. 14. Construction of the model for Lab color space: a – points of external shell (Fig. 11a), b – approximation of ellipsoids.

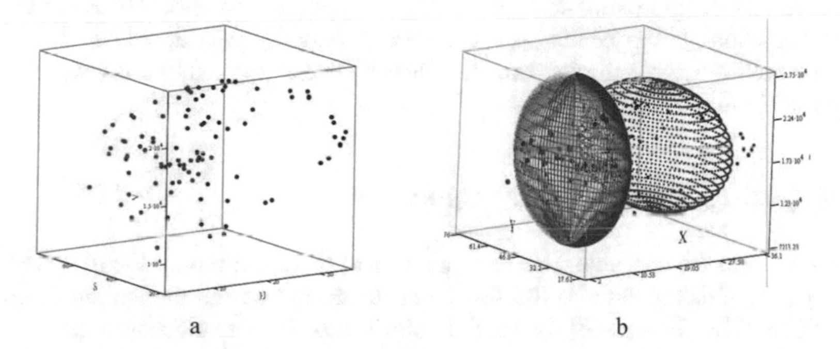

Fig. 15. Construction of the model for HSV color space: a – points of external shell (Fig. 12b), b – approximation of ellipsoids.

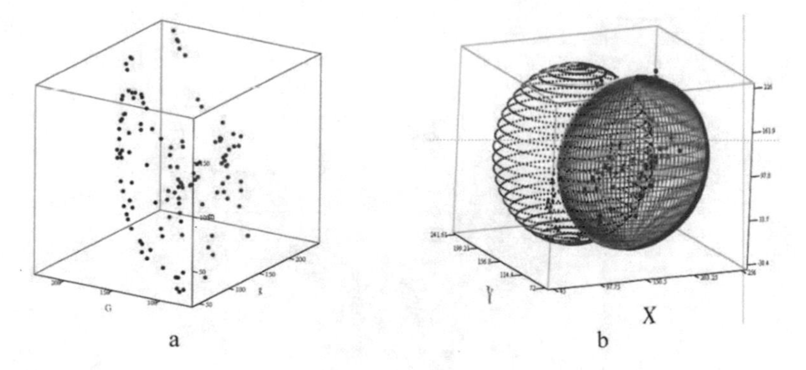

Fig. 16. Construction of the model for RGB color space: a – points of external shell (Fig. 12a), b – approximation of ellipsoids.

$$
\begin{aligned}
& a^{-2}x^2 - 2a^{-2}x \cdot x_0 + a^{-2}x_0^2 + b^{-2}y^2 - 2b^{-2}y \cdot y_0 \\
& + b^{-2}y_0^2 + c^{-2}z^2 - 2c^{-2}z \cdot z_0 + c^{-2}z_0^2 - 1 = 0
\end{aligned} \tag{7}
$$

Then using the least squares method [47], we find an approximation to the solution of this equation, which will minimize the sum of squared errors Eq. (7):

$$
\begin{aligned}
f(x_i, y_i, z_i) = & \hat{a}^{-2}x_i^2 - 2\hat{a}^{-2}x_i \cdot \hat{x}_0 + \hat{a}^{-2}\hat{x}_0^2 + \hat{b}^{-2}y_i^2 - 2\hat{b}^{-2}y_i \cdot \hat{y}_0 \\
& + \hat{b}^{-2}\hat{y}_0^2 + \hat{c}^{-2}z_i^2 - 2\hat{c}^{-2}z_i \cdot \hat{z}_0 + \hat{c}^{-2}\hat{z}_0^2 - 1, \quad i = 1 \ldots n,
\end{aligned} \tag{8}
$$

$$
\sum_{i=1}^{n} [f(x_i, y_i, z_i)]^2 \rightarrow \min. \tag{9}
$$

In this equation x_i, y_i, z_i – vertices of the shell clusters,x_0, y_0, z_0, a, b, c – the required parameters of the equation approximating, $f(x_i, y_i, z_i)$ – function approximation error.

The results of approximation shown in Figs. 13b, 14, 15 and 16b. As initial conditions approaching the center of the ellipsoid x_0, y_0, z_0 was appointed the gravity center of points the set for the original cluster. These approximations were used to justify the choice of color space.

7 Projecting Segmentation Algorithm

To build a segmentation algorithm as a measure of the proximity of the observed color to one of the clusters assume the likelihood function. As the elongation of clusters along different bright areas of the central color is uninformative because depending on lighting conditions, then we reduce the problem to a three-dimensional plane, display the selected color space in the plane of constant brightness (Fig. 17)

$$
A(x - x_0) + B(y - y_0) + C(z - z_0) = 0 \tag{10}
$$

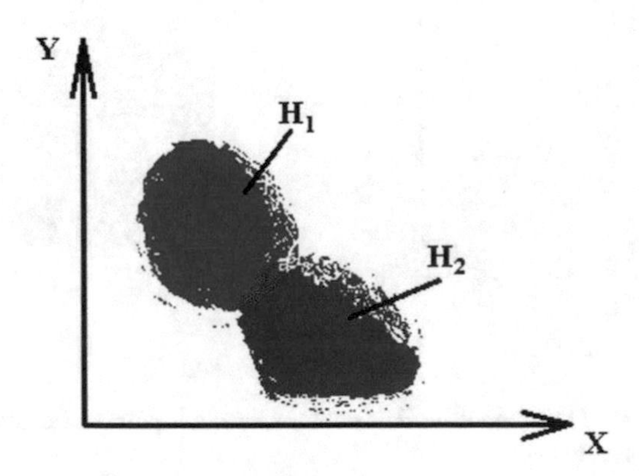

Fig. 17. The projection onto a plane equivalent to the brightness of selected laws for RGB model

The figure shows that the resulting projections are close to the centrally symmetric. A more detailed analysis shows that the density of the points of the two-dimensional clusters decreasing like the normal law.

In view of this as a measure of the proximity for the observed point to one of the clusters take two-dimensional function of the likelihood of alternative hypotheses. The calculation of the likelihood function of i-th hypothesis is performed by substituting the analyzed color in an approximation of conditional probability density of color

$$W(\mathbf{I}', i) = \frac{1}{2\pi\sigma_i^2}\exp\left\{-\frac{\|\mathbf{I}' - \mathbf{I}_i\|^2}{2\sigma_i^2}\right\} = \frac{1}{2\pi\sigma_i^2}\exp\left\{-\frac{(x - x_{0i})^2 + (y - y_{0i})^2}{2\sigma_i^2}\right\} \quad (11)$$

where $\mathbf{I}' = (x, y)$ – a two-dimensional projection of the color vector s analyzed the observed spatial element frame $s(n, m)$ with coordinates (n,m), i – number of color clusters (a testable hypothesis), $\mathbf{I}_i = (x_{0i}, y_{0i})$ – coordinates of gravity center for the respective distribution, $\|\cdot\|$ – norm of the vector, $\sigma_i = \sigma_{xi} = \sigma_{yi}$ – the standard deviation of the points from the center of the cluster (x_{0i}, y_{0i}) in the projection plane for each coordinate axis.

For the case of two alternative, $i = \{0, 1\}$, Bayesian optimal segmentation algorithm observed frame $s(n, m)$ with comparison of the likelihood ratio

$$\lambda(n, m) = \frac{W(\mathbf{I}'(n, m)|H_1)}{W(\mathbf{I}'(n, m)|H_2)} \quad (12)$$

with threshold $\lambda_{threshold}$ decision-making:

$$H(n, m) = \begin{cases} 1, & if \quad \lambda(n, m) \geq \lambda_{threshold}, \\ 0, & if \quad \lambda(n, m) < \lambda_{threshold}. \end{cases} \quad (13)$$

Figure 18a, likelihood ratio values are shown in the color plane $\lambda(x, y)$ at $\sigma_1 = 36.19$, $\sigma_2 = 33.5$.

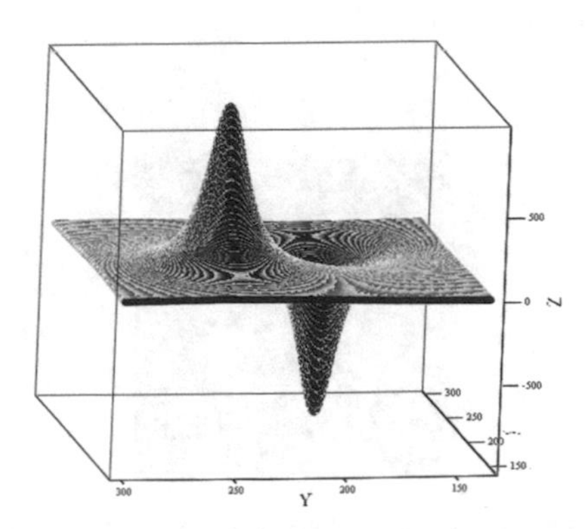

Fig. 18. The projection onto a plane equivalent to the brightness of selected laws for RGB model.

As a minimum sufficient statistics using log-likelihood ratio, for which, after collecting similar terms we obtain the expression for the corresponding algorithm:

$$H(n,m) = \begin{cases} 1, \ \mathit{if} \ u(n,m) > \ln \lambda_{threshold} - \ln \frac{\sigma_2^2}{\sigma_1^2}, \\ 0, \ \text{else}, \end{cases} \tag{14}$$

where $u(n,m) = (x - x_{02})^2 + (y - y_{02})^2 - (x - x_{01})^2 - (y - y_{01})^2$.

Accounting for the evolution of the object is performed by correcting estimates (x_{0i}, y_{0i}) for each iteration based on the results of segmentation. General notation of the algorithm then takes the form

$$H_k(n,m) = \begin{cases} 1, & \mathit{if} \quad \lambda_k(n,m) \geq \lambda_{threshold}, \\ 0, & \mathit{if} \quad \lambda_k(n,m) < \lambda_{threshold}, \end{cases} \quad k = 1, 2, 3, \ldots, K, \tag{15}$$

where k – the step number of object evolution (frame number $s_k(n,m)$), K – the number of evolution steps,

$$\lambda_k(n,m) = \frac{W(\mathbf{I}'(n,m)\,|\,H_1, x_{02}^{(k-1)}, y_{02}^{(k-1)}, \sigma_2^{(k-1)})}{W(\mathbf{I}'(n,m)\,|\,H_2, x_{01}^{(k-1)}, y_{01}^{(k-1)}, \sigma_1^{(k-1)})} \tag{16}$$

Estimates $x_{02}^{(0)}, y_{02}^{(0)}, \sigma_2^{(0)}, x_{01}^{(0)}, y_{01}^{(0)}, \sigma_1^{(0)}$ of the zero step formed at the stage of learning. For each subsequent iteration estimation of parameters conditional distributions of colors on the newly formed clusters object and background:

$$\left(x_{02}^{(k)}, y_{02}^{(k)}, \sigma_2^{(k)}, x_{01}^{(k)}, y_{01}^{(k)}, \sigma_1^{(k)}\right) = \Psi[s_k(n,m), H_k(n,m)] \tag{17}$$

Here, $\Psi[\bullet]$ – the operator of the parameter estimates forming the conditional distributions on the basis of the current frame segmentation

The work received frames illustrate the algorithm based on the minimum sufficient statistics are shown in Figs. 19 and 20.

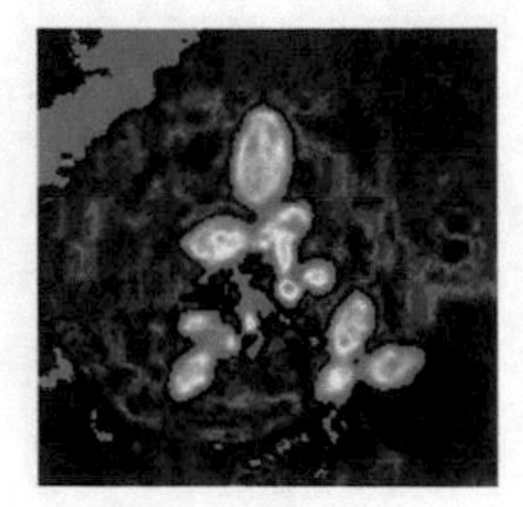

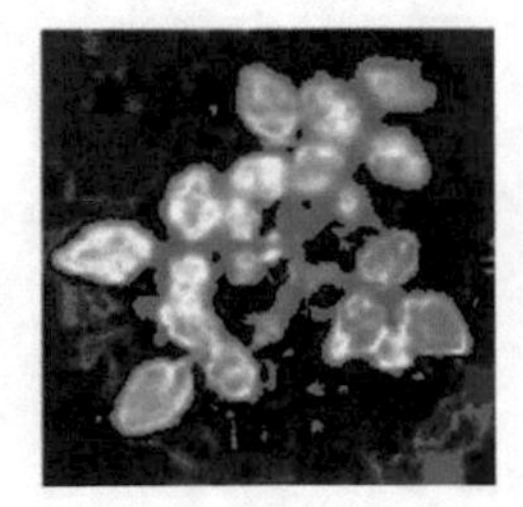

Fig. 19. The likelihood ratio of the scene $\lambda_1(n,m)$, $\lambda_2(n,m)$, $\lambda_3(n,m)$

  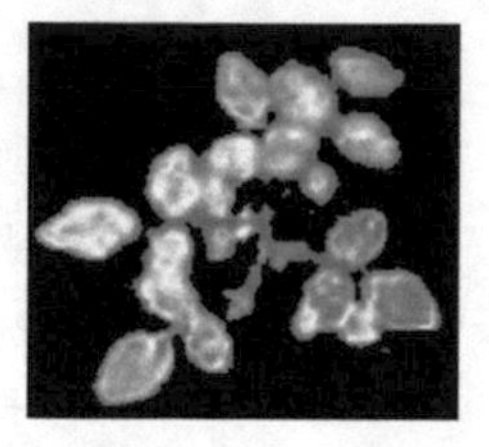

Fig. 20. Result after thresholding segmentation

Thus considered algorithm performs the mapping of the frame corresponding to the stages of evolution of objects in a set of image segmentation results

$$\mathbf{s}(n,m) = \{s_k(n,m)\}_{\overline{1,K}} \Rightarrow \mathbf{H}(n,m) = \{H_k(n,m)\}_{\overline{1,K}} \tag{18}$$

It should be noted that when an unknown background, such as switching to other observation conditions, the conditional probability distribution of the background color can be replaced by uniform. In such cases, the threshold value of the likelihood ratio appropriate to appoint the criterion of not exceeding a given probability of missing pixels with the color of the object.

8 Quality Assessment of the Algorithm Functioning

As estimates of the software implementation of the synthesized algorithm was implemented a series of experiments where the calculated estimates of the probabilities for false positives and false negatives. To calculate the wrong frame of reference pixels $s_k(n,m)$ for object and the background by expert assessments were set reference cards

$$g_k(n,m) = \begin{cases} 1, & \textit{for object}, \\ 0, & \textit{for background} \end{cases}, \ k = 1,2,3,\ldots,K. \tag{19}$$

Also, a comparison of the results (Table 3), according to estimates from known segmentation algorithms [21, 48]. Estimation error of the false positives was carried out on expression given by the formula

$$F = {}^{S_1}\!/_{S_2}, \tag{20}$$

where $S_1 = \sum_{k=1}^{K} \sum_{n,m} H_k(n,m) \cdot [1 - g_k(n,m)]$ – the total area of the background portion, falsely attributed to the type of the object of interest, $S_2 = \sum_{k=1}^{K} \sum_{n,m} [1 - g_k(n,m)]$ – the total area of the background on the reference cards.

The probability of false negatives

$$M = {}^{S_3}/_{S_4}, \tag{21}$$

where $S_3 = \sum_{k=1}^{K} \sum_{n,m} g_k(n,m) \cdot [\![H_k(n,m) - 1]\!]$ – the total area of the fragments of the object, referred to the background segmentation, $S_3 = \sum_{k=1}^{K} \sum_{n,m} g_k(n,m)$ – total area of the targets in the reference cards.

Table 3. Comparison of segmentation algorithms

Segmentation algorithm	$\overline{F}$	$\overline{M}$
Watershed segmentation algorithm	0.713	0.429
Segmentation algorithm k-means	0.479	0.043
Contour based segmentation algorithm	0.105	0.289
Created segmentation algorithm	0.021	0.087

From the comparison data segmentation algorithms (see Table 3) that the algorithm has generated minimum error of the false positives $F = 0.021$ and satisfactory value of the false negatives $M = 0.087$.

9 Conclusions

The problem of image segmentation is an important step in the process of extracting information from the observed image on the form and the geometric characteristics of objects. Despite the accumulated of the world set of different approaches to solving this problem due variety of purposes applications and classes images directly using published algorithms, it is not efficient enough for evolving biological objects.

During the investigation of this issue to the evolving plant objects revealed that the most informative feature for separating objects from the background image is a color characteristic that is invariant to brightness. In this paper a color space that is different from standard dimensions and at the best separability object color clusters and different backgrounds within the application.

It is shown that quasi-optimal Bayesian segmentation algorithm presented in the resulting color space pixel image is to convert the set of pictures stages of evolution object in the image likelihood ratios and binary quantization. Options segmentation algorithm on each evolutionary step of the algorithm are assigned recursively estimated parameters of the conditional probability color distributions of the previous stage. This allows more accurate assessment and adapt of the decision rule changes in the color characteristics of an object in the course of its evolution.

Synthesized segmentation algorithm provides the practical significance accuracy of the results and win the sum of the average probability of errors at the popular methods of watersheds and the method based on the edge contours.

As the observed image pixel color input segmentation algorithm can be assigned, and an integral characteristic of the neighborhood of the pixel, for example, the median or average in the 3 * 3. Potentially, this will reduce the likelihood functions blur and improve the selectivity of the algorithm. However, the average color of the neighborhood blurs the boundaries of the object and the background, it has the opposite effect. Median color neighborhood is devoid these disadvantages. This hardware and software system can serve as a tool for solving the original research, engineering and educational problems experienced.

Work supported by state order of the Ministry of Education and Science of the Russian Federation No. 2.31.35.2017/4.6, RFBR No. 16-01-00451a, state research 0073–2018–0003.

References

1. Pratt, W.K.: Introduction to Digital Image Processing. CRC Press, Boca Raton (2013)
2. Lukjanica, A.A., Shishkin, A.G.: Cifroavja obrabotka izobrazhenij. Aj-Jes-Jes Press, Moscow (2009)
3. Hartley, R., Zisserman, A.: Multiple View Geometry in Computer Vision. Cambridge University Press, Cambridge (2000)
4. Forsyth, D., Ponce, J.: Computer Vision: A Modern Approach. "Williams" Publishing House, Moscow (2004)
5. Gonzalez, R.C., Woods, R.E.: Digital Image Processing, 3rd edn. Prentice-Hall, New Jersey (2008)
6. Ajzerman, M.A., Braverman, J.M., Rozanojer, L.I.: Metod potencialnyh funkcij v teorii obuchenija mashin. Nauka, Moscow (1970)
7. Bakut, P.A., Kolmogorov, G.S., Vornovickij, I.J.: Segmentacija izobrazhenij. Metody porogovoj obrabotki. Zarubezhnaja radiojelektronika. (10), 6–24 (1987)
8. Vincent, P.S.: Watersheds in digital space: an efficient algorithms based on immersion simulation. IEEE Trans. Pattern Anal. Mach. Intell. **13**(6), 583–598 (1991)
9. Capaev, A.P., Kretinin, O.V.: Metody segmentacii izobrazhenij v zadachah obnaruzhenija defektov poverhnosti. Kompjuternaja optika **36**(3), 448–452 (2012)
10. Furman, Y.A.: Introduction to Contour Analysis and Its Applications in Image and Signal Processing. "Fizmatlit" Publisher, Moscow (2003)
11. Rogers, D.: Algorithmic Foundations of Computer Graphics. Mir, Moscow (1989)
12. Felzenswalb, P., Huttenlocher, D.: Efficient graph-based image segmentation. Int. J. Comput. Vis. **59**(2) (2004)
13. Manylov, I.V.: Ocenka tochnosti raspoznavanija klassov pri avtomatizirovannoj obrabotke ajerofotosnimkov. Izv. vuzov. Priborostroenie **54**(5), 35–39 (2011)
14. Statisticheskij analiz dannyh, modelirovanie i issledovanie verojatnostnyh zakonomernostej. Kompjuternyj podhod: monografija. Novosibirsk: Izd-vo NGTU, 888 p. (2011)
15. Palus, H.: The colour image processing handbook. In: Sangwine, S.J., Home, R.E.N. (eds.) Representations of Colour Images in Different Colour Spaces, pp. 67–90. Chapman & Hall, New York (1998)
16. Chochia, P.A.: Sglazhivanie cvetnyh izobrazhenij pri sohranenii konturov na osnove analiza rasstojanij v cvetovom prostranstve. Matematicheskie metody raspoznavanija obrazov, MMRO-13. Moskva, pp. 256–258 (2007)

17. Fisenko, V.T., Fisenko, T.: Metod avtomaticheskogo analiza cvetnyh izobrazhenij. Opticheskij zhurnal **70**(9), 18–23 (2003)
18. Goden, Zh.: «Kolorimetrija pri videoobrabotke» – M: Izdatel'stvo «Tehnosfera», 328 p. (2008)
19. Djuran, B., Odell, P.: Klasternyj analiz. In: Demidenko, E.Z. (ed.) Statistika (1977)
20. Ohta, N., Robertson, A.: Colorimetry: Fundamentals and Applications. Wiley, Hoboken (2005)
21. Goden, Zh.: Kolorimetrija pri videoobrabotke. Per. s fran. Tehnosfera, 328 p. (2008)
22. http://www.brucelindbloom.com/index.html?Eqn_RGB_XYZ_Matrix.html
23. Domasev, M.V., Gnatjuk, S.P.: Cvet, upravlenie cvetom, cvetovye raschety i izmerenija. SPb.: Piter, 224 p. (2009)
24. Shih, T.-Y.: The reversibility of six geometric color spaces. Photogram. Eng. Remote Sens. **61**, 1223–1232 (1995)
25. Munsell, A.: A Grammar of Color. Van Nostrand-Rcinhold, New York (1969)
26. Cheng, H.D., Sun, Y.: A hierarchical approach to color image segmentation using homogeneity. IEEE Trans. Image Process. **9**(12), 2071–2082 (2000)
27. CIE. Colorimetry. Official recommendation of the international commission on illumination: technical report 15.2, Bureau Central de la CIE, Vienna, Austria (1986)
28. Fairchild, M.D.: Color Appearance Models. Munsell Color Science Laboratory, Rochester (2004)
29. Fisenko, V.T., Fisenko, T.J.: Computer image processing and recognition. "SPbGU ITMO" Publisher, Saint-Peterburg (2008). 192 p.
30. Loughren, A.V.: Recommendations of the national television system committee for a color television signal. J. SMPTE **60**(321–326), 596 (1953)
31. Barber, C.B., Dobkin, D.P., Huhdanpaa, H.T.: The Quickhull Algorithm for Convex Hull, GCG53. The Geometry Center, Minneapolis (1993)
32. Tochechnye polja i gruppovye obekty/Ja. A Furman, A. A Rozhencov, R. G. Hafizov, D. G. Hafizov, A. V. Kreveckij, R. V. Eruslanov; pod obshh. red. prof. Ja. A. Furmana. FIZMATLIT, 440 p. (2014)
33. Fieguth, P.: Statistical Image Processing and Multidimensional Modeling. Springer, Cham (2011)
34. Ipatov, Y.A., Kreveckij, A.V.: Metody obnaruzhenija i prostranstvennoj lokalizacii grupp tochechnyh obektov. NB: Kibernetika i programmirovanie (6), S.17–S.25 (2014). https://doi.org/10.7256/2306-4196.2014.6.13642. http://e-notabene.ru/kp/article_13642.html
35. Rjabinin, K.B., Furman, J.A., Krasilnikov, M.I.: Provolochnaja model prostranstvennogo gruppovogo tochechnogo obekta. Avtometrija, (3), 3–16 (2008)
36. Kreveckij, A.V., Chesnokov, S.E.: Kodirovanie i raspoznavanie izobrazhenij mnozhestv tochechnyh obektov na osnove modelej fizicheskih polej. Avtometrija (3), 80–89 (2002)
37. Preparata, F.: Vychislitelnaja geometrija: Vvedenie. In: Preparata, F., Shepmos, M. (ed.) Mir, 478 p. (1989)
38. Graham, R.L.: An efficient algorithm for determining the convex hull of a finite planar set. Inf. Process. Lett. **1**, 132–133 (1972)
39. Cormen, T.H., Leiserson, C.E., Rivest, R.L., Stein, C.: Introduction to Algorithms, 2nd edn. MIT Press, Cambridge (2009)
40. Chan, T.M.: Optimal output-sensitive convex hull algorithms in two and three dimensions. Discret. Comput. Geom. **16**, 361–368 (1995)
41. Bronshtejn, E.M. Approksimacija vypuklyh mnozhestv mnogogrannikami. Sovremennaja matematika. Fundamentalnye napravlenija. **22**, 5–37 (2007)
42. Kamenev, G.K.: Optimalnye adaptivnye metody polijedralnoj approksimacii vypuklyh tel. Izd. VC RAN, 230 p. (2007)

43. Mikhlyaev, S.V.: Method for measuring the diameter of a growing crystal. Pattern Recogn. Image Anal. **15**(4), 690–693 (2005)
44. CVX: Matlab Software for Disciplined Convex Programming. http://cvxr.com/cvx/
45. Bedrincev, A.A., Chepyzhov, V.V.: Predstavlenie dannyh s pomoshhju jekstremalnyh jellipsoidov. Materialy konferencii «Informacionnye tehnologii i sistemy - 2013» 37-ja konferencija-shkola molodyh uchenyh i specialistov, Kaliningrad, Rossija, pp. 55–60 (2013)
46. Ovseevich, A.I., Chernous'ko, F.L.: Svojstva optimalnyh jellipsoidov, priblizhajushhih oblasti dostizhimosti sistem s neopredelennostjami. Izv. RAN «Teorija i sistemy upravlenija» (4), pp. 8–18 (2004)
47. Gill, P.E., Murray, W., Wright, M.H.: Practical Optimization, p. 402. Academic Press, New York (1982)
48. Deng, Y., Manjunath, B.S., Shin, H.: Color image segmentation. In: CVPR (1999)

Effectiveness of Recent Research Approaches in Natural Language Processing on Data Science-An Insight

J. Shruthi[1(✉)] and Suma Swamy[2]

[1] Department of Computer Science and Engineering, BMSITM, Bengaluru, India
shruthij.research@gmail.com
[2] Department of Computer Science and Engineering, Sir MVIT, Bengaluru, India

Abstract. With the exponentially increasing size and complexity of the data in present time, data quality has become a major concern with respect to data analytics. The potential capability of Natural Language Processing (NLP) is already known and being harnessed by various researchers to evolve up with some significant analytical process. However, there is less number of research works emphasizing on applying NLP over the data with complexity reported in current times in the area of big data. Therefore, the primary contribution of this manuscript is to review the most recent work towards NLP based approaches for data analysis where input data could be either text or non-textual too. The secondary contribution is to gauge the level of effectiveness from the existing research approach with NLP-based practices towards leveraging better data quality in data science.

Keywords: Data science · Natural Language Processing · Text mining Analytics · Big Data · Cloud · Clustering

1 Introduction

The sole purpose of Natural Language Processing (NLP) is to extract significant and highly logical information for any given textual data [1]. Basically, it comes in a category of artificial intelligence that offers potential capability of interactive communication bridge between humans and computers [2]. There are various applications of NLP e.g. summarization, fighting spam, machine translation, extracting information, answering question autonomously, etc. Although, it is a widely known fact that NLP is essentially meant for applying on data, but very least importance is given to the complexities associated with the data. In present time, there is an exponential rise of data to multi-fold that not only causes a problem in storage but also in performing analytical operation on it [3–5]. The presence of cloud environment assists in distributed data storage and retrieval as well as it is characterized by some good analytical tools too [6–10]. However, there are many pitfalls of existing analytics that are consistent reported by many researchers [11–14]. It will eventually mean that only a very less proportion of existing data is structured and majority of them are highly

R. Silhavy et al. (Eds.): CoMeSySo 2018, AISC 859, pp. 172–182, 2019.
https://doi.org/10.1007/978-3-030-00211-4_17

unstructured on which applying NLP is never possible. Hence, there is also a bigger impediment of applying NLP on existing form of complex data in the area of data science. Therefore, there are ongoing research work towards analyzing and obtaining information related to the text data and their possible processing capabilities over NLP. Reputed IT giants like Google and Siri have initiated NLP on their speech recognition-based application [15]. However, it is just a beginning and yet a long way to go to see proliferation of NLP based products and services inspite of various research works already carried out on it. Therefore, the prime motive of this paper is to discuss about the research progress in NLP. The organization of the paper is as follows: Sect. 2 discusses about the existing research approaches that are frequently used followed by Sect. 3 discussing about different approaches of NLP over text-based and non-text-based data. Section 4 discusses about the open research issues while Sect. 5 makes some concluding remarks about the paper.

2 Frequently Used Approaches of NLP

There are various mechanisms of applicability of NLP in order to perform text analysis. One of the frequently used approaches is *discourse analysis* that deals with establishing correlation among the isolated phrases and sentences [16, 17]. The second frequently used approach is *Word Sense Disambiguation* and *Co-reference Resolution* that deals with identification of sentences or word with equivalent meaning [18–21]. The third approach is called as *Named Entity Recognition* that deals with integrating different words as per some unique class function [22]. The approach of *Sentiment analysis* have been consistently researched on by various researcher that deal with extracting emotional-based information from given textual information [23]. It was also seen that conventional NLP-based approach makes usage of *chunking*, *part-of-speech*, and *stemming* more frequently. *Classification* is another frequently used approach in NLP which basically uses either rule-based or statistical-based methodology. It was also seen that *neural network* is predominantly utilized in classification and training operation while performing text analysis using NLP-based approach. *Deep Learning* and *Reinforcement Learning* techniques are usually found to be used by researchers while performing text analysis using NLP [24, 25]. Another frequently implemented research technique is that of *graph-based methodologies*. Some of them are (i) Clustering based classification [26], (ii) Usage of similarity measures [27], (iii) Textual features transformations to graphs [28], (iv) Conceptual and Semantic based graphical representation [29], (v) Ranking-based approaches [30], etc. It has been explored that similarity measures on the basis of multiple contextual parameters and clustering are some of the commonly used approaches seen in different forms in existing research approaches. However, it is not feasible to consider all the problems associated with text analysis in NLP, so existing research work considers discrete set of problems and case studies to carry out research work. At the same time, NLP application is not only limited to text-based contents but also for non-text based contents. The next section offers discussions about existing approaches where only recent literatures and their contributions are covered up.

3 Existing Research in NLP

At present, research work carried out towards NLP is basically in the direction of futuristic application with advance intelligence incorporation. It seeks consistent transformation in the communication technique between computer system and humans by harnessing the capabilities of cognitive as well as semantic technologies. A closer look into proportion of research work towards NLP is found basically in the direction of text analysis using different forms of application. It was also observed that existing approaches are reported to use semantics, linguistics, machine learning, and statistical methods in constructing various NLP-based solutions. This section briefs about signatory recent literatures of NLP with respect to different fields:

3.1 NLP Approaches Towards Text Analysis

Literature has witnessed some of the interesting and novel contribution of NLP towards text analysis in a very different way. Processing of unstructured text can be carried out using NLP as reported in the work of Fulda et al. [31] where discrete event information are extracted for better visual representation. The implementation assists in offering information about the time spent by the reader online for authoring specific set of documents. The complete prototype was designed using server scripts and is more of application form. Although, the outcome is benchmarked using error factor, but it was never evaluated by considering data complexity. The improved version of similar work towards visualization-based analysis was found in the work of Nafari and Weaver [32]. The study implements domain knowledge as the prominent key to extract information using NLP from the visualized information from various visual log files. However, the benefit was not found to be compared with any standards to retain the claim of this work. Research towards exploring visual context is also splitted in the form of analyzing textual content and image content discretely. Such form of approach was seen in the work of Ramisa et al. [33] where a canonical correlation analysis has been used for performing annotation of articles. However, the system doesn't behave in similar way for different dataset and this loophole is yet to be addressed. It was also seen that a significant adoption of distinct case study could further leverage the process of information extraction. The study carried out by Ki and Kim [34] have introduced a concept of contextual function for assisting in query processing using matrix-based methods considering the case study of analysis of data generated by nano-sensor. The technique also deals with elimination of noise from patent data. The extraction of technical information was carried out using conventional part-of-speech tagger using python script.

Such problems would have more addressed if linguistic-based approach was adopted. Literatures have reported the use of linguistic pattern as seen in the study of Poria et al. [35]. The study mainly uses sentiment analysis approach with polarity allocation for carrying out transformation of unstructured data. The authors have used computational intelligence-based approach for leveraging the accuracy. The study of Tang et al. [36] has used sentiment analysis for performing classification of both sentences as well as word. The authors have used classifiers e.g. neural network for carrying out training. Vioules et al. [37] have presented a significant application design of sentiment analysis where sentiments associated with suicide factor over social networking channels have been investigated.

Sentiment analysis was also investigated with respect to classification problems by Qiu et al. [38], where the authors have used fusion of features in order to perform sentiment classification. This is also one of the frequently adopted approaches found in the work of Yu et al. [39], Salas et al. [40], and Fang et al. [41]. Sahare and Dhok [42] have conventional k-classifier in order to perform segmentation of linguistic character right from document in the form of text to find 99% of accuracy. Figure 1 highlights the frequently adopted process flow using linguistic-based approach. Rodriqguez and Aguilar [43] have used semantic-based approach along with lexical to perform mining for unstructured data.

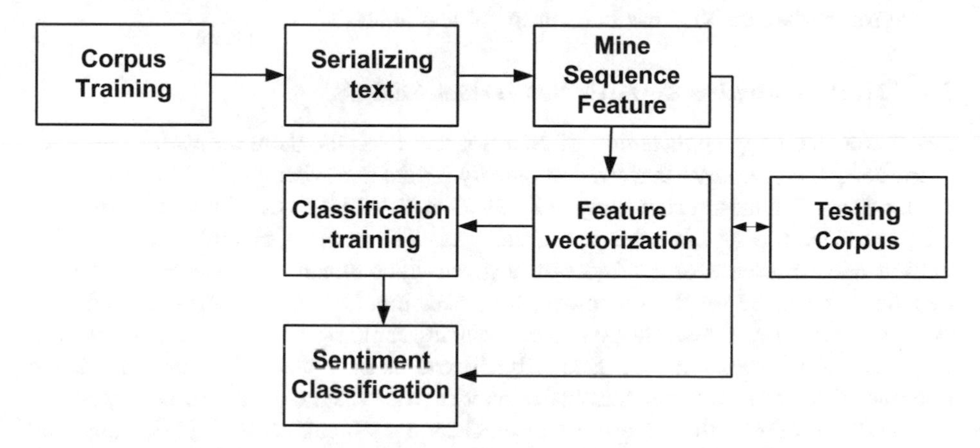

Fig. 1. Process flow of sentiment classification (Qiu et al. [38])

Usage of machine learning and its advanced form was also found in existing literature as it is suitable for extracting more intelligence from the unstructured data. The work carried out by Hassan and Mahmood [44] have combinedly used recurrent and convolution neural network for addressing the locality problems of different layers in it. Following expressions were deployed for implementing the above mentioned machine learning approach:

$$\mathrm{c} = [\mathrm{c}_1, \mathrm{c}_2, \mathrm{c}_3, \ldots \mathrm{c}_{\mathrm{n-h+1}}] \tag{1}$$

$$\mathrm{h_t} \rightarrow f(\mathrm{W_{xt}} + \mathrm{Uh_{t-1}} + \mathrm{b}) \tag{2}$$

$$p \rightarrow \exp(w_k^T x + b_k) \Big/ \sum_{k=1}^{k} \exp(w_k^T x + b_k) \tag{3}$$

The first and second expressions (1) and (2) were used for implementing convolution layer and recurrent layer respectively while the third expression is about implementing classification layer. The study outcome was also assessed with respect to different approaches of machine learning to find its higher accuracy. Usage of neural network of recursive form has been discussed to improve semantic operation by Wi and Chi [45]. According to author, a better form of memory management could be developed over complex structures of NLP using this technique. Thenmozhi and Aravindan [46] have used revised semantic-based approach over complex form of sentences for performing identification of paraphrase. The approach has also used integration of different similarity features for leveraging the identification process using machine translation attribute for carrying out extraction of feature. Whitehead et al. [47] have claimed that NLP could be used for analyzing system thinking process The authors have designed this model using statistical-based semantic approach using set of different conventional classifier. The work contributes to establish relationship between documents and system thinking. Therefore, it can be seen that there are different forms of approaches where NLP has been used for text analysis.

3.2 NLP Approaches Towards Non-textual Analysis

Apart from studying applicability of NLP for text analysis, there are various research work conducted to prove that NLP is equally productive when applied to non-textual contents too. The most recent study of Dilawari et al. [48] has considered implementing NLP on video file system using neural network. The target is mainly to obtain information about presence of a subject on a given video stream. An encoder as well as decoder is designed for the framework that takes input of video frames and offers an output in the form of text. Study towards ranking multimedia content was introduced by Etter and Domeniconi [49]. Similarly, literatures have also witnessed NLP based approach that extract text and formulates queries from video clips as claimed by Huang et al. [50]. Similarly, there are other researchers e.g. Kucuktuc et al. [51], Pouyanfar et al. [52], Wlodarczak et al. [53], and Zhang [54], who have dedicated research towards investigating applicability of NLP over multi-media file system as non-text contents using different forms of approach. All the approaches used have associated advantage as well as pitfalls too (Refer Table 1).

Table 1. Summary of existing research contribution in NLP

Author	Problems	Methodology	Advantage	Limitation
Fulda [31]	Event extraction from temporal reference	Prototyping, browser-based authoring	Better than manual curating process	Doesn't consider data complexity
Nafari and Weaver [32]	Visual query processing	Knowledge extraction from visual logs	Supports cross examination of queries	Less extensive analysis
Ramisa et al. [33]	Article annotation	Canonical correlation analysis	Supports extensive learning	No benchmarking

(*continued*)

Table 1. (*continued*)

Author	Problems	Methodology	Advantage	Limitation
Ki and Kim [34]	Technical information extraction	Semantic	Noise elimination	Doesn't consider heterogeneous text
Poria et al. [35], Qiu et al. [38], Tang et al. [36], Vioules et al. [37]	Flow analysis of sentiment data, word/sentence classification, identification of user behaviour	Sentiment analysis, neural network, text scoring	Supports contextual polarity	Less accuracy, less data complexity
Hassan and Mahmood [44]	Classification of sentence	Integrated convolution and recurrent neural network	Simplified process	Involves increased iterative steps, accuracy could be more increased
Sahare and Dhok [42]	Segmentation of multilingual character	k-nearest neighbor	Highest accuracy	Dimensional issues not addressed
Wu and Chi [45]	Memory management	Neural network	Increased classification accuracy	Involves highly iterative steps
Thenmozhi and Aravindan [46]	Identification of paraphrase	Machine translation attribute	Independent of sequence of sentences	dimensional issues not addressed
Whitehead et al. [47]	System thinking	Semantics, statistics	Supports multiword tokens	No benchmarking
Dilawari et al. [48]	Generating natural language from video	Neural network	Supports multiple events in video	Complexity issue no addressed
Eter [49]	Ranking of multimedia	Learning-based model	Large query processing	No benchmarking
Huang [50]	Auto quiz generation from video	Semantic-based approach	Simplified process	Narrowed scope of video dataset
Pouyanfar et al. [52]	Multimedia classification	Deep network	Supports multiple events in video	Lower accuracy
Wlodarczak et al. [53]	Mining video	Deep learning	Simplified implementation	Highly uncontrolled iteration
Zhang [54]	Video indexing, retrieval	Learning-by asking	Practical approach	Lower accuracy

4 Open Research Challenges

The previous section has discussed the existing mechanism of implementing NLP on both textual as well as non-textual contents. Irrespective of different forms of novel approaches, the beneficial features offered by existing system can be considered to be only marginal. Although, there are various studies carried out over unstructured data with NLP, but yet such approaches are not full-proofed against practical form of generated data in cloud computing environment. Hence, following are some of the open research challenges towards usage of NLP:

- **Lower Capability to Process Practical Data**: Majority of the existing research work has been carried out considering dataset, which doesn't necessarily represent the real-time data. Usually, the practical data is characterized by highly uncertain attribute, heterogeneity in its forms, ambiguity in its structure, and lacks any form of value within it. At present, there is no reported standard approach where NLP has been utilized to consider processing of such practical form of data with exponential complexity.
- **Marginal Identification Capabilities**: Existing approaches (whether text-based or non-text-based) uses lexical database to understand the alternative meaning of the word. However, human languages are comparatively complex in its structure which is quite a challenging task to precisely represent it in structured form. Therefore, there are less standard studies reported to offer more than 90% of accuracy in identifying true meaning of word/sentences considering complexity at the same time. The problem becomes much worst if transcribed documents are subjected to NLP for performing analysis.
- **Lower Cost-Efficient System**: Existing approaches have mainly used neural network, Support vector machine, deep learning, recurrent network, convolution network, as well as various revised forms of machine learning techniques. Owing to subjective target of the researchers towards solving one unique problem, the focus towards computational efficiency in terms of cost is completely missing. It is because there are various ubiquitous devices that not only generates the information but also assists in data processing. Hence, algorithms with more cost inclusion is highly impractical to be deployed in such resource-intensive devices.
- **Less Focus on Framework Modeling**: All the existing techniques are developed on the basis of a unique problem. However, it should be known that NLP is used for extracting valuable information from the complex data and hence there are various other aspects of the data that should be offered equal importance i.e. (i) dimensional problems of heterogeneous data, (ii) instance of generation of data, (iii) quality of the data bit-streams, (iv) degree of ambiguity associated with concurrent traffic flow of data, (v) presence of redundant data, etc. All these problems are highly interlinked with each other and hence it demands a formulation of a universal framework with more degree of applicability on various set of problems on various case studies as far as possible.

5 Conclusion

After reviewing the information related to existing system in this manuscript, it is now known that it is feasible to apply NLP in non-textual contents also unlike the conventional problems considering textual contents only. However, there is still a large scope for research to be carried out in text analytics itself. It is also known now, that NLP-based practices reported in literatures are not benchmarked efficiently as well as there is a less extensive study carried out with practical form of data. For this reason, majority of the existing approaches can be said to be more hypothetical and less applicable in real-world problems. It was also known that machine learning is another integrated part of all the NLP based practices, but all of the existing studies have almost no focus on discussing the computational complexity problems associated with the text analytics. Finally, it was known that existing system actually do not consider the standard form of data complexity and data dimensionality while applying NLP. Hence, our future work will be carried out in the direction of introducing a cost-effective computational model to address the problems associated with practical data.

References

1. Kurdi, Z.: Natural Language Processing and Computational Linguistics 2: Semantics, Discourse and Applications, vol. 2. Wiley, Hoboken (2018)
2. Lane, H., Howard, C., Hapke, H.: Natural Language Processing in Action. Manning Publications, Shelter Island (2018)
3. Ardagna, C.A., Ceravolo, P., Damiani, E.: Big data analytics as-a-service: issues and challenges. In: 2016 IEEE International Conference on Big Data (Big Data), Washington, DC, pp. 3638–3644 (2016)
4. Niño, M., Blanco, J.M., Illarramendi, A.: Business understanding, challenges and issues of Big Data Analytics for the servitization of a capital equipment manufacturer. In: 2015 IEEE International Conference on Big Data (Big Data), Santa Clara, CA, pp. 1368–1377 (2015)
5. Shuijing, H.: Big data analytics: key technologies and challenges. In: 2016 International Conference on Robots and Intelligent System (ICRIS), Zhangjiajie, pp. 141–145 (2016)
6. Barros, V.P., Notargiacomo, P.: Big data analytics in cloud gaming: players' patterns recognition using artificial neural networks. In: 2016 IEEE International Conference on Big Data (Big Data), Washington, DC, pp. 1680–1689 (2016)
7. Barga, R.S., Ekanayake, J., Lu, W.: Project Daytona: data analytics as a cloud service. In: 2012 IEEE 28th International Conference on Data Engineering, Washington, DC, pp. 1317–1320 (2012)
8. Schmid, S., Gerostathopoulos, I., Prehofer, C., Bures, T.: Self-adaptation based on big data analytics: a model problem and tool. In: 2017 IEEE/ACM 12th International Symposium on Software Engineering for Adaptive and Self-Managing Systems (SEAMS), Buenos Aires, pp. 102–108 (2017)
9. Makki, S., et al.: Fraud data analytics tools and techniques in Big Data era. In: 2017 International Conference on Cloud and Autonomic Computing (ICCAC), Tucson, AZ, pp. 186–187 (2017)
10. Schmid, S., Gerostathopoulos, I., Prehofer, C.: QryGraph: a graphical tool for Big Data analytics. In: 2016 IEEE International Conference on Systems, Man, and Cybernetics (SMC), Budapest, pp. 004028–004033 (2016)

11. Grolinger, K., Hayes, M., Higashino, W.A., L'Heureux, A., Allison, D.S., Capretz, M.A.M.: Challenges for MapReduce in Big Data. In: 2014 IEEE World Congress on Services, Anchorage, AK, pp. 182–189 (2014)
12. Jayasingh, B.B., Patra, M.R., Mahesh, D.B.: Security issues and challenges of big data analytics and visualization. In: 2016 2nd International Conference on Contemporary Computing and Informatics (IC3I), Noida, pp. 204–208 (2016)
13. Liu, Q., Ribeiro, B., Sung, A.H., Suryakumar, D.: Mining the Big Data: the critical feature dimension problem. In: 2014 IIAI 3rd International Conference on Advanced Applied Informatics, Kitakyushu, pp. 499–504 (2014)
14. Alam, A., Ahmed, J.: Hadoop architecture and its issues. In: 2014 International Conference on Computational Science and Computational Intelligence, Las Vegas, NV, pp. 288–291 (2014)
15. Hunckle, M., Article: This open-source AI voice assistant is challenging Siri and Alexa for market superiority. https://www.forbes.com/sites/matthunckler/2017/05/15/this-open-source-ai-voice-assistant-is-challenging-siri-and-alexa-for-market-superiority/#ed2d9e63ec01
16. Guiu, J.M.: Using latent semantic analyses and propositionalist methods in text comprehension. In: 2017 Computing Conference, London, pp. 187–191 (2017)
17. Geng, R., Jian, P., Zhang, Y., Huang, H.: Implicit discourse relation identification based on tree structure neural network. In: 2017 International Conference on Asian Language Processing (IALP), Singapore, pp. 334–337 (2017)
18. Punuru, J., Chen, J.: Learning taxonomical relations from domain texts using WordNet and word sense disambiguation. In: 2012 IEEE International Conference on Granular Computing, Hangzhou, China, pp. 382–387 (2012)
19. Cabezudo, M.A.S., Palomino, N.L.S., Perez, R.M.: Improving subjectivity detection for Spanish texts using subjectivity word sense disambiguation based on knowledge. In: 2015 Latin American Computing Conference (CLEI), Arequipa, pp. 1–7 (2015)
20. Shi, Z.: The design and implementation of domain-specific text summarization system based on co-reference resolution algorithm. In: 2010 Seventh International Conference on Fuzzy Systems and Knowledge Discovery, Yantai, Shandong, pp. 2390–2394 (2010)
21. Sleeman, J., Finin, T.: Type prediction for efficient coreference resolution in heterogeneous semantic graphs. In: 2013 IEEE Seventh International Conference on Semantic Computing, Irvine, CA, pp. 78–85 (2013)
22. Eletriby, M.R., Reynolds, T.L., Jain, R., Zheng, K.: Investigating named entity recognition of contextual information in online consumer health text. In: 2017 Eighth International Conference on Intelligent Computing and Information Systems (ICICIS), Cairo, pp. 396–402 (2017)
23. Yang, P., Chen, Y.: A survey on sentiment analysis by using machine learning methods. In: 2017 IEEE 2nd Information Technology, Networking, Electronic and Automation Control Conference (ITNEC), Chengdu, pp. 117–121 (2017)
24. Tang, Y., Wu, X.: Scene text detection using superpixel based stroke feature transform and deep learning based region classification. In IEEE Transactions on Multimedia
25. Zhu, F., Liu, Q., Zhang, X., Shen, B.: Protein interaction network constructing based on text mining and reinforcement learning with application to prostate cancer. IET Syst. Biol. **9**(4), 106–112 (2015)
26. Ali, I., Melton, A.: Semantic-based text document clustering using cognitive semantic learning and graph theory. In: 2018 IEEE 12th International Conference on Semantic Computing (ICSC), Laguna Hills, CA, pp. 243–247 (2018)
27. Tulu, C., Orhan, U.: PageRank based semantic similarity measure on a graph based Turkish WordNet. In: 2017 International Conference on Computer Science and Engineering (UBMK), Antalya, pp. 468–473 (2017)

28. Liu, H., Komandur, R., Verspoor, K.: From graphs to events: a subgraph matching approach for information extraction from biomedical text. In: Proceedings of the BioNLP Shared Task 2011 Workshop, pp. 164–172 (2011)
29. Al-Zaidy, R.A., Giles, C.L.: Extracting semantic relations for scholarly knowledge base construction. In: 2018 IEEE 12th International Conference on Semantic Computing (ICSC), Laguna Hills, CA, pp. 56–63 (2018)
30. Zhao, G., Zhang, X.: A domain-specific web document re-ranking algorithm. In: 2017 6th IIAI International Congress on Advanced Applied Informatics (IIAI-AAI), Hamamatsu, pp. 385–390 (2017)
31. Fulda, J., Brehmel, M., Munzner, T.: TimeLineCurator: interactive authoring of visual timelines from unstructured text. IEEE Trans. Visual Comput. Graph. **22**(1), 300–309 (2016)
32. Nafari, M., Weaver, C.: Query2Question: translating visualization interaction into natural language. IEEE Trans. Visual Comput. Graph. **21**(6), 756–769 (2015)
33. Ramisa, A., Yan, F., Moreno-Noguer, F., Mikolajczyk, K.: BreakingNews: article annotation by image and text processing. IEEE Trans. Pattern Anal. Mach. Intell. **40**(5), 1072–1085 (2018)
34. Ki, W., Kim, K.: Generating information relation matrix using semantic patent mining for technology planning: a case of nano-sensor. IEEE Access **5**, 26783–26797 (2017)
35. Poria, S., Cambria, E., Gelbukh, A., Bisio, F., Hussain, A.: Sentiment data flow analysis by means of dynamic linguistic patterns. IEEE Comput. Intell. Mag. **10**(4), 26–36 (2015)
36. Tang, D., Wei, F., Qin, B., Yang, N., Liu, T., Zhou, M.: Sentiment embeddings with applications to sentiment analysis. IEEE Trans. Knowl. Data Eng. **28**(2), 496–509 (2016)
37. Vioulès, M.J., Moulahi, B., Azé, J., Bringay, S.: Detection of suicide-related posts in Twitter data streams. IBM J. Res. Dev. **62**(1), 7:1–7:12 (2018)
38. Qiu, L., Lei, Q., Zhang, Z.: Advanced sentiment classification of tibetan microblogs on smart campuses based on multi-feature fusion. IEEE Access **6**, 17896–17904 (2018)
39. Yu, L.C., Wang, J., Lai, K.R., Zhang, X.: Refining word embeddings using intensity scores for sentiment analysis. IEEE/ACM Trans. Audio Speech Lang. Process. **26**(3), 671–681 (2018)
40. Salas, J.: Generating music from literature using topic extraction and sentiment analysis. IEEE Potentials **37**(1), 15–18 (2018)
41. Fang, Y., Tan, H., Zhang, J.: Multi-strategy sentiment analysis of consumer reviews based on semantic fuzziness. IEEE Access **6**, 20625–20631 (2018)
42. Sahare, P., Dhok, S.B.: Multilingual character segmentation and recognition schemes for indian document images. IEEE Access **6**, 10603–10617 (2018)
43. Rodriguez, T., Aguilar, J.: Knowledge extraction system from unstructured documents. IEEE Latin Am. Trans. **16**(2), 639–646 (2018)
44. Hassan, A., Mahmood, A.: Convolutional recurrent deep learning model for sentence classification. IEEE Access **6**, 13949–13957 (2018)
45. Wu, D., Chi, M.: Long short-term memory with quadratic connections in recursive neural networks for representing compositional semantics. IEEE Access **5**, 16077–16083 (2017)
46. Thenmozhi, D., Aravindan, C.: Paraphrase identification by using clause-based similarity features and machine translation metrics. Comput. J. **59**(9), 1289–1302 (2016)
47. Whitehead, N.P., Scherer, W.T., Smith, M.C.: Use of natural language processing to discover evidence of systems thinking. IEEE Syst. J. **11**(4), 2140–2149 (2017)
48. Dilawari, A., Khan, M.U.G., Farooq, A., Rehman, Z.U., Rho, S., Mehmood, I.: Natural language description of video streams using task-specific feature encoding. IEEE Access **6**, 16639–16645 (2018)
49. Etter, D., Domeniconi, C.: Multi2Rank: multimedia multiview ranking. In: 2015 IEEE International Conference on Multimedia Big Data, Beijing, pp. 80–87 (2015)

50. Huang, Y.T., Tseng, Y.M., Sun, Y.S., Chen, M.C.: TEDQuiz: automatic quiz generation for TED talks video clips to assess listening comprehension. In: 2014 IEEE 14th International Conference on Advanced Learning Technologies, Athens, pp. 350–354 (2014)
51. Kucuktunc, O., Gudukbay, U., Ulusoy, O.: A natural language-based interface for querying a video database. IEEE Multimed. **14**(1), 83–89 (2007)
52. Pouyanfar, S., Chen, S.C., Shyu, M.L.: An efficient deep residual-inception network for multimedia classification. In: 2017 IEEE International Conference on Multimedia and Expo (ICME), Hong Kong, pp. 373–378 (2017)
53. Wlodarczak, P., Soar, J., Ally, M.: Multimedia data mining using deep learning. In: 2015 Fifth International Conference on Digital Information Processing and Communications (ICDIPC), Sierre, pp. 190–196 (2015)
54. Zhang, D., Nunamaker, J.F.: A natural language approach to content-based video indexing and retrieval for interactive e-learning. IEEE Trans. Multimed. **6**(3), 450–458 (2004)

Data Gained from Smart Services in SMEs – Pilot Study

Lucie Kanovska[1(✉)] and Eva Tomaskova[2]

[1] Faculty of Business and Management, Brno University of Technology, Kolejni 2906/4, 612 00 Brno, Czech Republic
kanovska@fbm.vutbr.cz
[2] Faculty of Law, Masaryk University, Veveri 70, 611 80 Brno, Czech Republic
eva.tomaskova@law.muni.cz

Abstract. Nowadays, manufacturing companies increasingly invest in servitization by adopting 'smart services' enabled by connected product-service systems enabling data exchange between the customer and the service provider. Managing the transition toward smart services is not easy, especially among SMEs as many businesses struggle with lack of money, insufficient digital technologies, unskilled employees or gathering and using the proper data. The research presented in this paper is divided into two parts. The quantitative part focuses on researching possible correlations between business performance and the use of company IT systems among sixty Czech electrotechnical SMEs. The other part consists of a qualitative multi-case study and was conducted among seven Czech electrotechnical SMEs which have already started with smart service development. The findings indicate that companies gather and use the data in very different ways. They provide information to their customers, but also use information for themselves. The study is unique in highlighting the problems of smart services in small and medium manufacturers. Moreover, it investigates the gathering and the data usage gained from smart services in SMEs.

Keywords: Servitization · Smart services · SMEs
Electrotechnical manufacturers · Czech Republic

1 Introduction

In recent years, manufacturers have begun to offer products with services in inseparable formats to gain a competitive edge and survive in the contemporary market. More manufactures are providing their customers with an opportunity to obtain highly integrated services. The term servitization has been used in both academia and practice to capture this phenomena in which manufacturing companies provide services as an important strategy [1].

Smart service solutions in manufacturing companies include both hardware solutions as well as an essential service component. The focus shifts more and more from a product-centred focus towards a service component due to servitization [2]. Smart services can be seen as one of the enablers of servitization [3–5].

R. Silhavy et al. (Eds.): CoMeSySo 2018, AISC 859, pp. 183–200, 2019.
https://doi.org/10.1007/978-3-030-00211-4_18

The world has changed over the past years from a mainly physical to a more software-controlled economy with the information technology becoming an inseparable part of industries and society as a whole. Nowadays, it is no longer the product that matters, but it is the data that are generated by using the product or service. Those usage data collected and analysed serve as a starting point for new business models and services. The economic future of a company will much more rely on the ability to collect and use the data to generate smart services for their customers and transform from a simple product supplier to an entertainment provider.

Therefore, we focus on the topic related to gathering and using the data gained from smart products and related services. The aim of the research is to find out how electrotechnical SME's gather and use the data.

Service transformation has been studied in several studies, but not many studies have focused on the impact of digitalization on industrial services. Moreover, the novelty of this research lies in highlighting the problems of working with the data gained from smart services in SMEs in the Czech Republic, where the industrial sector is still dominant in comparison to other European countries. To merge the gap, this study explores how SMEs approach and work with the data gathered through smart services. To address the research objective, a qualitative multi-case study was conducted among seven electrotechnical companies which have already made some investments in smart service development. The study was conducted in the form of interviews. The findings reveal that companies gather and use the data gained from smart services in different ways. Further, the interpretation of results suggests possible approaches for managers to exploit the potential of the data stored in company IT systems as well as the data gained from smart services.

2 Theoretical Background

2.1 Smart Services

Today it is not sufficient for manufacturing companies to only provide tangible products to their customers. As a result, many manufacturers also include services and integrated solutions to their offer nowadays. Services and integrated solution can help to gain new sources of competitive advantage and value generation (e.g. [2, 5, 6]). Moreover, service revenues in manufacturing companies are rising – from 8.9% in 1990 to 42.2% in 2005 [7].

As Allmendinger and Lombreglia mention: "Soon, it will not be enough for a company to offer services; it will have to provide 'smart services' [8]. The term "smart services" has not been clearly defined and many conflicting definitions and terminologies are used [9]. Some researchers speak of 'teleservice' [10–12], 'telemaintenance' [13], 'telematics' [14], 'e-service' [15], 'e-maintenance' [16], or some alternatives combining the term 'remote', such as 'RRDM (Remote Repair, Diagnostics and Maintenance)' [17]. As Klein notes, the term 'smart service' has become popular nowadays [9].

Klein offers the following definition of smart services: "Smart services are technologically-mediated services actively delivered by the provider through accessing a remote asset and exchanging data through built-in control and/or feedback devices" [9]. A part of the definition focused on exchanging data through built-in control and/or feedback devices is based on [18] definition of remote services. Recording data via built-in devices negates aspects of on-site data capturing activities. The data can be exchanged in a form of automated read-outs, saving or analysis, or data transfer through one-way or two-way communication [10, 18]. "Smart service is the application of specialized competences, through deeds, processes, and performances that are enabled by smart products" [19].

Smart services are often considered one of the most important enablers of servitization (e.g. [3–5, 20]). According to [8], companies providing smart services derive more than 50% of revenue and 60% of margins from services rather than product sales as shown in benchmarks.

Smart services offer a variety of benefits for both manufacturing companies as service providers as well as their customers. Services become more competitive, can provide new sources of revenue, higher margins, and considerable cost savings [11]. In addition, smart services can offer a variety of non-monetary benefits, such as the opportunity to learn from their customers, establishing a basis for research and development, sales or marketing activities [21]. Moreover, they are gaining considerable strategic importance in B2B and B2C contexts [22]. Porter and Heppelmann summarize the importance of smart services: "[They] offer exponentially expanding opportunities for new functionality, far greater reliability, much higher product utilization, and capabilities that cut across and transcend traditional product boundaries" [23]. Further benefits listed by [23] include a potential broadening of value propositions, possibilities to gather valuable data, and enhance service offerings.

Smart services provide many benefits for customers as well. One of the most important benefit can be seen in "the value of removing unpleasant surprises from their lives" [8]. Lee, Kao and Yang give examples such as the form of reduction of machine downtimes, optimized scheduling of maintenance, more safety, improved information flow and transparency as well as a reduction of labour costs and creation of a better work environment [24]. Due to the considerable advantages of offering smart services, many companies increasingly adopt these novel services.

2.2 Data Gained from Smart Services

Companies have been offering common remote support via telephone connections for a long time. When internet connections became more available, the company's products' computerization was leveraged to enable data connection. Using this data connection, several different types of remote service became available: Remote diagnosis of errors, remote control and accompanying troubleshooting of machines, remote administration and installation of software updates as well as remote data backups of software and parameters.

Nowadays, smart services are characterized by including a component of IT, enabling transmission and analysis of data. Data as one of the key characteristics of smart services is leveraged by successful companies. Such smart service is based on two core properties of smart products: awareness and connectivity [8]. With these properties, smart products can provide data on how they are used back to the provider, who can use this information to offer their customers contextual and pre-emptive services that are based on the "hard field intelligence" the smart products provide [8]. This service enables providers to establish and cultivate close ties with their customers along the entire lifecycle of a smart product—from requirements analysis to disposal—and to inform their research and development processes.

In addition, service providers might grow their businesses by taking over adjacent activities from customers. As a result, they may evolve from being pure providers of hardware to solution-providers that optimize their product's entire lifecycle. They may even evolve into service aggregators who manage the channel between customers and third-party providers of complementary services that are based on the data that is retrieved from their smart products [8].

Historical data can serve as a database of cases that can be used for statistical analysis to better diagnose incidents and predict future issues. It can also be used to develop automated alert algorithms for detecting problems. "Historical data are thus an important enabling factor which greatly complements the skills and knowledge of individual experts. It supports a speedy diagnosis process through recommendations and suggestions built from past cases. In a way, it enables and facilitates quick pattern recognition and pattern matching in the domain of faults and root causes. These data are built over many years" [3]. However, other researchers note that the information itself might not be as valuable as the opportunity in its own [8]. This is acknowledged by informants in this research's cases who state that data should be leveraged to become truly valuable. Westergren advises that companies begin to facilitate more and more agreements with external partners that represent the expert voice in order to leverage data and make sense of increasing complexity [25].

According to Klein, what follows is the kind of data that can be transferred from the connected installed base: only data about components of the product, data about the whole product, data about the system environment or data about the whole customer process [9]. Furthermore, Klein suggests activities in which can the gathered data be successfully used: optimization of service solutions, development of new service solutions, optimization of hardware products, development of new hardware products or improvement of internal processes [9].

As smart services are quite different from conventional services, they are not perceived by some companies as business opportunities. The use of data in novel ways can open new business areas. Particularly "product-centric companies are often unable to identify these chances" [9]. Several authors mention that additional problems are caused by the invisible and continuous information exchange that is essential for smart services, and spark concerns among many customers regarding privacy violations and data security [22, 26].

The fear of loss of control over information, or even production, can lead to customers' unwillingness to transfer data in the first place. Once data transfer is taking place, the legal status of data ownership is often unclear. However, it is important to own the data, or at least secure the usage rights. As Porter and Heppelmann explain: "As a company chooses which data to gather and analyse, it must determine how to secure rights to the data and manage data access. The key is who actually owns the data. The manufacturer may own the product, but product usage data potentially belongs to the customer. […] Although we are seeing the early stages of a movement toward more transparency in data gathering across industries, data disclosure and ownership standards often have yet to be established. […] Customers and users want a say in these choices" [23]. Ensuring security to alleviate customer concerns about data connections and usage also create the need for modernized processes and infrastructure: "Smart, connected products create the need for robust security management to protect the data flowing to, from, and between products; protect products against unauthorized use; and secure access between the product technology stack and other corporate systems" [23].

3 Methodology

Existing research into the unique traits of smart services, however, is mostly limited to insufficient evidence in case studies (e.g. [22, 27]. Literature often describes mostly generic process steps regarding the operation in a smart service environment [2]. Theoretical understanding of smart services is at the very beginning despite the advanced practical knowledge. Consequently, there is a lack of knowledge about how smart services are used in praxis by manufacturing companies [3]. As Wünderlich et al. summarize, "*Despite the accelerating development of these smart services, academic research is still in its* infancy. We see the need to further explore the effect that smart service has on organizations, customers and the evolving service landscape" [22].

In addition, Dachs et al. notes that most of the existing research is based on case studies [28] (e.g. [2, 4, 5, 28, 29] and concludes that these case studies should be complemented by quantitative, survey based analysis across a sample of manufacturing companies [4, 28, 30].

Therefore, the research (including both a quantitative and a qualitative part) was prepared and conducted to discover more about services and smart services in manufacturing. Benkenstein et al. identifies the field of IT-driven services as a promising topic for further research, highlighting the digitalization of services, process management of digital services, information systems for services, mobile devices for services, value co-creation in digital services and big data in service industries [31].

The first part of the research included sixty electrotechnical SMEs in the Czech Republic, South Moravian Region (CZ-NACE 26 and CZ-NACE 27). The research aims to contribute to the existing knowledge base by offering insights into the service offerings of manufacturing companies, especially electrotechnical ones. The first part of the research was important not only to learn more about the current situation of service offerings in manufacturing companies, but also to find electrotechnical companies which also provide smart services to their customers. The companies providing smart services were included in the second part of the research.

The second part of the research process was conducted as a multi-case study among seven electrotechnical SMEs in the South Moravian Region of the Czech Republic. The investigation focused on unveiling the gathering and use of data gained from smart services in manufacturing SMEs. The multi-case approach provided analytical benefits over a single-case by enabling comparison of the results to find the distinction of case specific findings along with discovering some general patterns. This paper mainly presents the part of the study focused on the gathering and usage of the data gained from smart services. The kind of data transferred from the connected installed base could be the following ones: only data about components of the product, data about the whole product, data about the system environment or data about the whole customer process [9]. The gathered data can be successfully used in the following activities: optimization of service solutions, development of new service solutions, optimization of hardware products, development of new hardware products or improvement of internal processes [9].

3.1 Context of the Research

In past decades, manufacturers have begun to offer services inseparable from the products to gain a competitive edge to survive in the current market. More manufactures are providing their customers with an opportunity to obtain highly integrated services. Manufacturing companies have started to invest in servitization by delivering smart services, which enable a data exchange between their customers and a service provider via connected product-service systems. Therefore, research was prepared and conducted focusing on the area related to service delivery, i.e. on the usage of the company's IT system in the 1st part and the gathering and the usage of the data related to smart services in the 2nd part.

The 1st Part of the Research

A questionnaire was used to gather information to investigate the relationship between service delivery (particularly to the company IT system) and business performance for the 1st part of the research. The questionnaire focusing on service delivery is a part of a larger questionnaire focusing on both the interfunctional coordination and the services provided by electrical engineering companies. The questionnaire uses the Likert scale ranging from 1 (No, I don't agree) to 5 (Yes, I agree). The data related to service delivery and particularly to company IT system were used for the purposes of this paper. The part related to services was based on the previous research held in 2005 among the sector of saw and band saw SMEs in the Czech Republic [32]. The rest of the items are new and were developed based on (a) the study of the literature, mainly [33–35]; (b) interviews with manufacturers; (c) current information about sale and service support in manufacturing companies; (d) the study of information contained in the periodicals targeting this sphere.

The 2nd Part of the Research

All of the companies included in the second part of the research are SMEs from the same industry – electrotechnical producers. Although they operate in one industry, they provide a wide range of products and services to their customers with varying degrees of service orientation. All case companies have been implementing smart services to their companies in different levels and ranges. Case companies mostly provided the following smart services: remote monitoring, control and diagnostics, remote repairs, preventive and predictive maintenance.

The different level and wide range of smart services provides valuable insights into smart services in SMEs in different contexts. The goal was to select companies from the same industry, but in different maturity phases in their service transformation journey. The case companies were selected based on purposive sampling [36]. Four companies providing smart services were detected from the first part of the research and were included in the second part. Another three companies were found as the members of Electrotechnical Association of the Czech Republic (https://www.electroindustry.cz/). The details of the case companies are described in Table 1.

Table 1. Case company description (Source: Authors)

Firm	Respondents	Number of employees	The length of smart service provision in years
A	Owner	15	1
B	Product manager	50	1
C	Owner	10	2
D	Owner	4	2
E	Owner	25	2
F	Owner	148	2
G	Product manager	170	More than 2

3.2 Data Collection

The 1st Part of the Research
The respondents participating in the research were directors or managers of electrical engineering companies in the Czech Republic. The data were collected from February to November 2014. The research focused on the following industry classifications belonging to CZ-NACE 26 (Manufacturing of computer, electronic and optical products): CZ-NACE 26.1, CZ-NACE 26.3, CZ-NACE 2651, CZ-NACE 266, and CZ-NACE 27 (The production of electric equipment): CZ-NACE 27, CZ-NACE 271, and CZ-NACE 273. According to the Czech Statistical Office, the total number of these SMEs reaches 107. A total of 60 valid questionnaires were processed. Therefore, the research study covered a representative sample (56%) of existing companies. Incomplete questionnaires were discarded.

The 2nd Part of the Research
The empirical data of this case study consists of in-depth interviews with owners or with experienced senior managers in the case organizations. The interviews were carried out between April 2017 and January 2018. Each interview was between 50 to 100 min in length and was performed on site, which gave a chance to tour each company and get a sense of the work environment. The sample size matches the recommendations for exploratory research [37]. To enable relaxed communication, the informants' anonymity was guaranteed through the assurance that the results would be released without any identifying information. Each interview was recorded and transcribed. The transcriptions were crosschecked by both authors.

After selecting the case companies, semi-structured interviews with predefined themes were conducted. Interview contents were continuously adapted on the basis of previous interviews [38]. The interview consisted of open-ended questions, initially crafted based on the literature review and then modified and adjusted during the research process. All interviews were conducted face-to-face, and the transcribed data obtained from the interviews were analysed continuously during the whole research process in order to exclude or include the predefined themes that did not seem to resonate with the initial interview structure.

3.3 Data Analysis

The 1st Part of the Research

The data were analysed using the statistical software package Minitab, version 17. Descriptive statistics (minimum, maximum, mean and standard deviation) were applied to describe the characteristics of the organizational performance of the sample. Both the Spearman's rank correlation coefficient and the Pearson's chi square test can be used to measure the correlation of two variables.

The 2nd Part of the Research

The data analysis followed an abductive analysis process, and the understanding of the phenomenon based on the study of literature laid the foundation for early interviews, which then used evolving themes to track important issues as the interviews progressed and the understanding of smart services in the real-life setting increased [39]. In practice, the literature informed about the importance of smart services and their benefits and also barriers and working with data (e.g. the kinds of data transferred from the connected installed base and how to successfully use the gathered data) [9].

The interviews explored the following categories: the type of smart products and smart services, the length and the type of smart service provision, customer perception of smart services, the reasons for starting with smart service provision, the benefits gained from smart services, barriers connected with smart service provision, the data gathering and use of the data gained from smart services, the specifics of the Czech industrial market, collaboration with other firms and "learnings" for other firms which want to begin offering smart services.

Open coding was used to organize and to convert the data to discrete thematic blocks. As the qualitative case research is sensitive to researchers' subjective interpretations, some checks and peer debriefing to reduce researcher bias were conducted to increase the objectivity of the study. A rich set of direct interview quotations to demonstrate interpretations was added to support the transparency and conformability of the findings.

4 Findings

The 1st Part of the Research

Firstly, mean value and standard deviation were processed to discover whether the respondents have and use their own company IT system to collect and store data related to services (e.g. repairs, claims, complaints), and if they use the data for further decision-making. The results are presented in Table 2. Two items were used in the questionnaire as a closed question with a five-level scale to measure the degree of importance in a particular area, where 1 means "No, I don't agree" and 5 means "Yes, I agree".

Table 2. Company IT systems (Source: Authors)

Item	Mean value	Standard deviation
Company has IT system to collect and store data related to services (e.g. repairs, claims, complaints)	3.86	0.99
IT system evaluates the data related to services and management uses the data for further decision-making	3.02	0.86

The results reveal that SMEs gather and store data in their IT systems (mean value 3.86). However, the figure depicting further use of the data in the decision-making process is much lower (3.02). These figures illustrate the fact that even though the companies have a great pool of data at their disposal, they fail to use them. The question arises whether the companies perceive that there is no point in further usage of the data, or whether they do not have the capacity to work with the information.

Secondly, a hypotheses H was designed resulting from the part related to measuring business performance and company IT systems. It was formulated in the following statement: H: IT systems related to services have a positive influence on business performance. Five different items were chosen to measure the business performance. The first three items are related to the market business performance; the last two items are related to the financial business performance. The findings show a relationship between the five items related to business performance and the two items related to company IT systems (see Table 3). Verification of the relationship between company IT systems and the business performance was analysed using the Spearman correlation coefficient. The first value is Spearman's rank correlation: Spearman's rho, the second value is p-value. If $p < 0.05$, then we reject the null hypothesis (H0: items are independent), i.e. accept that the sample gives reasonable evidence to support the alternative hypothesis (H: items are dependent). The correlations are shown in Table 3.

The findings presented in the Table 3 show that two indicators of business performance (The number of warranty claims decreases and Production effectiveness increases) have a positive correlation with both the two selected indicators related to IT systems of Electrotechnical companies ($p < 0.05$). Furthermore, there is also another positive correlation between the indicator of business performance (ROA increases year-on-year) and the one selected indicator related to data. No other positive correlations have been found ($p > 0.05$).

The 2nd Part of the Research

The analysis of the part related to gathering and using data identified one research question RQ, which was formulated in the following statement: RQ: How today's

Table 3. Correlation analysis: business performance and service delivery by using the Spearman's rank correlation (Source: Authors)

Item	Company registers the sales volume increase by current customers	The number of new customer increases year-on-year	The number of warranty claims decreases	ROA increases year-on-year	Production effectiveness increases
Company has IT system to collect and store data related to services (e.g. repairs, claims, complaints)	0.169 0.218	0.220 0.106	0.365 **0.006**	0.128 0.358	0.390 **0.004**
IT system evaluates the data related to services and management use the data for further decision-making	0.131 0.339	0.116 0.399	0.371 **0.005**	0.433 **0.001**	0.388 **0.004**

electrotechnical companies gather and use the data which are gained from their smart services? To address this RQ, the approach employed in data collection was organized into five sets of themes and related questions as follows: Frequency of data gathering (How often do you gather the data?), Customer perception of data gathering (How do the customers perceive the data gathering process?), The usage of the data gathered (How do you use the gathered data?), The kinds of data for evaluation (Which kinds of transferred data do you use/evaluate?) and Legal framework of data gathering (How do you deal with/approach the data security according to the legal regulations?). All five themes are discussed next.

The analysis revealed that smart services are provided by current companies, but in very different ranges and conditions. It was also found that the gathering and the usage of the data gained from smart services differs from one company to another (2nd part). Below are some quotes that illustrate the research findings. However, all case companies unanimously agreed that smart services are the future of manufacturing. Smart services are still perceived as a possible competitive edge in some industries, but in a couple of years, smart services will become a necessity. It is likely that a sustainable competitive edge may be achieved through complex combinations of interconnected products and services found within manufacturers, customers and intermediaries, if needed.

Frequency of Gathering Data

There are differences among the case companies regarding the data gathering. While some gather the data randomly, others collect the data regularly, e.g. once a day, and other respondents download data constantly. The frequency is very much dependent on the individual customer's requirements and therefore can vary significantly. For some customers, a remote online data transfer is impossible due to data transfer security concerns.

> "The data collection frequency is unique to each of our customers. One of our customers requires daily collection and processing of data along with a more in-depth analysis monthly. Naturally, should a problem arise, we need to collect the data immediately and promptly find a solution to the problem."

> "We collect the data based on the customer's requirements. While some customers wish to download the data every 15 min or every 30 min, some are content with an hourly data gathering."

> "We now collect 80% of the data in person via our technicians who attend to the customer in a problematic situation on site, and 20% is transferred online. However, we are planning to introduce an online remote data transfer in the second half of the year."

Customer Perception of Data Gathering

Great differences among the respondents' customer perception regarding the data collection can also be seen in this part. Some customers are immediately open to data collection from the very beginning of their cooperation or product use, perceive no problems and welcome such possibilities. Some even require this option to gather additional information about a product. On the other hand, other customers are extremely sceptical, and more time is needed to see the benefits. The third part of customers do not appreciate this at all as they do not see great benefits, fear lack of data protection security, or are prevented from the data usage by internal regulations.

> "Most of our customers' reactions show no sign of problems. However, some customers require a contract to limit the feedback. Frequently, a unique solution is necessary. In such cases, we collect the data only at the time of installations based on the customer's explicit permission."

> "It varies greatly. Some customers do not perceive it as necessary, so they are not pressed to deal with it. Furthermore, not all customers see the benefits, it is a slow process…."

> "Very positively. They require it!"

The Usage of Gathered Data

There were many differences in answers in this part as well. Some respondents do not use any data, but rather share the data with the customer, who works with it as needed. They use the data only in the case of a technical problem, with the company's active participation. Such an approach to data use seems to be marginal. The majority of respondents use the data either for meeting customer needs, remote monitoring, diagnostics, or for remote repairs. The information is also further used by the respondents, especially for a faster and cheaper service. Three respondents already use the data for predictive maintenance and two are trying to use the processed data for innovation planning and development. The two latter areas are for many a plan or a dream. Overall, it is currently an operational data processing. On the other hand, respondents agree on the use of data not only for monitoring and remote management, but also for predictive maintenance, innovations and new product development.

"A data analysis follows a claim, diagnostics are used in a failure, change control in new product development."

"The data are used by the customer, very often a distributor, and not processed except for repairs and distant processing."

"We collect and evaluate the data, but we select only the data relevant to finding a solution to the problem."

The Kind of Data Used for Evaluation
The respondents' answers were unique and differed according to the product types, market conventions and customer requirements. The type of data inspected also varies from product to product even within one company, from the data following the use of only individual parts of the product to data collected for the whole product or the system environment, or to data collected in the entire customer process.

"The use of data is dependent on the automatization type of our products. While we only process individual product parts, data collected from the datalog of the system unit and system environment, with others we use the data gathered for the whole product and the customer process. In the case of cloud services, we record the entire volume of data including the communication data according to the product release phase."

"We most frequently process the data collected from the individual product parts related to our products, but it can differ based on customer requirements."

Legal Framework for Data Gathering
The majority of respondents agreed that the legal framework for data gathering is an inevitable part of purchasing agreements, collaboration agreements and other documents. This area needs to be well defined and is perceived as a very sensitive topic by customers. Each agreement is complemented with a Non-Disclosure Agreement (NDA). The NDA is an agreement between the two parties who wish to mutually share information and also wish to limit the information provided to third parties. Such agreements are composed based on consultations with lawyers. The manufacturers further treat the information provided by customers according to such agreements. Nevertheless, handling information is a sensitive issue and the protocol is outlined in the agreement.

To sum up, the analysis revealed five basic themes of gathering and using data in current Electrotechnical SMEs.

5 Discussion

The focus of this research has been an exploration of aspects related to data, which a company can store in their IT system (1st part of the research) and also data gained from smart services (2nd part of the research). One hypotheses for the 1st part and one research question were formulated in this research to discover how companies gather and use the data.

The aim of the first part of the research among 60 SMEs was to ascertain whether the businesses dispose of their own IT systems, where they record information collected from their customers and whether they further use this information in the managerial decision-making process. Further, the aim was to discover whether any relation exists between the IT department and the items related to the company performance. The research results show that majority of businesses have their own IT systems, where they gather data about their customers, and that more than half further use this data. The company IT systems do not influence all of the company performance items except for two: The number of warranty claims decreases and Production effectiveness increases. Therefore, company's preferences play the most important part; whether it prefers profitability increases or effectiveness increases. Two items of the company's performance show a positive influence, specifically the warranty claims decrease and effectiveness increases.

Most of the business managers consider computers to be an important productivity tool which can be used in their daily activities. Besides that, the importance of computer-based information systems used to make decisions, decision support systems, are gaining increased popularity as one of the unique sides of businesses [40]. Therefore, management information systems can help managers to find a necessary solution for given problems since problem identification and decision-making analysis is an important function of decision support systems. Therefore, management information systems are crucial to grouping different sort of information into relevant phases and to creating good sources of information which can be useful for decision-making purposes [40].

These results can serve as a motivational tool for the businesses to implement such systems. At the same time, the independence of the increases of customer numbers and purchases made by existing customers is understandable. Customers do not choose a business according to the IT system in use and similarly, the IT system is not a reason for the customers to increase their purchases.

The question of profitability increase needs to be investigated in more depth as one of the items is dependent and the other independent. Therefore, it is recommended to explore the area of profitability increase (ROA) in relation to IT in further research.

The case study investigates the gathering and use of the data gained from smart service among seven companies. Manufacturing service providers are responding to the changing needs of customers in order to always be "in control" and provide also smart services, which can be useful for both customer and manufacturers. Data can be leveraged in offerings to provide additional value for the customer. It can also be used by the service provider to verify if its value proposition matches customer needs [23]. However, companies are often unsure which smart service value proposition appeals to their customers [11, 27]. A fundamental reason for this can be a general lack of knowledge of customer needs. Therefore, understanding customer needs is one of the most important tasks for service managers [2, 41]. Respondents frequently mentioned that their customers play a key role in deciding which data should be processed, as well as how and when the data should be processed. In addition, the respondent frequently mentioned a wide range of differences of customer needs. Moreover, the data are an important source of information about customers and products. In order to understand the data process properly, a thorough analysis of the data was performed. The analysis identified five main themes related to the gathering and use of the smart services data within one industry.

5.1 Theoretical Implications

The 1st part of the research confirmed the positive influence of IT systems on selected areas of the company's performance. The company's performance can also be inspected from various perspectives and some items show positive influence. Therefore, the recommendation for future research is to focus more on the analysis of rentability (ROA) as one item showed a positive influence whereas another item was independent.

The topical issue of smart services based on smart products discussed in *the 2nd part of the research* has not been thoroughly researched yet and provides an opportunity for further investigation. As smart services are quite different from conventional services, companies sometimes do not perceive them as business opportunities. The use of data in novel ways can open new business areas. Especially product-centric companies are often unable to identify these opportunities [9]. Additional problems caused by the invisible and continuous information exchange that is indispensable for smart services spark concerns among many customers regarding privacy violations and data security [22, 26].

As the case studies revealed, every company is unique in its approach to providing smart services both from the perspective of data collection frequency, and the use of data and further processing. As mentioned previously, the differences in customer needs and wishes related to the use of data play an important part. Therefore, it is very difficult to provide general theoretical framework for both the approach to smart service delivery, and to the data gathered from the customers and their further use. The ways respondents approach this area vary greatly. However, the majority of companies try to process the data or plans to do so not only for the sake of their customer, but also for their own benefit. Boyd and Crawford noted that companies can analyze customer data to understand patterns of customer behavior [42]. This could help them to learn more about their customers, e.g. why they make certain decisions or behave in certain ways [43] and how to design new services or improve existing ones [44]. Respondents in our research use the gathered data solely for optimization of service solutions, hardware products and internal processes. Development of new service solutions and new hardware products is quite rare.

Based on the findings, a simple scheme of data use among respondents was designed (Scheme 1). The points 1 through 3, which present methods of data processing and use, is naturally offered by the majority of respondents. However, the points 4 and 5 are now provided infrequently. According to the respondents, these areas will become a part of the data processing in their companies, will definitely gain significance, and become not only a competitive practice, but a commonplace one. Three companies are mostly focused on data monitoring for their customers and also on very general data analytics. The rest of companies in the research not only monitor data, but also try to analyse information much deeper to be able to use the data for more smarter and informed decisions.

The interview results reveal that some customers refuse to download data remotely or agree to it to a limited extent based on predefined conditions. However, continuous service interactions that are based on a continuous exchange of data provide new opportunities for routinizing interactions in service systems [45]. Routinizing

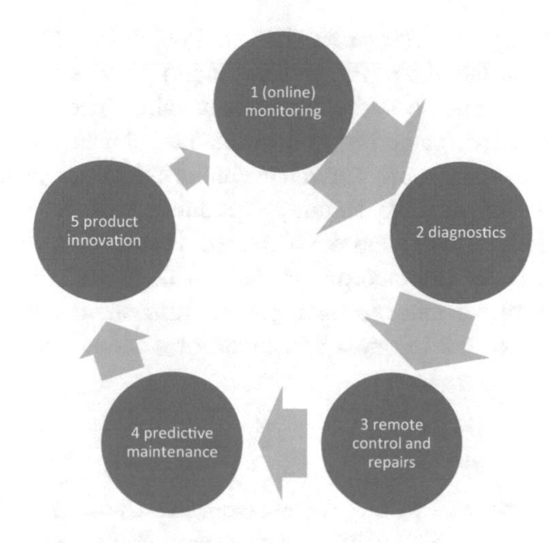

Scheme 1. Data use in SMEs (Source: Authors)

interactions can help to coordinate and control activities, legitimate the actions performed by the actors involved, economize on resources, reduce uncertainty, increase stability, and store information about the interactions [46] in a smart service system.

5.2 Managerial Implications

Many companies today manage their own IT systems and gather data, the majority of which is further evaluated. Even if the influence of IT systems on business performance has not been proven in the 1st part of the research with the exception of two items (The number of warranty claims decreases and Production effectiveness increases), it would be useful for the companies to further process the gathered data as they are already at their disposal. Naturally, this can lead to increases in the use of data gathered through smart services. This provides the companies with an incentive to change their perception of IT systems as cost increase with no further utilization. As confirmed in research, the company IT systems can have an influence on a lower number of warranty claims and on increases of production effectiveness. It may be concluded that it is up to the company to decide its preferences and how it wishes to use the company IT system in place.

The findings of the 2nd part illustrate the impact of services as well as smart services on industrial services and contribute to the discussion of smart servitization of industrial companies. Some authors, e.g. [8, 23] named possible benefits for customers, which can be seen in smart services. However, despite the considerable advantages of using smart services, some customers do not perceive any benefits, and are prudent and unwilling to share the data gained from smart product and smart services with manufacturers. For practicing managers, the implications are rather straightforward: to achieve the potential of the data gained from smart services, the managers need to persuade their customers to be more open to sharing the data from their products with them as the information gained from the products can help both sides be more efficient and competitive in today's fierce market. The findings further illustrate the importance of acknowledging that the data gained from smart services not only monitor the products by customers but also add an important value. Therefore, managers should also be more interested in gathering and using the relevant data, which should be carefully evaluated to learn more about customers and their products, but also for managing predictive maintenance leading to reduction of costs and providing key information for product research and development. Progressive companies need to use the gained data effectively and incorporate them in their strategic vision.

The case studies prove that the data gained from smart services are considered crucial and beneficial for the business and future of such companies.

6 Conclusion

This study tried to contribute to the understanding of the role of data in present day manufacturing companies which also provide smart services to their customers. Data are seen as a key business asset; data management and analysis are business capabilities of high importance. Moreover, the data-centred convergence of essential digital technologies has started to focus on data science and data-mining procedures [47]. Based on the case studies findings, implications for practice and theory are drawn. Therefore, other researchers can build on the findings mentioned in the paper. However, the study presented in this paper is still in the initial phases. Therefore, the limited number of cases and examples in both parts of the research collected seriously discourage an overgeneralization of the findings achieved so far which need to be confirmed by further investigations. Nevertheless, the authors believe that the research has a valuable potential impact both on theory and practice. Different problems that firms face during the data procession were highlighted in the paper. Even if these problems are not completely new to servitization literature, their weight and intensity are being redefined in a new business environment, i.e. the electrotechnical industry, small and medium enterprises and the Czech Republic.

Future research could be extended to other companies from other industries, which have already started to provide smart services with their products. Future studies should also investigate further how to process the management of digital services and information systems for smart services.

References

1. Vandermerwe, S., Rada, J.: Servitization of business: adding value by adding services. Eur. Manag. J. **6**(4), 314–324 (1988)
2. Brax, S., Jonsson, K.: Developing integrated solution offerings for remote diagnostics: a comparative case study of two manufacturers. Int. J. Oper. Prod. Manag. **29**(5), 539–560 (2009)
3. Grubic, T., Peppard, J.: Servitized manufacturing firms competing through remote monitoring technology: an exploratory study. J. Manuf. Technol. Manag. **27**(2), 154–184 (2016)

4. Neu, W.A., Brown, S.W.: Forming successful business-to-business services in goods-dominant firms. J. Serv. Res. **8**(1), 3–17 (2005)
5. Oliva, R., Kallenberg, R.: Managing the transition from products to services. Int. J. Serv. Ind. Manag. **14**(2), 160–172 (2003)
6. Wise, R., Baumgartner, P.: Go downstream. Harvard Bus. Rev. **77**, 5 (1999)
7. Fang, E., Palmatier, R.W., Steenkamp, J.B.E.: Effect of service transition strategies on firm value. J. Market. **72**(5), 1–14 (2008)
8. Allmendinger, G., Lombreglia, R.: Four strategies for the age of smart services. Harvard Bus. Rev. **83**(10), 131 (2005)
9. Klein, M.M.: Design rules for smart services: overcoming barriers with rational heuristics. Doctoral dissertation, Universität St. Gallen (2017)
10. Borgmeier, A.: Schlußbetrachtung. In: Teleservice im Maschinen-und Anlagenbau. Deutscher Universitätsverlag, pp. 209–217 (2002)
11. Küssel, R., Liestmann, V., Spiess, M., Stich, V.: "TeleService" a customer-oriented and efficient service? J. Mater. Process. Technol. **107**(1–3), 363–371 (2000)
12. Pfeiffer, S.: Teleservice im Werkzeugmaschinenbau. Arbeit **9**(4), 293–305 (2000)
13. Garcia, E., Guyennet, H., Lapayre, J.C., Zerhouni, N.: A new industrial cooperative tele-maintenance platform. Comput. Ind. Eng. **46**(4), 851–864 (2004)
14. Chatterjee, A., Greenberg, J., Jones, M., Kaas, H.W., Wojcik, P.: Telematics: decision time for detroit. Lond. Bus. School Rev. **12**(2), 21–38 (2001)
15. Rowley, J.: An analysis of the e-service literature: towards a research agenda. Internet Res. **16**(3), 339–359 (2006)
16. Levrat, E., Iung, B., Crespo Marquez, A.: E-maintenance: review and conceptual framework. Prod. Plan. Control **19**(4), 408–429 (2008)
17. Biehl, M., Prater, E., McIntyre, J.R.: Remote repair, diagnostics, and maintenance. Commun. ACM **47**(11), 100–106 (2017)
18. Wünderlich, N.V.: Acceptance of Remote Services, 1st edn. Gabler, Wiesbaden (2009)
19. Beverungen, D., Matzner, M., Janiesch, C.: Information systems for smart services
20. Jonsson, K., Holmström, J.: Ubiquitious computing and the double immutability of remote diagnostics technology: an exploration into six cases of remote diagnostics technology use. In: Designing Ubiquitous Information Environments: Socio-Technical Issues and Challenges, pp. 153–167. Springer (2005)
21. Laine, T., Paranko, J., Suomala, P.: Downstream shift at a machinery manufacturer: the case of the remote technologies. Manag. Res. Rev. **33**(10), 980–993 (2010)
22. Wünderlich, N.V., Heinonen, K., Ostrom, A.L., Patricio, L., Sousa, R., Voss, C., Lemmink, J.G.: "Futurizing" smart service: implications for service researchers and managers. J. Serv. Mark. **29**(6/7), 442–447 (2015)
23. Porter, M.E., Heppelmann, J.E.: How smart, connected products are transforming competition. Harvard Bus. Rev. **92**(11), 64–88 (2014)
24. Lee, J., Kao, H.A., Yang, S.: Service innovation and smart analytics for Industry 4.0 and big data environment. Proc. CIRP **16**, 3–8 (2014)
25. Westergren, U.H.: Opening up innovation: the impact of contextual factors on the co-creation of IT-enabled value adding services within the manufacturing industry. IseB **9**(2), 223–245 (2011)
26. Rixon, L., Hirani, S.P., Cartwright, M., Beynon, M., Selva, A., Sanders, C., Newman, P.S.: What influences withdrawal because of rejection of telehealth—the whole systems demonstrator evaluation. J. Assist. Technol. **7**(4), 219–227 (2013)
27. Grubic, T.: Servitization and remote monitoring technology: a literature review and research agenda. J. Manuf. Technol. Manag. **25**(1), 100–124 (2014)

28. Dachs, B., Biege, S., Borowiecki, M., Lay, G., Jäger, A., Schartinger, D.: Servitisation of European manufacturing: evidence from a large scale database. Serv. Ind. J. **34**(1), 5–23 (2014)
29. Davies, A., Brady, T., Hobday, M.: Organizing for solutions: systems seller vs. systems integrator. Ind. Mark. Manage. **36**(2), 183–193 (2007)
30. Gebauer, H., Kowalkowski, C.: Customer-focused and service-focused orientation in organizational structures. J. Bus. Ind. Market. **27**(7), 527–537 (2012)
31. Benkenstein, M., Bruhn, M., Büttgen, M., Hipp, C., Matzner, M., Nerdinger, F.W.: Topics for service management research – a European perspective. SMR J. Serv. Manag. Res. **1**(1), 4–21 (2017)
32. Kanovska, L.: Customer services and their importance for company prosperity. Vutium, Brno (2005)
33. Gebauer, H., Worch, H., Truffer, B.: Absorptive capacity, learning processes and combinative capabilities as determinants of strategic innovation. Eur. Manag. J. **30**(1), 57–73 (2012)
34. Kindström, D., Kowalkowski, C.: Service innovation in product-centric firms: a multidimensional business model perspective. J. Bus. Ind. Market. **29**(2), 96–111 (2014)
35. Baines, T., Lightfoot, W.H.: Servitization of the manufacturing firm: exploring the operations practices and technologies that deliver advanced services. Int. J. Oper. Prod. Manag. **34**(1), 2–35 (2013)
36. Eisenhardt, K.M., Graebner, M.E.: Theory building from cases: opportunities and challenges. Acad. Manag. J. **50**(1), 25–32 (2007)
37. Corbin, J., Strauss, A., Strauss, A.L.: Basics of Qualitative Research. Sage, Thousand Oaks (2014)
38. Silverman, D.: Interpreting Qualitative Data. Sage, Thousand Oaks (2015)
39. Dubois, A., Gadde, L.E.: "Systematic combining"—a decade later. J. Bus. Res. **67**(6), 1277–1284 (2014)
40. Aina, A.A.M., Hu, W., Noofal, A.N.: Use of management information systems impact on decision support capabilities: a conceptual model. J. Int. Bus. Res. Market. **1**(4), 27–31 (2016)
41. Brax, S.: A manufacturer becoming service provider – challenges and a paradox. Manag. Serv. Qual. Int. J. **15**(2), 142–155 (2005)
42. Boyd, D., Crawford, K.: Critical questions for big data: provocations for a cultural, technological, and scholarly phenomenon. Inf. Commun. Soc. **15**(5), 662–679 (2012)
43. Huang, M.H., Rust, R.T.: IT-related service: a multidisciplinary perspective. J. Serv. Res. **16** (3), 251–258 (2013)
44. Lim, C.H., Kim, M.J., Heo, J.Y., Kim, K.J.: Design of informatics-based services in manufacturing industries: case studies using large vehicle-related databases. J. Intell. Manuf. **29**, 1–12 (2015)
45. Becker, J., Beverungen, D., Knackstedt, R., Matzner, M., Müller, O., Pöppelbuß, J.: Bridging the gap between manufacturing and service through IT-based boundary objects. IEEE Trans. Eng. Manag. **60**(3), 468–482 (2013)
46. Becker, M.C.: Organizational routines: a review of the literature. Ind. Corp. Change **13**(4), 643–677 (2004)
47. Fajszi, B., Cser, L., Fehér, T.: Business Value in an Ocean of Data. Alinea Kiadó with TSystems, Budapest (2013)

Enhancing Concurrent ETL Task Schedule with Altruistic Strategy

Li Tang[1,2(✉)], Hui Li[1,2], Mei Chen[1,2], Zhenyu Dai[1,2], and Ming Zhu[3]

[1] Department of Science and Technology,
Guizhou University, Guiyang 550025, China
yntanglil993@126.com, {cse.HuiLi,gychm}@gzu.edu.cn
[2] Guizhou Engineering Lab of ACMIS,
Guizhou University, Guiyang 550025, China
[3] National Astronomical Observatories,
Chinese Academy of Sciences, Beijing 100016, China

Abstract. In representative ETL software such as Informatica and DataStage, their ETL task scheduler are only supports timing scheduling, meanwhile, they neither take the resource consumption into consideration nor make the user to be easy to configure resource utilization strategy, which often make concurrent task schedule to be inefficient. In this paper, we propose a long-term altruism strategy based concurrent ETL task schedule approach named Aetsa to solve this problem. In order to make critical jobs are have the needed resource to execute efficiently, Aetsa approach can pause certain jobs temporarily, and then schedule other jobs execute in a higher priority. After that, once there available appropriate resource for the aforementioned paused jobs, Aetsa will resume it for further execution. We evaluate the efficiency of Aetsa in our real medical data integration scenarios. It involved medical datasets from more than 1600 primary health care institution of GuiZhou province, China. Our experimental result show that Aetsa's average waiting time is very close to the well-known scheduling solution SJF, meanwhile, compared with FCFS, both the average response time and the efficiency are achieved satisfactory improvement.

Keywords: ETL task · Schedule · Altruistic

1 Introduction

With the development of the Internet, data in all areas has grown exponentially. Even the internal departments of the company also have their own databases, these data are generally inconsistent, non-standard, and redundant, which making it difficult for enterprises to conduct integrated data management and obtain application-specific data [16]. Due to the information and insights inside the data can bring huge value to the enterprise, internal data often need to be integrated for further processing (e.g., multidimensional analysis, data mining, etc. [12–14]) to support decision-making [14]. Data integration [15] often includes three sub-processing procedures, data extraction, transformation and loading (ETL) [1, 3]. First, data is extracted from the data source,

R. Silhavy et al. (Eds.): CoMeSySo 2018, AISC 859, pp. 201–212, 2019.
https://doi.org/10.1007/978-3-030-00211-4_19

then, the data is cleaned and converted to make it becomes consistent and standardized data, and finally loaded the data into the target analytical store [1].

Generally, we'd like to use ETL toolkit for data integration. This procedure often produce a large number of ETL tasks. Since it is inefficient to run the ETL tasks manually, especially when the number and volume of tasks is very large, and a proper strategy is urgently needed to scheduling ETL tasks for efficient execution and better QoS metrics. There are two kinds of ETL task schedule methods: simple timing schedule and period timing schedule. Simple timing schedule is cyclic execution of tasks within a specified time. Period timing schedule can perform tasks periodically at specified time point. However, both of these ETL tasks scheduling methods did not consider the issues of resource utilization and efficiency.

In this study, in order to improving resources utilization and task execution efficiency, we propose an altruism based ETL Tasks Schedule Algorithm (Aetsa) to schedule ETL tasks. Altruistic [2] strategy is a kind of long-term altruistic, resource scheduling strategy, in this approach, job can be paused to make other jobs to be executed until there have appropriate resource to start it, meanwhile it doesn't affect the response time and job isolation. Therefore, altruistic strategy based ETL task can improve the resource utilization, shorten the average job response time and makespan[1] while reducing resource competition.

2 Background

ETL toolkits such as Informatica [7] and DataStage [8] have done well in data integration. However, Informatica only support timing schedule. DataStage have a more complex schedule strategy, it take resources consumption into consideration during task schedule, but it requires users to set many resource thresholds themselves (e.g., CPU, memory and IO), which will always make users feel difficult and confusing.

Other professional batch job scheduling system, include TASKCTL [4], TBSchedule [5] and Control-M [6]. TASKCTL and TBSchedule only support timing schedule. The use of Control-M is very complicated, and users must have higher professional skills, and a huge amount of training costs must be paid.

With the increasing amount of data, resources needed for ETL tasks often to be limited, and the consideration of the resource utilization and efficiency during ETL tasks scheduling to be critical.

Most batch task scheduling systems are using timing schedule and can only handle one task at a time. Figure 1 illustrates how a timing scheduler schedules a new task: the task start to execute when it arrives at the time point set by the timing scheduler, due to set aside enough execution time for each task is needed before task execution, user must set start time for each task separately to ensure that the tasks can have enough resources to be executed correctly, this not only make it too troublesome for usage, but also result in enormous waste of resources.

[1] The completion time of all jobs within one scheduling period.

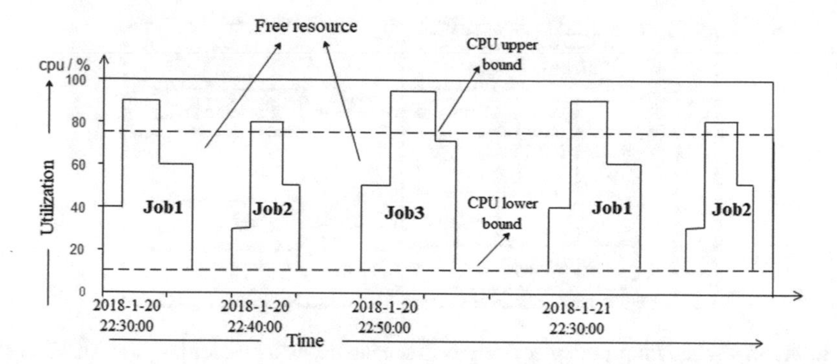

Fig. 1. Timing schedule

The FCFS [11] scheduling strategy is based on the order in which the tasks are submitted, and the task that is submitted first is executed first until the task queue is empty and the scheduling is completed. This kind of strategy is usually used initially in the process scheduling of the operating system [17, 19, 20], and is rarely used in the ETL task scheduling.

The Shortest Job First (SJF) [10] policy refers to the prioritization of jobs with the shortest job execution time. It can theoretically achieve the best average job waiting time and job response time, but it always takes the longest overall job completion time (makespan).

Unfortunately, all the above-mentioned scheduling strategies are easy to perform poorly with the resource utilization and efficiency constraints, and often do not lead to significant long-term benefits. In this paper, we use the FCFS strategy and SJF strategy as comparative experiments. The user only needs to specify a start time for all tasks with the same execution cycle, and these tasks will be executed automatically when the time point arrives. Obviously, the resource utilization and efficiency of these two strategies have been greatly improved compared to timing schedule.

3 Aetsa Approach

Our Aetsa (Altruism based ETL Tasks Schedule Algorithm) approach is a kind of batch scheduling. Users only need to set a start time for all tasks with the same period. After the task is added to the scheduling plan, all tasks are suspended. When the specified time point is reached, the Aetsa scheduler will automatically schedule the most suitable task execution, which is transparent to the user.

Figure 2 gives an overview of the Aetsa scheduler approach. First, the jobs and tasks are submitted to the Scheduling plan, and then Aetsa scheduler pause the tasks. Second, all tasks are iterated to calculate their respective total IO costs and sorted by IO cost. Thirdly, the scheduling policy captures real-time CPU utilization to determine which task is best for execution, and then scheduler resumes the task, until all the tasks in the cycle are executed.

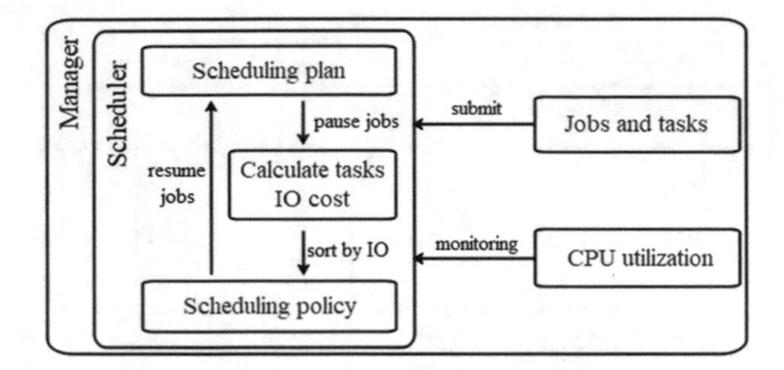

Fig. 2. Aetsa's scheduling policy revise the execution sequence according to task IO cost and CPU utilization, when CPU utilization is low, the scheduler schedules small IO cost task execution, otherwise it schedules large IO cost task execution.

Assume that all jobs can be executed at the same time. Consider the issue of scheduling ETL jobs J ($J = \{J_1, J_2, \ldots, J_n\}$), first adding jobs to the scheduling plan. We need to iterate through all jobs to calculate the IO cost of jobs IO ($IO = \{IO_1, IO_2, \ldots, IO_n\}$), then sort all jobs by IO cost ascending to get J_{sorted}. When the specified execution time point arrives, the job is scheduled according to the scheduling policy. It is easy to know that the higher the IO cost, the longer the execution time and the more CPU time required. Figure 3 shows that in the appropriate range of CPU utilization (by default, CPU upper bound $CPU_{up} = 0.75$, CPU lower bound $CPU_{down} = 0.05$, which can be set by the user), when the CPU utilization is high, the small IO job will be scheduled, and when the CPU utilization is low, large IO jobs will be scheduled. The position L of the job to be executed can be expressed as following:

$$L = \left\lceil 1 - (CPU - CPU_{down}) \bullet \frac{100}{CPU_{up}} \right\rceil \bullet n \quad (1)$$

(CPU is the current CPU utilization, n is the number of jobs).

As shown in Figs. 1 and 4, our altruistic scheduling theoretically improves resource utilization and shortens the overall job completion time (makespan).

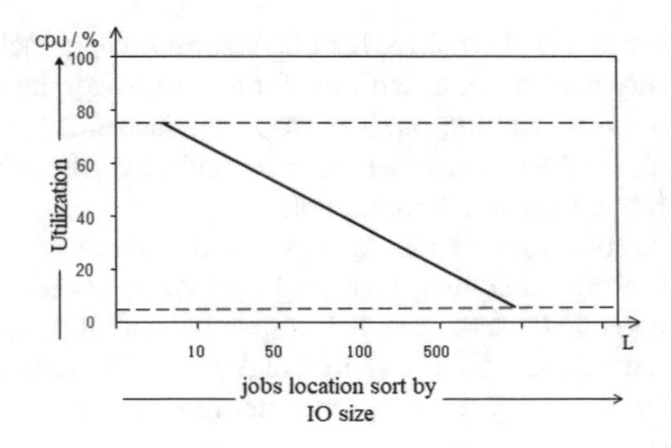

Fig. 3. Scheduling policy. Maps to the job location based on the current CPU utilization, and schedules the corresponding job to execute.

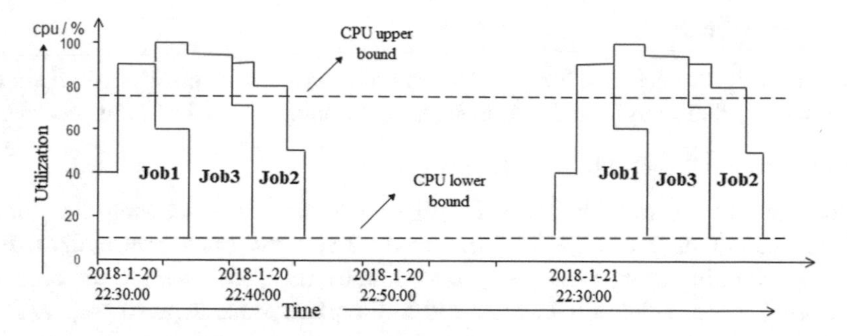

Fig. 4. Altruistic scheduling. When the CPU utilization is within the feasible range, the scheduler will choose the most appropriate task to perform.

4 Design Details

In this section, we describe how to enable and implement our altruism based Aetsa approach in ETL task scheduling based on Quartz [18], and discuss how to estimate resource requirements for tasks.

4.1 Enable Altruistic in ETL Task Schedule

In order to improve resource utilization and efficiency, we designed an altruistic strategy based scheduler (Pseudocode 1) in ETL task scheduling, which can reduce resource contention and meanwhile shorten the average response time of jobs. After receiving the task, Aetsa calls Pseudocode 1 to schedule tasks with the same execution cycle. First, add the ETL tasks to the schedule plan. Second, evaluate task's IO cost and sort them by the cost. Finally, use timer to monitor jobs start time, when specified time point is arrived, call another timer to capture the current CPU utilization, and according the CPU utilization to proceed atruistic schedule.

Pseudocode 1 Atruistic Scheduler

1: **procedure** SCHEDULE(Jobs J)
2: AddToSchedulePlan(J)
3: **for all** $J_k \in$ J **do**
4: IO_k = CalculateIOExpense(J_k)
5: **end for**
6: sort Jobs by IO, get J_{sorted}
7: **Timer** to set jobs start time
8: cpu = **Timer** (eg., every five second) read cpu info
9: AltruisticSchedule(J_{sorted}, cpu)
10: **for all** $J_k \in$ J **do**
11: pause J_k
12: **end for**
13: **end procedure**

4.2 Aetsa System

In the following, we describe how we implement Aetsa to offer intrinsic altruism. From Pseudocode 1, there are roughly three steps to complete our scheduling process.

- Add jobs to the scheduling plan

We use Quartz to help us with this process. After receiving the job, the AddToSchedulePlan procedure in Pseudocode 2 sets the executionConfiguration of each task according to the task type (Trans or Job), then adds them to the scheduling plan through the scheduleJob J_k, after add to the plan, pause them to prepare for our altruistic schedule.

Pseudocode 2 Add jobs to the scheduling plan

1: **procedure** AddToSchedulePlan(Jobs J)
2: **for all** $J_k \in$ J **do**
3: set executionConfiguration
4: scheduleJob J_k
5: pauseJob J_k
6: **end for**
7: **end procedure**

Since the execution of ETL tasks is pipelined, data will not reside in memory for a long time after being added to the schedule, so we do not consider their memory usage. Therefore we choose to evaluate the task IO cost to estimate their CPU requirements. From the CalculateIOCost procedure in Pseudocode 3, for each task, we need to read all the steps to determine which step needs to calculate the IO cost. If a step is an input step or has the function to read/write disk, calculate its stepIO as following:

$$stepIO = \frac{datasize}{blocksize} \tag{2}$$

It is proportional to the size of the read/write data, so we must read the data source information to get the data size of the step. After calculating the IO cost for the job, the job is sorted by its IO cost to get J_{sorted}. After that, call the timer to set the start time of the schedule (e.g., 22:30:00 every day), and when the specified time point is reached, another timer is called to capture the real-time CPU utilization of the system (by default, every 5 s).

Pseudocode 3 Calculate tasks IO cost

```
1: procedure CalculateIOCost (Job J_k)
2:   read all steps in J_k
3:   for all steps in J_k
4:     if(step is input step || step has read/write disk)
5:       IO_k += stepIO
6:   end for
7:   return IO_k
8: end procedure
```

The altruistic Schedule procedure in Pseudocode 4 responsible for altruistic scheduling. As long as the length of jobs *J* is not equal to 0, and the CPU utilization is within the feasible range, the position (*L*) of the job most suitable for scheduling is found according to function (*1*), then, restore the *Lth* job, and it will be executed. After execution, remove it from *J*. All jobs are scheduled until *J* has no jobs rest. Finally, suspend all jobs to wait for the next cycle to execute.

Pseudocode 4 Altruistic schedule

```
1: procedure AltruisticSchedule(Jobs J,CPU cpu)
2:   while J.length!=0 do
3:     if(CPU_down < cpu < CPU_up)
4:       L = [1 - (CPU - CPU_down) • 100/CPU_up] • n
5:       resume J_L
6:       delete J_L
7:   end while
8: end procedure
```

5 Result and Evaluation

5.1 Experimental Setup

Workloads. Our workload is composed of five different ETL tasks that we designed. The data comes from the medical data of more than 1600 primary medical institutions of GuiZhou Province. We choose a hospitalized registration table, a hospital doctor's advice table, a patient information table, a hospital expenses settlement table and a hospitalization expenses detail table to design five different ETL jobs to load the validated data into the target data warehouse. The Job details are shown in Table 1, the job data size and IO cost are shown in Table 2. Job4 has a SortRows step, which requires write/read disks, that makes its IO cost greater than Job5.

Table 1. ETL Job detail

Job num	Job description	Source data description
Job1	Verification of the developer code and the medical institutions code. Finally, load the correct data into the data warehouse, and the wrong data is logged into the error table	The hospitalized registration table is store in Mysql's qzj database, the developer table, medical institution table are store in Mysql's sjck database
Job2	The hospital doctor's advice table is store in Mysql's qzj database, and category detail table is store in Mysql's sjck database	Verification of the medical advice category code and the type of medical advice code. Finally, load the correct data into the data warehouse, and the warn data into the error table
Job3	The patient information table is store in Mysql's qzj database	Full extraction of patient information table, verify address classes code, if address classes code is wrong, set to "9". Finally, load to the target data warehouse
Job4	Full extraction of hospital expenses settlement table, and sort by total amount of expense settlement. Finally, load to the target data warehouse	The hospitalization expense settlement table is store in Mysql's qzj database
Job5	Partial field extraction of hospitalization expenses detail table, and set to 0 if the actual amount of the itemized project is null. Finally, load to the target data warehouse	The details of hospitalization expenses table is store in Mysql's qzj database

Table 2. Job data size

Job number	Data size	IO cost
Job1	52600 line	3980
Job2	101200 line	5780
Job3	152200 line	6288
Job4	202800 line	15012
Job5	252600 line	13720

Environment. Our experiments are run over a 64-bit Windows 10 operating system with 16 GB of memory and a 600 GB disk size. Both the data warehouse and the source database use MySQL 5.6, and the metadata is stored in H2 database.

Compared Scheduler Baselines. We mainly compare Aetsa with the following two methods:

FCFS: The first-come-first-servicing scheduler schedules jobs based on job submission order.

SJF: A shortest-job-first scheduler [10], it uses IO cost to infer job execution time and schedule jobs in the order of shortest IO first. We use SJF as the upper limit of Aetsa's average job waiting time and average job response time.

Metrics. Our metrics to quantify performance and efficiency are the improvement in the average response time and makespan, computed as:

$$Factor\ of\ Improvement = \frac{Duration\ of\ an\ Approach - Duration\ of\ Aetsa}{Duration\ of\ an\ Approach} \quad (3)$$

Factor of improvement greater than 0 means Aetsa is performing better, and vice versa.

Additionally, we use average waiting time to measure job performance.

5.2 Experiment Evaluation

In this section, we compare Aetsa with multiple metrics for two solutions (FCFS and SJF) to evaluate its impact on overall workloads and individual jobs.

Performance vs. Efficiency. Figure 5 depicts the average response time, average waiting time, and makespan for five different ETL tasks in our experiment. For each CPU upper bound (70%, 75%, 80%, 85%) and each method, we take the average of three experiments.

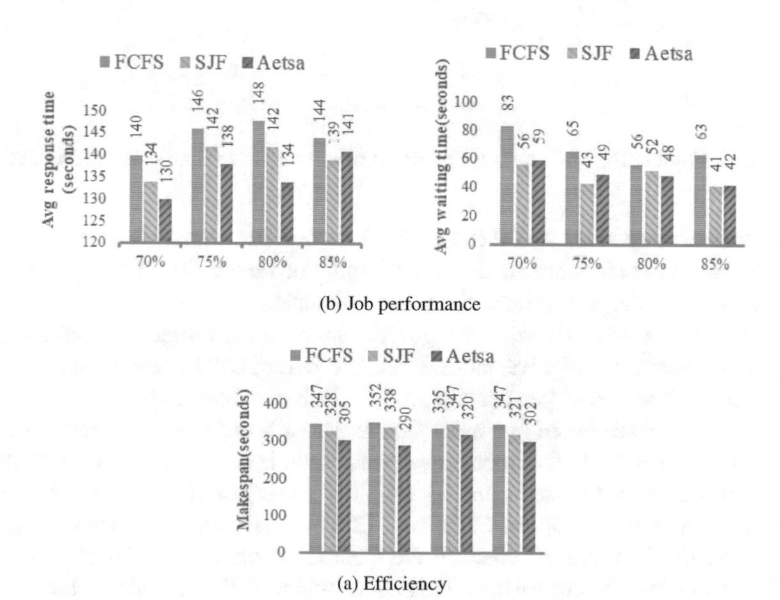

Fig. 5. Aetsa's performance meets or exceeds SJF, always better than FCFS, and achieves the best efficiency at different CPU upper bound, the lower the pillar, the better.

From Fig. 5 We observe that Aetsa's average response time is best. When the CPU upper bound is 85%, it is only slightly worse than SJF. In this case, the CPU is always busy and the task is running. Due to fierce competition in resources, it takes longer time for execution. For the average waiting time, SJF behave best because short jobs are executed first and then the CPU is quickly released for other jobs. Even so, our Aetsa approach is close to SJF in average waiting time, and FCFS has a long way to go. Finally, Aetsa offers the highest efficiency (minimum makespan), because it chooses the most suitable job to execute at any time, rather than equally handling all runnable tasks.

Improvement of Individual Job's Response Time. Figure 6 shows the cumulative distribution of job response time for FCFS and Aetsa under our workload. Obviously, Aetsa has more than 45% of the job's response time far exceeds FCFS, about 45% of the job's response time is slightly better, and less than 10% of the job's response time is up to 1.1 times worse than using FCFS. We see that about 40% of long jobs have a response time similar to that of FCFS. This is because when there are few long jobs left, their CPU is no longer a bottleneck, but it takes a lot of time to write databases, so their response time is not much different.

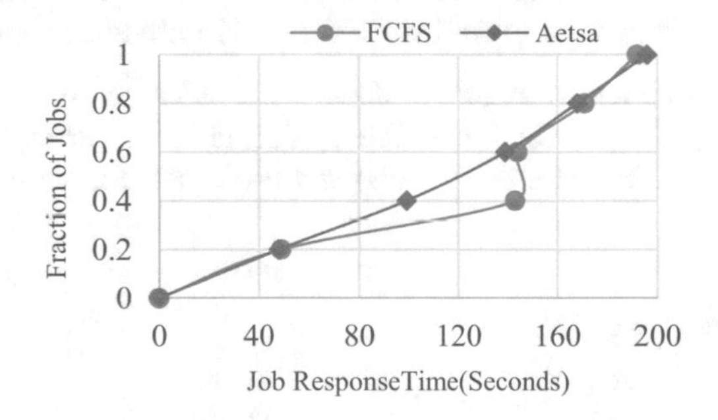

Fig. 6. CDF (Cumulative Distribution Function) of job response time using Aetsa and FCFS.

Two Factors of Improvement Across the Entire Workloads. Although Aetsa performs well for different metrics, the most important metrics from the perspective of ETL users is the average response time and makespan.

Figure 7 shows the factors of improvement of the average response time and makespan of Aetsa versus FCFS and SJF under different CPU upper bound. For Fig. 7 (a), we observe that Aetsa performs significantly better than FCFS, and the average response time increases by an average of more than 6% under the different CPU upper bound, At CPUup = 85%, the average response time is only reduced by less than 2% compared to SJF. It can be seen that as the CPU upper bound increases, the factor of improvement on average response time on FCFS and SJF tend to decrease as a whole. This is due to the fact that the load on the CPU is becoming heavier and the competition for resources is increasing. However, when CPUup = 80%, the factor of improvement is high because CPU overloads is not so severe at this time, and resources are fully utilized for Aetsa scheduling. For Fig. 7(b), we observe that Aetsa's makespan

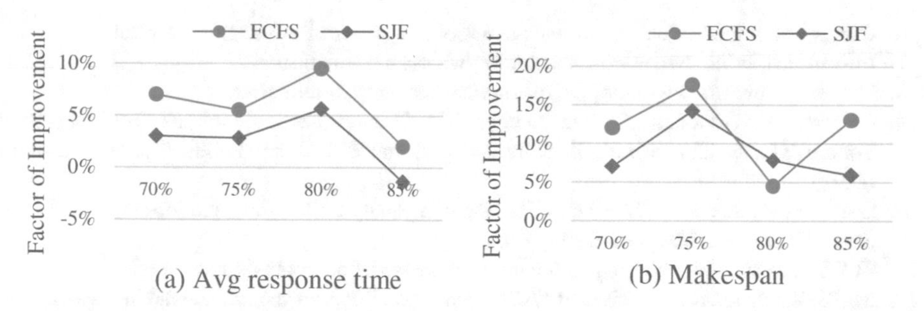

Fig. 7. Factors of improvement in average response time and makespan using Aetsa relative to FCFS and SJF.

has improved significantly both with respect to FCFS and SJF. Under the different CPU upper bound, the factor of improvement relative to FCFS has increased by more than 12%, and which is the most concerned issue for ETL users.

6 Conclusion

Existing timing based ETL schedule techniques only need user set the corresponding start times for every task to insure them execute correctly. The intrinsic mechanism of timing schedule can rarely take the performance and efficiency of the tasks into consideration. In order to achieve a better overall performance metrics, balance between usage, performance, resource utilization and efficiency are heavily needed. In this paper, we designed and implemented an altruism strategy based schedule approach named Aetsa. Benefit from the altruistic strategy, most appropriate job can be chosen to execute first, while it can significantly improve job performance, resource utilization and efficiency, meanwhile, all of these are transparent to users, i.e., without need they to make the corresponding configuration before tasks scheduling. The real scenarios based experimental result showed that Aetsa is very close to SJF in terms of average waiting time. Compared with FCFS, both the average response time and the efficiency are achieved satisfactory improvement.

References

1. Kimball, R., Ross, M.: The Data Warehouse Toolkit: The Definitive Guide to Dimensional Modeling, 3rd edn. Wiley Publishing, Inc., Toronto (2013)
2. Grandl, R., Chowdhury, M., Akella, A., Ananthanarayanan, G.: Altruistic scheduling in multi-resource clusters. In: Proceedings of the 12th USENIX Symposium on Operating Systems Design and Implementation (OSDI 2016), pp. 65–80 (2016)
3. Kimball, R., Caserta, J.: The Data Warehouse ETL Toolkit: Practical Techniques for Extracting, Cleaning, Conforming and Delivering Data. Wiley Publishing, Inc., Toronto (2004)
4. TASKCTL 5.0 Product White Paper. http://www.taskctl.com
5. TBSchedule. https://github.com/taobao/TBSchedule

6. Control-M. http://www.bmcsoftware.de/it-solutions/control-m-integrations.html
7. Informatica. https://www.edureka.co/blog/what-is-informatica/
8. DataStage. http://www-03.ibm.com/software/products/zh/ibminfodata
9. Casters, M., Bouman, R., van Dongen, J.: Pentaho Kettle Solutions: Building Open Source ETL Solutions with Pentaho Data Integration. Wiley Publishing, Inc., Toronto (2010)
10. Garey, M.R., Johnson, D.S., Sethi, R.: The complexity of flowshop and jobshop scheduling. Math. Oper. Res. **1**(2), 117–129 (1976)
11. FCFS. https://www.techopedia.com/definition/23455/first-come-first-served-fcfs
12. Multi-Dimensional Analysis. http://www.learn.geekinterview.com/data-warehouse/data-analysis/multi-dimensional-analysis.html
13. Han, J., Jian, P.: Data Mining, 3rd edn. Machinery Industry Press, Beijing (2012)
14. Vitt, E., Luckevich, M., Misner, S.: Business Intelligence: Making Better Decisions Faster. Microsoft Press, New York (2008)
15. Doan, A., Halevy, A., Ives, Z.: Principles of Data Integration. Machinery Industry Press, Beijing (2014)
16. Fei, Z.: Master Data Management MDM Detailed Explanation and Practice Based on the Whole Life Cycle. Tsinghua University Press, Beijing (2014)
17. Silberschatz, A., Galvin, P., Gagne, G.: Operating System Concepts. Wiley Publishing, Inc., Toronto (2005)
18. Cavaness, C.: Quartz Job Scheduling Framework. Prentice Hall, London (2006)
19. Coffman, E.G., Bruno, J.L.: Computer and Job-Shop Scheduling Theory. Wiley, Toronto (1976)
20. Chowdhury, M., Stoica, I.: Efficient coflow scheduling without prior knowledge. In: Proceedings of the 2015 ACM Conference on Special Interest Group on Data Communication – SIGCOMM 2015, pp. 393–406 (2015)

Determination of Stationary Points and Their Bindings in Dataset Using RBF Methods

Zuzana Majdisova(✉), Vaclav Skala, and Michal Smolik

Department of Computer Science and Engineering, Faculty of Applied Sciences, University of West Bohemia, Univerzitní 8, 30614 Plzeň, Czech Republic
{majdisz,smolik}@kiv.zcu.cz
http://www.vaclavskala.eu

Abstract. Stationary points of multivariable function which represents some surface have an important role in many application such as computer vision, chemical physics, etc. Nevertheless, the dataset describing the surface for which a sampling function is not known is often given. Therefore, it is necessary to propose an approach for finding the stationary points without knowledge of the sampling function.

In this paper, an algorithm for determining a set of stationary points of given sampled surface and detecting the bindings between these stationary points (such as stationary points lie on line segment, circle, etc.) is presented. Our approach is based on the piecewise RBF interpolation of the given dataset.

Keywords: Stationary points · RBF interpolation
Shape parameter · Shape detection · Nearest neighbor

1 Introduction

Stationary points of the given explicit function $f(\boldsymbol{x})$ are points where the gradient of the function $f(\boldsymbol{x})$ is zero in all directions, i.e. all partial derivatives are zero:

$$\begin{aligned} \nabla f(\boldsymbol{x}) &= \boldsymbol{0} \qquad \boldsymbol{x} \in \mathbb{E}^n, \text{ i.e.} \\ \frac{\partial f(\boldsymbol{x})}{\partial x_k} &= 0 \qquad k = 1, \ldots, n, \end{aligned} \tag{1}$$

where n denotes the dimension of space. The knowledge of stationary points is required in many areas that are used a multidimensional data analysis, e.g. [1–5]. The significant features of the given dataset can be determined using the set of stationary points. This properties can be further used for improving the quality of the RBF approximation [6,7], etc. In the technical applications, the sampling function is not often known and only the dataset describing the given surface is specified. Therefore, it is necessary determining the stationary points without knowledge of the sampling function. Moreover, for a higher dimension

R. Silhavy et al. (Eds.): CoMeSySo 2018, AISC 859, pp. 213–224, 2019.
https://doi.org/10.1007/978-3-030-00211-4_20

of space $n \geq 2$, it is possible that the stationary points of given surface are not only isolated but they can be formed into line segments, circles or some other shapes. A new approach for searching of bindings between stationary points will be described in this paper. Knowledge of these bindings is suitable, for example, for pruning purposes.

In the following sections, the fundamental the RBF interpolation will be described. The finding of stationary points of surface using the RBF interpolation will be described in Sect. 3. Moreover, the method, how the bindings between stationary points are searching, is introduced in this section. In the section Sect. 4, the results of our proposed algorithm will be presented. Finally, a final discussion of results will be performed.

2 RBF Interpolation

In this section, the RBF interpolation method, recently introduced, e.g. in [8,9], and its properties are described.

We assume that we have an unordered dataset $\{\boldsymbol{x}_i\}_1^N \in \mathbb{E}^n$, where n denotes the dimension of space and N is the number of given points. Further, each point $\boldsymbol{x}_i$ from the dataset is associated with a vector $\boldsymbol{h}_i \in \mathbb{E}^p$ of the given values, where p is the dimension of the vector, or a scalar value, i.e. $h_i \in \mathbb{E}^1$. In the following, we will deal with scalar data interpolation, i.e. the case when each point $\boldsymbol{x}_i$ is associated with a scalar value h_i is considered. Our goal is determined the unknown function which is sampled at given points $\{\boldsymbol{x}_i\}_1^N$ by values $\{h_i\}_1^N$. For these purposes, it can be used the RBF interpolation which is based on the distance computation between two points $\boldsymbol{x}_i$ and $\boldsymbol{x}_j$ from the given dataset.

The interpolated value can be determined as:

$$f(\boldsymbol{x}) = \sum_{j=1}^{N} c_j \phi(r_j) = \sum_{j=1}^{N} c_j \phi\left(\|\boldsymbol{x} - \boldsymbol{x}_j\|_2\right), \tag{2}$$

where the interpolating function $f(\boldsymbol{x})$ is represented as a sum of N RBFs, each centered at a different data points $\boldsymbol{x}_j$ and weighted by an appropriate weight c_j which has to be determined, see Fig. 1.

Applying (2) for all data points $\boldsymbol{x}_i, i = 1, \ldots, N$, we get a linear system of equations:

$$h_i = f(\boldsymbol{x}_i) = \sum_{j=1}^{N} c_j \phi\left(\|\boldsymbol{x}_i - \boldsymbol{x}_j\|_2\right) \quad i = 1, \ldots, N. \tag{3}$$

The linear system of equations can be represented in a matrix form as:

$$\boldsymbol{A}\boldsymbol{c} = \boldsymbol{h}, \tag{4}$$

where the matrix $\boldsymbol{A} = \{A_{ij}\} = \{\phi\left(\|\boldsymbol{x}_i - \boldsymbol{x}_j\|_2\right)\}$ is $N \times N$ symmetric square interpolation matrix, the vector $\boldsymbol{c} = (c_1, \ldots, c_N)^T$ is the vector of unknown weights and $\boldsymbol{h} = (h_1, \ldots, h_N)^T$ is a vector of values in the given points. This

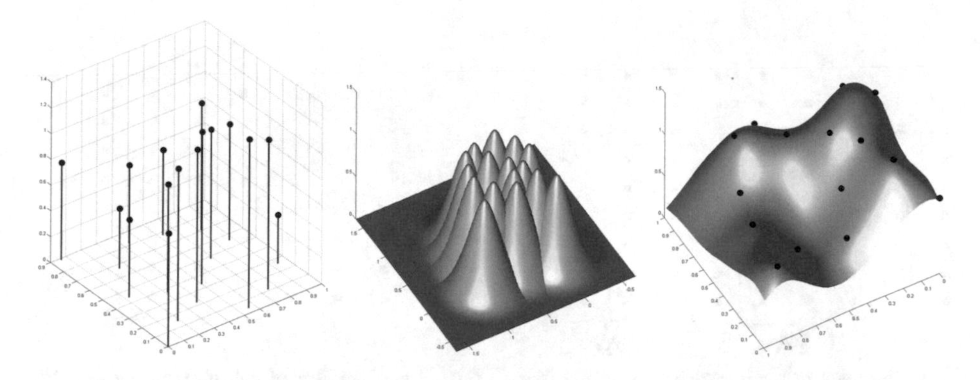

Fig. 1. Data values, the RBF collocation functions, the resulting interpolant.

linear system of equations can be solved by the Gauss elimination method, the LU decomposition, etc.

From the above, it can be seen that, in order to solve the interpolation problem, the distance matrix and a radial basis expansion are used.

3 Proposed Approach

In this section, determination of stationary points of the given dataset is described. Moreover, the approach includes the method for searching of bindings between stationary points because whole shape of stationary points may lie on the sampled surface.

3.1 Piecewise Approach for Determination of Stationary Points

For simplicity we assume that we have given dataset $\{\boldsymbol{x}_i\}_1^N \in \mathbb{E}^2$ and each point $\boldsymbol{x}_i$ from this dataset is associated with a scalar value $h_i \in \mathbb{E}^1$. Further, for purposes of determination of stationary points, we assume that the given dataset contains the points on a $N_x \times N_y$ regular grid, where Δx and Δy are real numbers representing its grid spacing. Moreover, the row-major ordering of the given data is performed at first. After that, the piecewise approach is applied on the given data.

The process which is performed at each step of the piecewise approach is following. Every sixteen points $\{\boldsymbol{x}_m\}_1^{16} = \{\boldsymbol{x}_{i,j}, \ldots, \boldsymbol{x}_{i,j+3}, \ldots, \boldsymbol{x}_{i+3,j}, \ldots, \boldsymbol{x}_{i+3,j+3}\}$ from the given dataset, where $i \in \{1, \ldots, N_y - 3\}$ denotes the row index and $j \in \{1, \ldots, N_x - 3\}$ denotes the column index, are interpolated by the RBF interpolation (2), i.e. the linear system (4) has to be solved and the vector of weights $\hat{\boldsymbol{c}} = (c_1, \ldots, c_{16})$ is computed. It mean that during one step of proposed approach, the RBF interpolation for $3\Delta x \times 3\Delta y$ area, where Δx and Δy are real numbers representing the input grid spacing, is performed, see Fig. 2a.

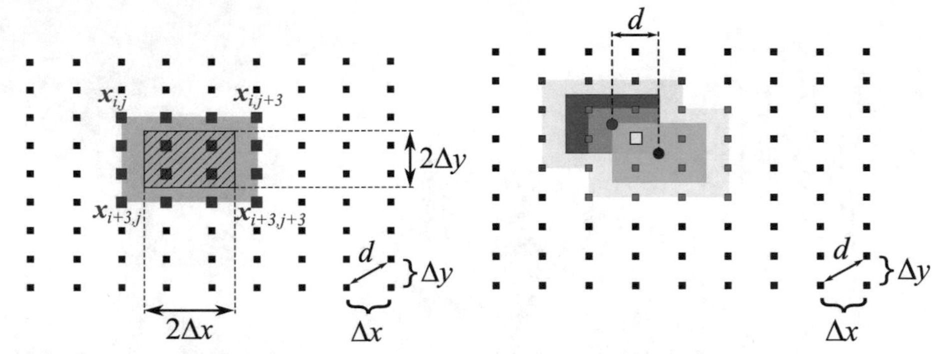

(a) The grey area shows the all points from the given dataset which are interpolated by the RBF method during one step of piecewise approach. The hatched area illustrates the domain for which the stationary points of the given dataset are determined from the obtained RBF interpolation.

(b) Visualization of the stationary points reduction which is performed if the two points are identical or very close to identical. The green circle and red circle mark the stationary points which were determined from two different RBF interpolations and which were merged to one stationary point marked by yellow square.

Fig. 2. Proposed piecewise approach

Then, the stationary points $\{\boldsymbol{s}_q\}$ of this interpolation function are determined using (1). Specifically, for stationary points of the RBF interpolation function the nonlinear system of equations:

$$\mathbf{0} = \sum_{m=1}^{16} c_m \frac{\phi'\left(\|\boldsymbol{x} - \boldsymbol{x}_m\|_2\right)}{\|\boldsymbol{x} - \boldsymbol{x}_m\|_2} * \left(\boldsymbol{x} - \boldsymbol{x}_m\right), \tag{5}$$

where $\phi'(r)$ is the derivation of RBF function ϕ with respect to variable r, $*$ denotes the element-wise multiplication and $\hat{\boldsymbol{c}} = (c_1, \ldots, c_{16})$ is the vector of weights, has to be solved. The solution of (5), i.e. the stationary points $\{\boldsymbol{s}_q\}$ of the RBF interpolation, is searched for the domain defined as:

$$\begin{gathered} \boldsymbol{x}_{i,j} + \boldsymbol{\varepsilon}_{min} \leq \boldsymbol{s}_q \leq \boldsymbol{x}_{i+3,j+3} - \boldsymbol{\varepsilon}_{max}, \\ \boldsymbol{\varepsilon}_{min} = \begin{cases} \left[\frac{\Delta x}{2}, 0\right] & \text{if } i = 1 \\ \left[0, \frac{\Delta y}{2}\right] & \text{if } j = 1 \\ \left[\frac{\Delta x}{2}, \frac{\Delta y}{2}\right] & \text{otherwise} \end{cases} \quad \boldsymbol{\varepsilon}_{max} = \begin{cases} \left[\frac{\Delta x}{2}, 0\right] & \text{if } i = N_y - 3 \\ \left[0, \frac{\Delta y}{2}\right] & \text{if } j = N_x - 3 \\ \left[\frac{\Delta x}{2}, \frac{\Delta y}{2}\right] & \text{otherwise} \end{cases} \end{gathered} \tag{6}$$

where Δx and Δy are real numbers representing the input grid spacing, N_x indicates the number of grid column and N_y is the number of grid rows, see Fig. 2a, and the resulting set is added to the set of stationary points $\{\boldsymbol{s}_l\}$. It should be noted, that the values $\boldsymbol{\varepsilon}_{min}$ and $\boldsymbol{\varepsilon}_{max}$ include the correction for the boundary areas.

The determination of stationary points of a function corresponds to the problem of finding critical points of the vector field, where the vector field is defined by Eq. (5) for our purposes, and, therefore the method for determining critical points [10, 11] may be used for obtaining the result.

The advantage of the above mentioned process is that the matrix $\boldsymbol{A}$ of the linear system (4) for the RBF interpolation, is not dependent on the position of the given points (the matrix is dependent only on the distances between given points) and, therefore, this matrix is constant for all steps of piecewise approach. It should be noted that the approximation by a quadric surface could be used instead of the RBF interpolation, but the experimental results proved that this variant returns worse results in terms of stationary point locations.

The set of stationary points in the current form $\{\boldsymbol{s}_l\}$ may contain two identical points or points very close to identical. This problem is caused by the fact that one stationary point can be obtained from more RBF interpolations. The situation is illustrated in Fig. 2b. However, this problem can be solved by reduction of the set of stationary points. Then, the final set of stationary points $\{\boldsymbol{\sigma}_u\}$ of the given data is determined as follows. The subset S_u of stationary points is removed from the unreduced set of stationary points $\{\boldsymbol{s}_l\}$. The points in the subset S_u meet relation:

$$S_u = \{\boldsymbol{s}_k : \|\boldsymbol{s}_k - \boldsymbol{s}_1\| \leq d\}, \tag{7}$$

where $d = \sqrt{(\Delta x)^2 + (\Delta y)^2}$ is the diagonal step in the regular grid, and the new stationary point is determined as a centroid of points from subset S_u:

$$\boldsymbol{\sigma}_u = \frac{\sum \boldsymbol{s}_k}{|S_u|}, \tag{8}$$

where $|S_u|$ is a number of points in the subset S_u. The process is repeated until the unreduced set of the stationary points is not empty.

The whole algorithm for determining the stationary points of the given dataset is summarized in Algorithm 1.

3.2 Estimation of Shape Parameter for RBF Interpolation

The piecewise RBF interpolation is used during the process of the determining the stationary points of the given dataset. Nevertheless, the quality of the resulting RBF interpolation strongly depends on the choice of the shape parameter α. Therefore, in this section, the determination of suitable shape parameter α will be performed.

For the above mentioned process, the surface with the least possible tension is required, i.e. the surface must contain as little wavy as possible if the interpolated points allow it. It means that the shape parameter α has to be sufficiently large.

Therefore, for these purposes, we proposed and experimentally verified that shape parameter α is chosen so that the radius of circle of non-stationary inflection points of used RBF function $\phi(r)$ corresponds to the maximum distance of the interpolated points, within one step of proposed piecewise approach, which is $r = 3d$, where $d = \sqrt{(\Delta x)^2 + (\Delta y)^2}$ is the diagonal step in the regular grid.

Algorithm 1. Determination of the stationary points $\{\boldsymbol{\sigma}_u\}_1^{N_S}$.

Input: given points $\{\boldsymbol{x}_i\}_1^N$ and their associated scalar values $\{h_i\}_1^N$, size of grid $N_x \times N_y$, grid spacing Δx and Δy, used RBF ϕ and its shape parameter α.
Output: stationary points $\{\boldsymbol{\sigma}_u\}_1^{N_S}$

1: Row-major ordering the given points $\{\boldsymbol{x}_i\}_1^N$.
2: $d = \sqrt{(\Delta x)^2 + (\Delta y)^2}$.
3: Compute matrix $\boldsymbol{A}$ of linear system (4) for the set of points $\{\boldsymbol{x}_1, \dots, \boldsymbol{x}_4, \boldsymbol{x}_{N_x+1}, \dots, \boldsymbol{x}_{N_x+4}, \boldsymbol{x}_{2N_x+1}, \dots, \boldsymbol{x}_{2N_x+4}, \boldsymbol{x}_{3N_x+1}, \dots, \boldsymbol{x}_{3N_x+4}\}$.
4: **for** $i = 1, \dots, N_y - 3$ **do**
5: **for** $j = 1, \dots, N_x - 3$ **do**
6: $\hat{\boldsymbol{x}} = \left\{\boldsymbol{x}_{(i-1)N_x+j}, \dots, \boldsymbol{x}_{(i-1)N_x+j+3}, \dots, \boldsymbol{x}_{(i+2)N_x+j}, \dots, \boldsymbol{x}_{(i+2)N_x+j+3}\right\}$
7: $\hat{\boldsymbol{h}} = \left\{h_{(i-1)N_x+j}, \dots, h_{(i-1)N_x+j+3}, \dots, h_{(i+2)N_x+j}, \dots, h_{(i+2)N_x+j+3}\right\}$
8: Compute the vector of unknown weights $\hat{\boldsymbol{c}}$, eq. (4), where $\boldsymbol{h} = \hat{\boldsymbol{h}}$.
9: Compute the coefficients ε_{min} and ε_{max}, eq. (6).
10: Determine the stationary points $\{\boldsymbol{s}_q\}$ from eq. (5) in the domain (6).
11: $\{\boldsymbol{s}_l\} = \{\boldsymbol{s}_l\} \cup \{\boldsymbol{s}_q\}$
12: **while** the set $\{\boldsymbol{s}_k\}$ is not empty **do**
13: Find $S_u = \{\boldsymbol{s}_k : \|\boldsymbol{s}_k - \boldsymbol{s}_1\|_2 \leq d\}$ in the set $\{\boldsymbol{s}_l\}$.
14: Add the stationary point $\frac{\sum \boldsymbol{s}_k}{|S_u|}$ to the final set of stationary points $\{\boldsymbol{\sigma}_u\}$.
15: Delete all points $\boldsymbol{s}_k \in S_u$ from the set $\{\boldsymbol{s}_l\}$.

From this assumption, the following expression for shape parameter α was derived:

$$\alpha = \frac{\omega}{3d}, \tag{9}$$

where $d = \sqrt{(\Delta x)^2 + (\Delta y)^2}$ is the diagonal step in the regular grid and ω is a constant parameter depending on the type of used RBF, see Table 1.

Table 1. Different RBFs, their derivation $\phi'(r)$ and their parameter ω, Eq. (9).

RBF	$\phi(r)$	$\phi'(r)$	ω
Gaussian RBF	$e^{-(\alpha r)^2}$	$-2\alpha^2 r e^{-(\alpha r)^2}$	$1/\sqrt{2}$
Inverse quadric	$\left(1 + (\alpha r)^2\right)^{-1}$	$-2\alpha^2 r \left(1 + (\alpha r)^2\right)^{-2}$	$1/\sqrt{3}$
Wendland's $\phi_{3,1}$	$(1 - \alpha r)_+^4 (4\alpha r + 1)$	$-20\alpha^2 r (1 - \alpha r)_+^3$	$1/4$

3.3 Searching of Bindings Between Stationary Points

It is possible that the given surface does not contain only isolated stationary points, but the curves of stationary points, such as line segments, circles, parabolas or some other shapes, can lie on the given surface. Therefore, the method for searching of bindings between stationary points will be described.

At the beginning, the maximal possible distance δ_{max} of two stationary points for which these stationary points still lie on the same curve has to be established. The situation of the worst case is illustrated in Fig. 3. In this figure, it can be

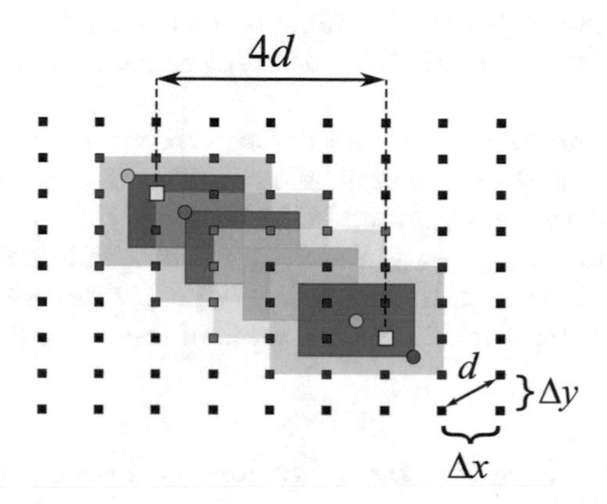

Fig. 3. The figure shows the worst case in which two stationary points (yellow squares) still lie on the same curve of stationary points, i.e. the distance between two stationary points is maximal possible distance. Moreover, the reduction of stationary points is again shown.

seen four subdomains of the piecewise approach and for each of them, the one stationary point is indicated using circle mark. Based on Eq. (7), the stationary points of blue and red subdomains are reduced and are replaced by their centroid. The same case occurs for the green and purple subdomain. New stationary points which are obtained after the reduction are indicates by yellow squares in the figure. It is also obvious that the distance of these two stationary points, which is also the maximum possible distance δ_{max}, is:

$$\delta_{max} = 4d,$$

where $d = \sqrt{(\Delta x)^2 + (\Delta y)^2}$ is the diagonal step in the regular grid.

Now, the stationary points $\{\boldsymbol{\sigma}_u\}$ of the given dataset are sequentially processed by following. For the current stationary point $\boldsymbol{\sigma}_u$, the all stationary points $\{\boldsymbol{\sigma}_w\}$ which lying in the distance δ_{max} are determined:

$$\{\boldsymbol{\sigma}_w\} = \{\boldsymbol{\sigma}_w : \|\boldsymbol{\sigma}_w - \boldsymbol{\sigma}_u\| \leq \delta_{max}\}. \tag{10}$$

If no stationary point is found, then the stationary point $\boldsymbol{\sigma}_u$ is isolated. Otherwise, the binding $f_v = \{\boldsymbol{\sigma}_w\} \cup \{\boldsymbol{\sigma}_u\}$ is obtained and newly added stationary points are processed in the same way. Finally, the result of this approach is the set of points described the curve of stationary points. This procedure is repeated until the all stationary points $\{\boldsymbol{\sigma}_u\}$ are processed.

One of the possible solution of this problem is the kd–tree which can be simply applied for purposes of searching of bindings between stationary points.

4 Experimental Results

In this section, the experimental results for our proposed approach will be presented and their comparison with the exact stationary points which were determined analytically from the sampling function will be made. The implementation was performed in Matlab. In addition, different radial basis functions have been used, see Table 1.

For purposes of our experiments, a uniform distribution of points on a rectangular domain was used for the testing data. Thus, the given dataset contains 120×120 points uniformly distributed in the interval $[x_{min}, x_{max}] \times [y_{min}, y_{max}]$, where the values x_{min}, x_{max}, y_{min} and y_{max} are chosen based on the used sampling function, see (11a) – (11b) and (12a) – (12d). Moreover, each point from this dataset is associated with a function value of the selected sampling function at this point.

4.1 Comparison of Determined Stationary Points with Exact Stationary Points

In this section, the results for datasets whose stationary points do not contain mutual bindings, i.e. all stationary points are isolated, will be presented. The sampling functions f_1 (11a) and f_2 (11b), which were defined in [12], fulfill these properties.

$$\begin{aligned} f_1(\boldsymbol{x}) = & \frac{3}{4} e^{-\frac{(9x_1-2)^2}{4} - \frac{(9x_2-2)^2}{4}} + \frac{3}{4} e^{-\frac{(9x_1+1)^2}{49} - \frac{(9x_2+1)}{10}} \\ & + \frac{1}{2} e^{-\frac{(9x_1-7)^2}{4} - \frac{(9x_2-3)^2}{4}} - \frac{1}{5} e^{-(9x_1-4)^2-(9x_2-7)^2} \end{aligned} \qquad \boldsymbol{x} \in [0,1] \times [0,1] \quad (11a)$$

$$f_2(\boldsymbol{x}) = \sin(3 \cdot x_1) \cdot \cos(3 \cdot x_2) \qquad \boldsymbol{x} \in [-2,2] \times [-2,2] \quad (11b)$$

Figure 4a presents the results for the dataset in which each point is associated with a value from the f_1 function (11a) when the Gaussian RBF has been used for the piecewise RBF interpolation. Using our proposed approach, five isolated stationary points which are marked by white circles were found for this dataset. The exact stationary points of f_1 function (11a) are shown using the red asterisks ($*$).

The results for the dataset in which each point is associated with a value from the f_2 function (11b), when the Gaussian RBF has been used for the piecewise RBF interpolation, are presented in Fig. 4b. Twenty four isolated stationary points which are represented by white circles were found for this dataset using our proposed approach. The exact stationary points of f_2 function (11b) are again shown using the red asterisks ($*$).

It can be seen that obtained results for both datasets correspond to the stationary points calculated analytically from the sampling functions. Moreover, it should be noted, that the same results were obtained even when other RBF function, see Table 1, were used for the piecewise RBF interpolation.

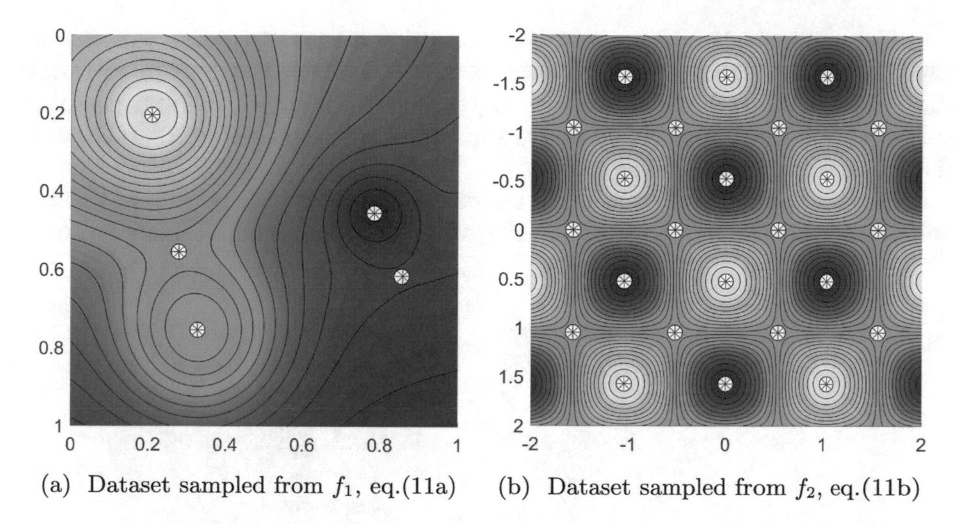

(a) Dataset sampled from f_1, eq.(11a) (b) Dataset sampled from f_2, eq.(11b)

Fig. 4. The white circles indicate the stationary points of the given dataset that are obtained using the proposed approach when the RBF interpolation used the Gaussian RBF. The tested dataset contains 120×120 points. The red asterisks ($*$) denote the exact positions of the stationary points of the appropriate function. Furthermore, the contour map of given dataset is shown.

4.2 Bindings Between Stationary Points

In this section, the results for datasets whose stationary points contains mutual bindings will be presented. The four following sampling functions (12a) – (12d) fulfill these properties.

$$f_{11}(\boldsymbol{x}) = -\left(x_1 - x_2\right)^2, \qquad \boldsymbol{x} \in [-1,1] \times [-1,1] \qquad (12a)$$

$$f_{12}(\boldsymbol{x}) = \sin\left(x_1 + x_2^2\right), \qquad \boldsymbol{x} \in [-3,3] \times [-2,2] \qquad (12b)$$

$$f_{13}(\boldsymbol{x}) = \sin\left(3\pi\left(\sqrt{x_1^2 + x_2^2} + 0.25\right)\right), \qquad \boldsymbol{x} \in [-1,1] \times [-1,1] \qquad (12c)$$

$$f_{14}(\boldsymbol{x}) = -2 \cdot \left(x_1^2 - x_2^2\right)^2 + 1, \qquad \boldsymbol{x} \in [-1,1] \times [-1,1] \qquad (12d)$$

At the beginning, it should be noted that the white solid line indicates the curve of stationary points obtained for the given dataset using our proposed approach and the isolated stationary point obtained using our proposed approach is marked by the white circle. The red dashed line indicates the curve of stationary points calculated analytically from the given sampling function and the isolated stationary point calculated analytically from the given sampling function is represented by the red asterisk ($*$).

Figure 5a presents the results for the dataset in which each point is associated with a value from the f_{11} function (12a) when the Gaussian RBF has been used for the piecewise RBF interpolation. Using our proposed approach, one curve of stationary points, specifically the line segment, were found for this dataset. This result coincides with the result obtained using analytically approach.

The results for the dataset in which each point is associated with a value from the f_{12} function (12b), when the Gaussian RBF has been used for the piecewise RBF interpolation, are presented in Fig. 5b. For this dataset, the four curves of stationary points, specifically two parabolas and two segments of parabola, were found using our proposed approach.

Figure 5c presents the results for the dataset in which each point is associated with a value from the f_{13} function (12c) when the Gaussian RBF has been used

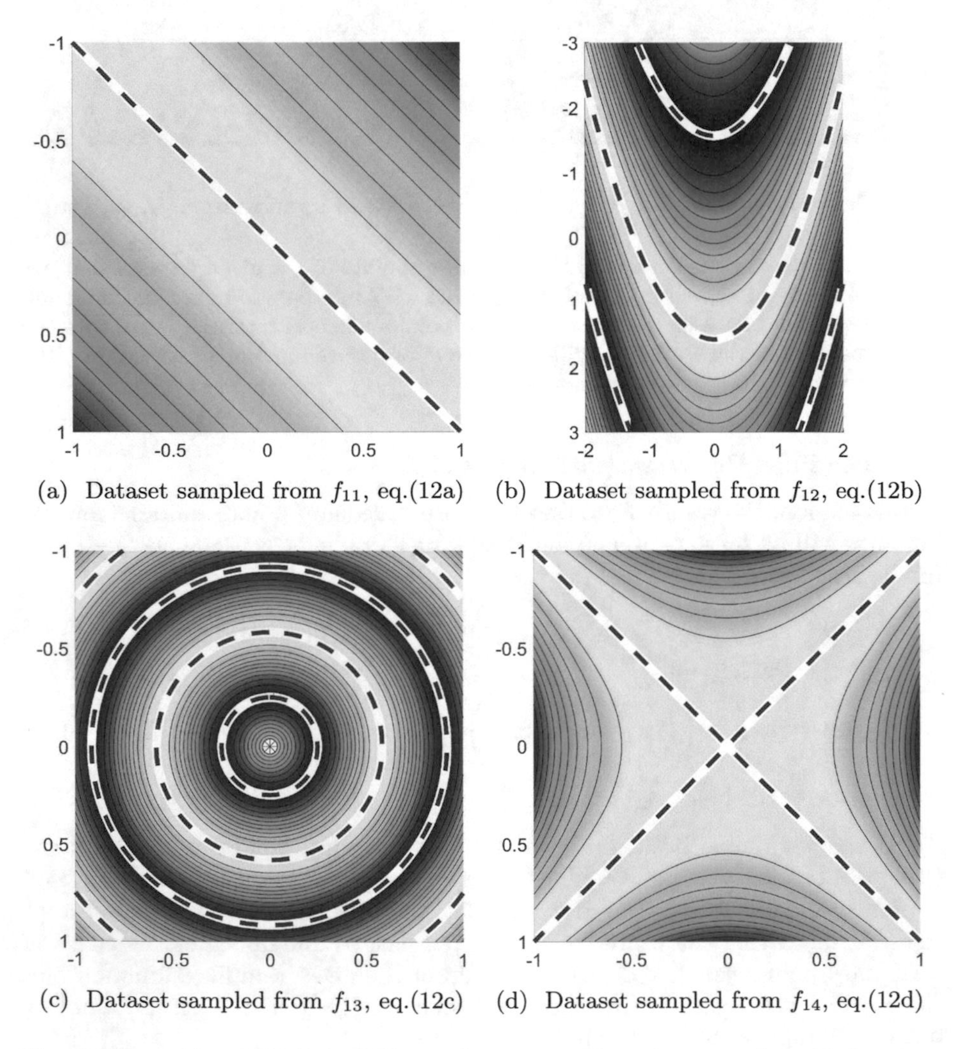

(a) Dataset sampled from f_{11}, eq.(12a) (b) Dataset sampled from f_{12}, eq.(12b)

(c) Dataset sampled from f_{13}, eq.(12c) (d) Dataset sampled from f_{14}, eq.(12d)

Fig. 5. The white solid lines indicate the curves of stationary points of the given dataset that are obtained using the proposed approach when the RBF interpolation used the Gaussian RBF. The tested dataset contains 120×120 points. The red dashed lines denote the exact curves of the stationary points of the appropriate function. Furthermore, the contour map of given dataset is shown.

for the piecewise RBF interpolation. Using our proposed approach, seven curves of stationary points, specifically three circles and four arcs, and one isolated stationary point were found for this dataset.

The results for the dataset in which each point is associated with a value from the f_{14} function (12d), when the Gaussian RBF has been used for the piecewise RBF interpolation, are presented in Fig. 5d. For this dataset, the two curves of stationary points, specifically two line segments, were found using our proposed approach.

For all mentioned experiments, it can be seen that the results obtained using our proposed approach correspond to results obtained using analytically approach. Moreover, it should be again noted that the same results were obtained even when other RBF function, see Table 1, were used for the piecewise RBF interpolation.

5 Conclusion

In this paper, a new approach for determination of stationary points of given sampled surface without knowledge of the sampling function is presented. The proposed method is based on the piecewise RBF interpolation of the given dataset. Moreover, the proposed approach includes the method of detecting the bindings between the found stationary points, i.e. the approach is able to associate the points from the same curve of stationary points.

The experiments proved that the stationary points determined by our proposed approach coincide with the exact stationary points which were determined analytically form the sampling function.

The results of the proposed approach can, for example, be used for determination of the set of reference points for the RBF approximation which enable appropriate compression of given dataset. The knowledge of the bindings between stationary points is possible to use for pruning the subset of related stationary points to the required number of points on the appropriate curve of stationary points.

In the future work, the proposed approach can be generalized for scattered data using the k-nearest neighbors algorithm.

Acknowledgments. The authors would like to thank their colleagues at the University of West Bohemia, Plzeň, for their discussions and suggestions, and the anonymous reviewers for their valuable comments. Special thanks belong to Jan Dvorak, Lukas Hruda and Martin Červenka for their independent experiments and valuable comments. The research was supported by the Czech Science Foundation GAČR project GA17-05534S and partially supported by the SGS 2016-013 project.

References

1. Banerjee, A., Adams, N., Simons, J., Shepard, R.: Search for stationary points on surfaces. J. Phys. Chem. **89**(1), 52–57 (1985)
2. Tsai, C.J., Jordan, K.D.: Use of an eigenmode method to locate the stationary points on the potential energy surfaces of selected argon and water clusters. J. Phys. Chem. **97**(43), 11227–11237 (1993)
3. Comaniciu, D., Meer, P.: Mean shift: a robust approach toward feature space analysis. IEEE Trans. Pattern Anal. Mach. Intell. **24**, 603–619 (2002)
4. Strodel, B., Wales, D.J.: Free energy surfaces from an extended harmonic superposition approach and kinetics for alanine dipeptide. Chem. Phys. Lett. **466**(4), 105–115 (2008)
5. Liu, Y., Burger, S.K., Ayers, P.W.: Newton trajectories for finding stationary points on molecular potential energy surfaces. J. Math. Chem. **49**(9), 1915–1927 (2011)
6. Majdisova, Z., Skala, V.: Radial basis function approximations: comparison and applications. Appl. Math. Model. **51**, 728–743 (2017)
7. Majdisova, Z., Skala, V.: Big geo data surface approximation using radial basis functions: a comparative study. Comput. Geosci. **109**, 51–58 (2017)
8. Skala, V.: RBF interpolation with CSRBF of large data sets. Procedia Comput. Sci. **108**, 2433–2437 (2017). International Conference on Computational Science, ICCS 2017, 12–14 June 2017, Zurich, Switzerland
9. Smolik, M., Skala, V.: Large scattered data interpolation with radial basis functions and space subdivision. Integr. Comput.-Aided Eng. **25**(1), 49–62 (2018)
10. Bhatia, H., Gyulassy, A., Wang, H., Bremer, P.T., Pascucci, V.: Robust detection of singularities in vector fields. In: Topological Methods in Data Analysis and Visualization III, pp. 3–18. Springer (2014)
11. Wang, W., Wang, W., Li, S.: Detection and classification of critical points in piecewise linear vector fields. J. Vis. **21**, 147–161 (2018)
12. Franke, R.: A critical comparison of some methods for interpolation of scattered data. Technical report NPS53-79-003, Naval Postgraduate School, Monterey, CA (1979)

A Novel Approach of the Approximation by Patterns Using Hybrid RBF NN with Flexible Parameters

Mariia Martynova(✉)

University of West Bohemia, 306 14 Pilsen, Czech Republic
martynov@kiv.zcu.cz

Abstract. This paper describes a new solution to the optimization task of approximation by radial-basis-function (RBF) neural network. The proposed method is addressed to the problem of variable shape parameters for data using the RBF. It involves the max-min algorithm, the RBF neural network, the algorithm for placing new neuron's centers and the structure of the future deep learning complex neural network (NN).

Keywords: Approximation · Big data · Complex function · Deep learning Neural network · Radial basic function · RBF · Variable shape parameters

1 Introduction

1.1 Solution Resources

The problem of the data approximation has not lost its relevance for decades, as the approximation solves the problem of incompleteness and noisy data, and as a result - it helps to improve the quality of data, which is important for a further processing. The approximation helps to study the numerical characteristics and qualitative properties of the object, reducing the problem with the study of simpler or more convenient objects which properties are known or calculated easily, which simplifies sometimes rather complicated and time-expensive calculations [1].

The application value of the data approximation is quite large since it can be used wherever any kind of the data set, which supposedly follows a certain mathematical law, is. For example, to calculate values of complex functions or to process experimental and natural data. With the growing popularity of the Big Data, a data mining, the Internet of things and other technologies based on large volumes of information, the question of the qualitative approximation becomes more distinct.

For the scattered data approximation radial basis function (RBF) was proposed, as the RBF formulation does not actually depend on data dimensionality and leads to a system of linear equations $\mathbf{Ax} = \mathbf{b}$. However, there are problems with numerical stability, i.e. ill conditioned systems [2–4], memory requirements for large data sets [5, 6]. Also, problems of data compression using RBF and Least Square Error application have been solved recently [7, 8].

R. Silhavy et al. (Eds.): CoMeSySo 2018, AISC 859, pp. 225–235, 2019.
https://doi.org/10.1007/978-3-030-00211-4_21

RBF approximation methods use fixed shape parameters with some "experimental estimation". However, it means that non-fixed shape parameters are needed to obtain better interpolation or approximation.

On the other hand, neural network (NN) technologies are well-applicable tool in problems of finding parameters and dependencies, and the variety of architectures and methods of learning neural networks allows choosing the best option for a particular type of problem. Neural networks are also good due to their efficiency of small datasets or poorly structured data given at the beginning.

The NNs are widely used for solutions of approximation problems [9]. The analysis of existing solutions is placed in Chap. 4 of this paper. Figure 1 presents the results of a data processing by different types of neural networks. It is easy to see that the results are quite different, each of them has its own shortcomings and advantages.

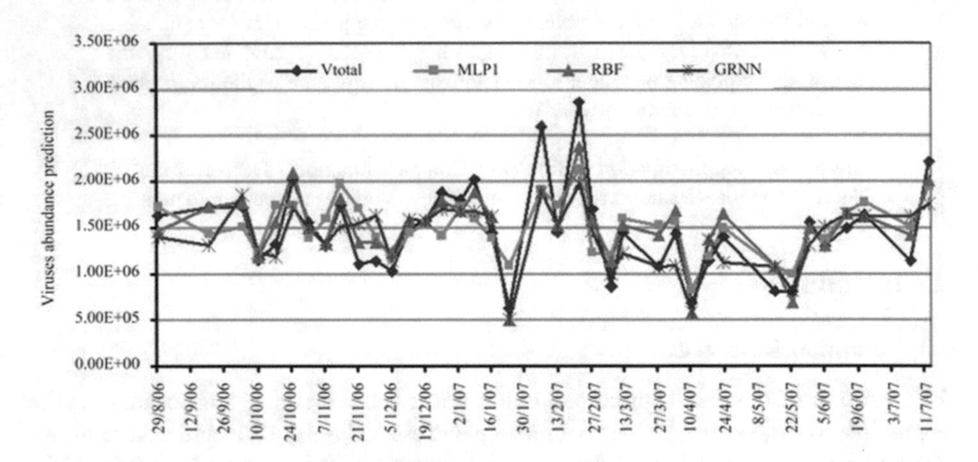

Fig. 1. Approximation results by some types of NN (taken from [10])

1.2 Curve Approximation with RBF NN

In the following the proposed hybrid RBF NN is explained on a curve approximation in E^2, i.e. y = f(x). In this solution, the radial basis function neural network (RBF NN) was chosen, since they possess certain undeniable advantages, such as:

- the simplicity of structure,
- the flexible and versatile adjustment of parameters,
- it allows to significantly reduce the neurons number, necessary for a separation of different classes/descriptions of complex curves.

Figure 2 presents the classical structure of a radial basis neural network [11]. Due to its features, this network has only one hidden layer, which allows bypassing the task of selecting the number of hidden layers.

The network itself consists of:

N – a number of training data,
$\boldsymbol{X} = [x_1 \ldots x_N]$ - a vector of input data,
k – a number of RBF neurons,
$\boldsymbol{\varphi} = [\varphi_1 \ldots \varphi_k]$ - a radial basis function as an activation function,
• 1 – bias,
$\boldsymbol{\omega} = [\omega_0 \ldots \omega_k]$ - the weights of each of these functions,
$\sum$ - the output adder neuron,
$\boldsymbol{y} = [y_1 \ldots y_N]$ - the vector of output values of the neural network.

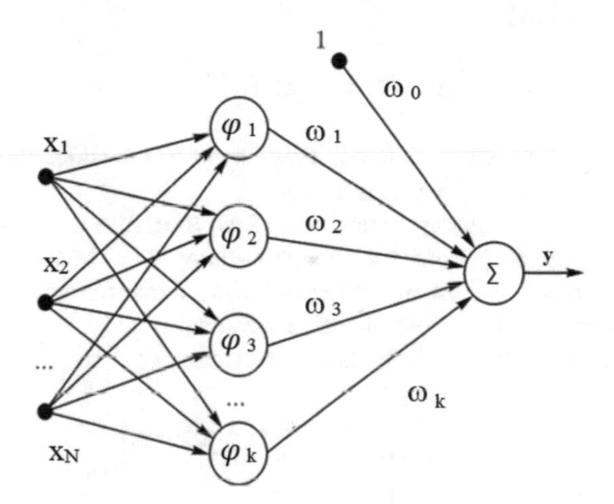

Fig. 2. The typical structure of RBF NN

The processing of data in this network is divided into 2 types: linear and nonlinear algorithm. The difference between these methods is that the shape parameters of the activation functions are known before the training begins, and in the process of nonlinear processing these parameters are changed by the network itself [12].

The Gauss function is the most common and used radially symmetric function [13]. It has two settings: center $\boldsymbol{c} = [c_1 \ldots c_k]$ and width $\boldsymbol{\sigma} = [\sigma_1 \ldots \sigma_k]$. It can be written as:

$$\varphi_j(\mathbf{x}) = \varphi(||\mathbf{x} - \mathbf{c}_j||) = exp\left(\frac{((\mathbf{x}_i - \boldsymbol{c}_j))^2}{2\sigma_j^2}\right), \quad i = 1, \ldots, N, \quad j = 1, \ldots, k. \quad (1)$$

In the case of a linear algorithm, a vector of input parameters enters to the input layer, after which the radial Green's matrix $\boldsymbol{G}$, i.e. the matrix of the reaction values of each neuron for each data example or interpolation matrix, is calculated:

$$G = \begin{bmatrix} \varphi_1(\mathbf{x}_1) & \cdots & \varphi_k(\mathbf{x}_1) \\ \vdots & \ddots & \vdots \\ \varphi_1(\mathbf{x}_N) & \cdots & \varphi_k(\mathbf{x}_N) \end{bmatrix} = \begin{bmatrix} exp\left(\frac{((\mathbf{x}_1-\mathbf{c}_1))^2}{2\sigma_1^2}\right) & \cdots & exp\left(\frac{((\mathbf{x}_1-\mathbf{c}_k))^2}{2\sigma_k^2}\right) \\ \vdots & \ddots & \vdots \\ exp\left(\frac{((\mathbf{x}_N-\mathbf{c}_1))^2}{2\sigma_1^2}\right) & \cdots & exp\left(\frac{((\mathbf{x}_N-\mathbf{c}_k))^2}{2\sigma_k^2}\right) \end{bmatrix} \tag{2}$$

After that, the problem of finding the vector of weights leads to a solution of an overdetermined system of linear equations.

$$\boldsymbol{G}(\boldsymbol{\omega}) = \boldsymbol{y} \tag{3}$$

Then, it is possible to calculate the weights using the least square error method, e.g. by pseudo inversion [14]:

$$\boldsymbol{\omega} = \boldsymbol{G}^{+}\boldsymbol{y} = \left(\boldsymbol{G}^{T}\boldsymbol{G}\right)^{-1}\boldsymbol{G}^{T}\boldsymbol{y} \tag{4}$$

In the case of a large error value, additional neurons are added to the network and re-computed with the new network structure.

In the case of the non-linear algorithm in 2-D, at first, a linear algorithm is performed and linear network parameters – weights – have to be chosen. Then a correction of nonlinear parameters - widths and centers - is carried out. In this case, the number of neurons in the hidden layer is determined in advance.

The objective function for minimization is defined as:

$$E = \frac{1}{2}\sum\nolimits_{i=1}^{N}\left[\sum\nolimits_{j=1}^{k}\omega_j\varphi(\mathrm{x}_i) - y_i\right]^2 \tag{5}$$

After differentiation:

$$\frac{\partial E}{\partial c_j} = \sum\nolimits_{i=1}^{N}\left[\left(\sum\nolimits_{j=1}^{k}\omega_j\varphi(x_i) - y_i\right)\omega_j \exp\left(-\frac{(x_i - c_j)^2}{2\sigma_j^2}\right)\left(\frac{x_i - c_j}{2\sigma_j^2}\right)\right] \tag{6}$$

$$\frac{\partial E}{\partial \sigma_j} = \sum\nolimits_{i=1}^{N}\left[\left(\sum\nolimits_{j=1}^{k}\omega_j\varphi(\mathrm{x}_i) - y_i\right)\omega_j\, exp\left(-\frac{(x_i - c_j)^2}{2\sigma_j^2}\right)\left(\frac{(x_i - c_j)^2}{2\sigma_j^3}\right)\right] \tag{7}$$

Then a gradient method is used to change the values of the widths and centers according to the following formulas [15]:

$$\boldsymbol{c}_j(t+1) = \boldsymbol{c}_j(t) - \mu\frac{\partial E(t)}{\partial \boldsymbol{c}_j(t)}, \quad j = 1,\ldots,k, \tag{8}$$

$$\boldsymbol{\sigma}_j(t+1) = \boldsymbol{\sigma}_j(t) - \mu\frac{\partial E(t)}{\partial \boldsymbol{\sigma}_j(t)}, \quad j = 1,\ldots,k, \tag{9}$$

where:

t - the step of training,

μ - coefficient of training, selected individually.

Thus, in the linear algorithm, the choice of weights and the number of neurons takes place, while in the nonlinear solution the problem of selecting linear (weight) and nonlinear (centers and widths) parameters is solved.

2 Hybrid RBF Neural Network

The main goal of the work is to create a neural network based on the RBF functions, which will be able to recreate the curve with optimal parameters and accuracy. There is the design of the deep training neural network (Fig. 3), which has been developed.

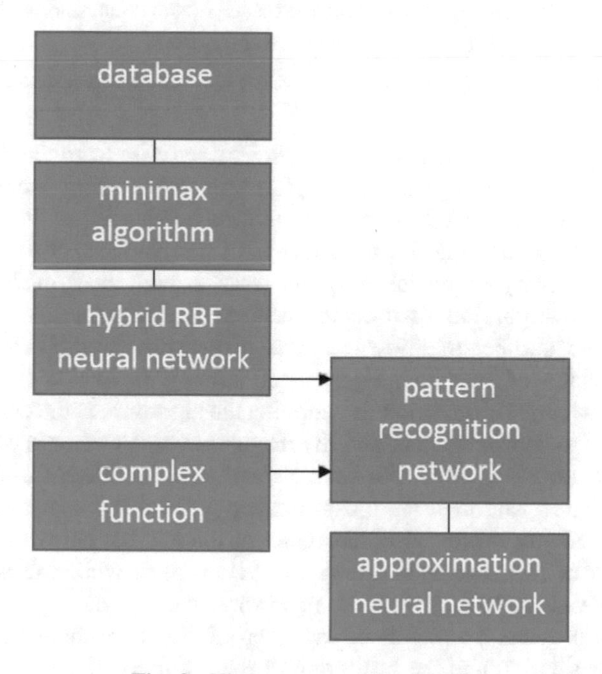

Fig. 3. The structure of result NN

At first, a database of simple functions was created. The database of functions is a kind of constantly updated set of functions that will later be used as a reference for the pattern recognition neural network. Using the algorithm for finding extremes, the expected centers of future neurons of the hybrid network and their number (with the addition of neurons to the ends of functions) are found.

Currently, a hybrid RBF neural network has been developed, which is based on the both of algorithms (linear and non-linear) and represents a combination of them using some additional methods. This network allows changing all possible parameters within the work of one algorithm.

Algorithm 1. The algorithm of creating new neurons

Require: Centers, Sigmas, NumN, x, y

Ensure: TLSE

Until the approximation accuracy is not insufficient **do**

  For each element and each center **do**

    Calculate the Euclidean distances between Centers and examples of data

  End for

  Find a point that is as far away as possible from all the centers

  Find the nearest center for this point

  Create a new RBF neuron center between the found point and the nearest center

  Recalculate the linear parameters ω using a linear algorithm

End until

As starting parameters, the neural network receives data from the database created in the previous step, which includes the function, the starting number of neurons (NumN) and the values of the Centers and the widths (Sigmas) of the functions. Then the calculation of the missing linear parameters - weights - is performed and if the approximation accuracy - calculated by total least square error method (TLSE) - is unsatisfactory, the Algorithm 1 for adding neurons is activated.

Values of the number of known centers of neurons (numN), their coordinates (Centers) and widths of functions (Sigmas) and training dataset (x, y) are given as the input of the algorithm. The criterion for stopping the algorithm is the correspondence of the accuracy of the approximation, and the magnitude of the Error is calculated by the method of least squares (TSLE). For each element, the Euclidean distances from each center are calculated, and then the most distant point and the nearest Center to it are located. A new neuron is created in the center of the interval between them, added to the hidden layer of NN scheme and then the cycle repeats while the stop condition is executed. The result of RBF NN work is shown in the Fig. 4.

The Fig. 4a presents the results of each step of the neural network algorithm, the graph of the distribution and the histogram of the absolute error are presented on the Fig. 4b, c and d is the fall in the error value calculated by the TLSE method.

Then it is needed to optimize the number of neurons and width. This approach has the risk of duplication or proximity of the centers. To solve this problem, neurons are fused with establishing a single center and changing the width in such a way as to cover the activation Gaussian curves of all group neurons.

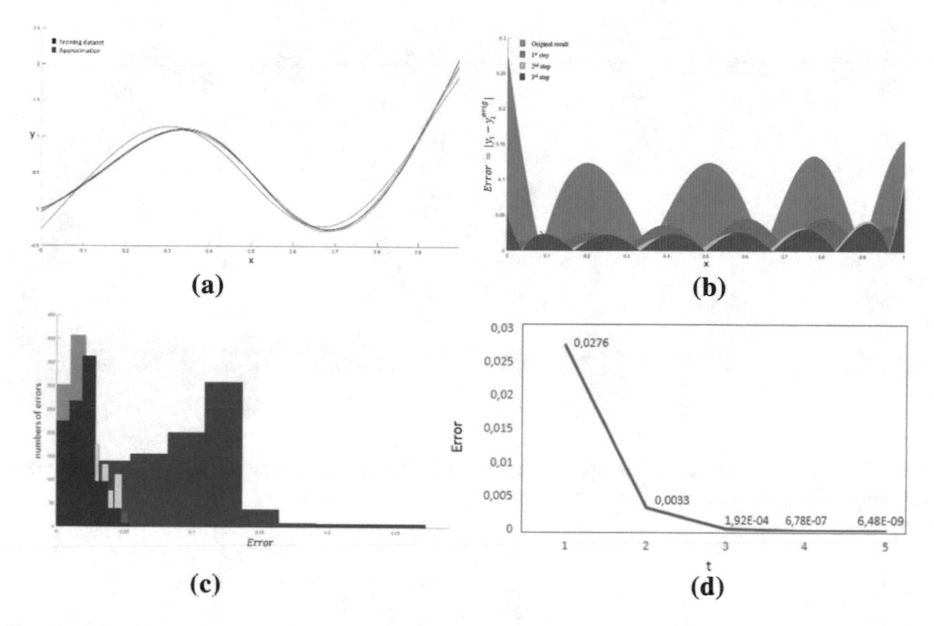

Fig. 4. The example of the method. (a) Each step off approximation RBF NN, (b) absolute error distribution for each algorithm step, (c) histograms of absolute errors, (d) the error graph

Such a simple method makes possible to optimize the network. The advantage of the method is an adding new center not only in the approximation interval of the function but beyond the ultra-centers of the approximation interval if the method was started only with the part of the function, increasing the accuracy of the network.

3 Advantages and Problems

The presented approach has been applied to several testing datasets with similar results. It should be noted that this method gives us:

- variable shape parameters of RBF approximation,
- analytical description of the approximated dataset in the RBF form, i.e. $f(x_n) = \sum_{j=1}^{k} \omega_j \varphi_j(x_n)$.

The proposed method has its drawbacks, for the most part it concerns some types of functions, such as symmetric ones (Fig. 5).

In the case of this type (Fig. 5a), the neural network is "confused" in local extras and loses symmetry, with different error rates on parts of the function (those zones are marked on Fig. 5b).

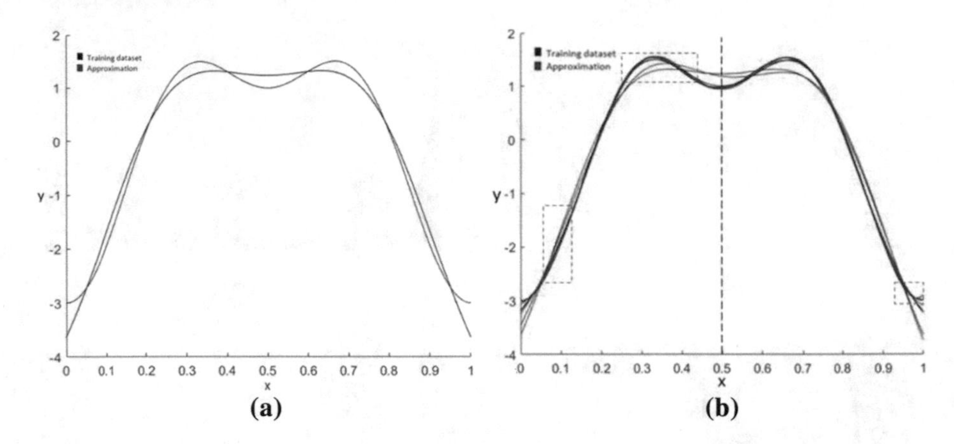

Fig. 5. The symmetric function approximation. (a) Result approximation with start parameters, (b) approximation process

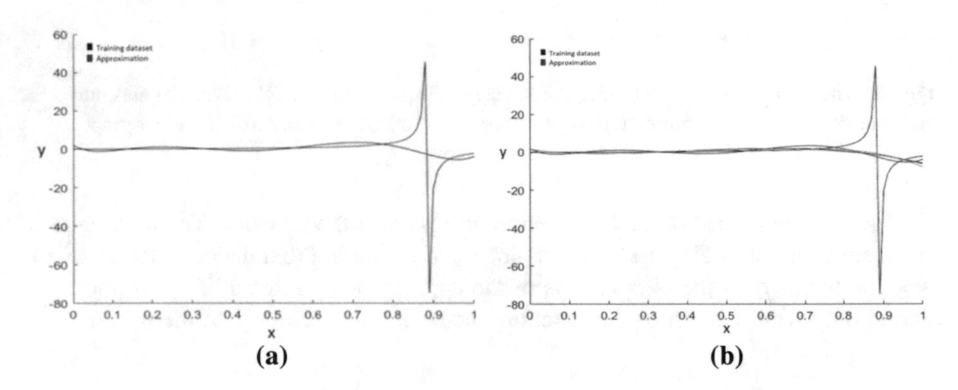

Fig. 6. The symmetric function approximation. (a) Result approximation with start parameters, (b) approximation process

In the case of areas with sufficiently painful emissions from the main trend (Fig. 6), a typical problem for RBF, neuron activation functions are not "sharp enough" to properly cover acute single emissions. Especially for this case post-processing algorithm is provided. In case of failure to achieve criteria accuracy, an algorithm for optimizing zones based on the absolute error values is started.

4 Related Works

As mentioned above, the approximation problem keeps importance its importance and relevance. Every year there are new increasingly sophisticated algorithms and increasingly complex systems for solving increasingly complex approximation problems. This paper describes the creation of a modular neural network of in-depth training for approximating complex data sets with highly noisy or sparse data. Therefore, the implementation, the RBF neural network was chosen, since this type of network is mostly used for approximation problems [16, 17].

In [18] authors propose a new improved K-means Algorithm to optimizing the structure of RBF NN for reaching the better approximation ability. This algorithm is designed to effectively remove noise and outliers, based on the density of the point. Using the algorithm of subtractive clustering, the K-means algorithm is initialized and the final clustering centers are obtained. As a result, the centers of hidden RBF neurons are clustering centers. The authors try to solve a number and positions of RBF-neuron centers, but that system is not flexible enough and it is needed to use another algorithm as post-processing.

The algorithm proposed in [19] is based on the method of subtractive clustering and the methods of generating the centers of the neural network RBF and uses the method of orthogonal least squares to calculate the weights for the output layer.

An evolutionary approach for learning parameters based on genetic algorithms was presented in [20]. It consists of three well-defined feed-forwarding phases and uses a Power function as a fitness evaluation method. This function is highly related to the distribution of the dataset and the structure of the RBF. Therefore, the Power function may be used for choosing good parameters and it will be computationally more efficient than evaluating the MSE. This method is really good for choosing parameters, but it was needed to have the number of RBF neurons before starting algorithm. Depends on the efficient pruning of the parameters and employing scaled basis functions based on Power function are problems of this algorithm.

The positions of hidden layer RBF centers using modified particle swarm clustering algorithm is discussed in [21], then the RBF neural network was trained and the output layer weights were gotten by the least square method. The HIPRBF is the RBF neural network based on improved particle cluster, which has less calculation and better accuracy compared with other RBF neural network based clustering algorithm.

The problem of local approximation is solved in [22], initial weights of RBF neural network and the base width were globally optimized and identification precision of RBF NN on the nonlinear system was improved. The testing results showed that RBF NN based on proposed PSO algorithm had better precision than ones based on GA algorithm or the traditional RBF NN.

5 Conclusion

The neural network proposed in this paper has certain advantages over existing architectures. It makes possible to calibrate all existing parameters and reduce the number of neurons without a loss in an accuracy of the approximation. The proposed solution is a modular part of the complex neural network of in-depth training for approximating of complex functions. The hybrid RBF neural network has some problematic cases discussed above, which require further solutions to improve data preprocessing. In the further, the hybrid RBF neural network algorithm creates an optimal way for managing of all possible parameters in auto-mode and next problems will be solved:

- incomplete data,
- incorrectly spaced centers,
- duplicated neurons,
- functions, which is difficult to approximate (symmetric functions, functions with peaks and ejections, with relatively fast changes at the ends of the curve.

In the future, the complex deep-learning neural network will be created. It is based on described in this paper the radial basis functions with an algorithm module and composed by the neural network, which detects curve patterns and makes a compound solution for many-parts approximation and sufficient mathematical accuracy.

Acknowledgement. The author would like to thank their colleagues at the University of West Bohemia, Plzen, for their discussions and suggestions and anonymous reviewers for their valuable comments and hints provided. The research was supported by projects Czech Science Foundation (GACR) No. 17-05534S and SGS 2016-013.

References

1. Collins English Dictionary. https://www.collinsdictionary.com/dictionary/english/approximation. Accessed 22 May 2018
2. Skala, V.: RBF Interpolation and approximation of large span data sets. In: Corfu, Greece, to Appear in IEEE. MCSI 2017, pp. 212–218 (2018). https://doi.org/10.1109/mcsi.2017.44
3. Skala, V.: Least square error method robustness of computation: what is not usually considered and taught. In: FedCSIS 2017, pp. 537–541. IEEE (2017)
4. Skala, V.: RBF interpolation with CSRBF of large data sets. In: ICCS 2017. Proceedia Computer Science, vol. 108, pp. 2433–2437. Elsevier (2017)
5. Majdisova, Z., Skala, V.: Big geo data surface approximation using radial basis functions: a comparative study. Comput. Geosci. **109**, 51–58 (2017). https://doi.org/10.1016/j.cageo.2017.08.007. ISSN 0098-3004
6. Majdisova, Z., Skala, V.: Radial basis function approximations: comparison and applications. Appl. Math. Model. **54**, 728–743 (2017)
7. Majdisova, Z., Skala, V.: A radial basis function approximation for large datasets. In: SIGRAD 2016, Sweden, pp. 9–14 (2016)
8. Majdisova, Z., Skala, V.: A new radial basis function approximation with reproduction. In: CGVCVIP 2016, Portugal, pp. 215–222 (2016). ISBN 978-989-8533-52-4
9. Filatova, T.V.: Application of neural networks for data approximation. Cybernetics, 121–125 (2004)
10. Pereira, G., Oliveira, M., Ebecken, N.: Genetic optimization of artificial neural networks to forecast Virioplankton abundance from cytometric data. J. Intell. Learn. Syst. Appl. **5**(1), 57–66 (2013)
11. Han, H.G., Wang, L.D., Qiao, J.F.: A spiking-based mechanism for self-organizing RBF neural networks. In: International Joint Conference on Neural Networks (IJCNN), Beijing, China (2014)
12. Nekipelov, N.: Introduction to RBF networks. Data Analysis Technologies. https://basegroup.ru/print/218. Accessed 11 May 2018
13. Xie, T., Yu, H., Hewlett, J., Rózycki, P.: Fast and efficient second-order method for training radial basis function networks. IEEE Trans. Neural Netw. Learn. Syst. **23**, 609–619 (2012)
14. Soldatova, O.P.: Neuroinformatics. In: Lecture Course, Samara, pp. 43–62 (2013)

15. Osovskiy, S.: Neural networks for information processing/trans. In: Finance and Statistic, p. 145. IDM, Rudinsky (2002)
16. Broomhead, D., Lowe, D.S.: Radial basis functions, multivariable functional interpolation and adaptive networks. Technical report, DTIC Document (1988)
17. Park, J., Sandberg, I.W.: Universal approximation using radial-basis-function networks. Neural Comput. **3**(2), 246–257 (1991)
18. Yuqing, S., Junfei, Q., Honggui, H.: Structure design for RBF neural network based on improved K-means algorithm. In: CCDC 2016, Yinchuan, Chinese (2016)
19. Xiao, D., Li, X., Lin, X., Shi, C.: A time series prediction method based on self-adaptive RBF neural network. In: ICCSNT 2016, Harbin, China (2016)
20. Pazouki, M., Wu, Z., Yang, Z.: An efficient learning method for RBF neural networks. In: International Joint Conference on Neural Networks (IJCNN), Killarney, Ireland (2015)
21. Guo, Y., Wang, H.: Hybrid learning algorithm based modified particle swarm clustering for RBF neural network. In: 8th International Symposium on Computational Intelligence and Design (ISCID), Hangzhou, China (2015)
22. Guoqiang, Y., Weiguang, L., Hao, W.: Study of RBF neural network based on PSO algorithm in nonlinear system identification. In: 8th International Conference on Intelligent Computation Technology and Automation (ICICTA), Nanchang, China (2015)

Evaluation of Machine Learning Algorithms in Predicting CO_2 Internal Corrosion in Oil and Gas Pipelines

Wan Mohammad Aflah Mohammad Zubir(✉), Izzatdin Abdul Aziz, and Jafreezal Jaafar

Centre for Research in Data Science, Universiti Teknologi PETRONAS, Bandar Seri Iskandar, 32610 Perak, Malaysia
wanmohammadaflah@gmail.com, {izzatdin,jafreez}@utp.edu.my

Abstract. Over recent years, a lot of money have been spent by the oil and gas industry to maintain pipeline integrity, specifically in handling CO_2 internal corrosion. In fact, current solutions in pipeline corrosion maintenance are extremely costly to the companies. The empirical solutions also lack intelligence in adapting to different environment. In the absence of a suitable algorithm, the time taken to determine the corrosion occurrence is lengthy as a lot of testing is needed to choose the right solution. If the corrosion failed to be determined at an early stage, the pipes will burst leading to high catastrophe for the company in terms of costs and environmental effect. This creates a demand of utilizing machine learning in predicting corrosion occurrence. This paper discusses on the evaluation of machine learning algorithms in predicting CO_2 internal corrosion rate. It is because there are still gaps on study on evaluating suitable machine learning algorithms for corrosion prediction. The selected algorithms for this paper are Artificial Neural Network, Support Vector Machine and Random Forest. As there is limited data available for corrosion studies, a synthetic data was generated. The synthetic dataset was generated via random Gaussian function and incorporated de Waard-Milliams model, an empirical determination model for CO_2 internal corrosion. Based on the experiment conducted, Artificial Neural Network shows a more robust result in comparison to the other algorithms.

Keywords: Artificial Neural Network · Random Forest
Support vector machine · de Waard-Milliams · Corrosion · Pipeline

1 Introduction

Pipelines are an integral part of the oil and gas industry. They are mainly used for transportation of hydrocarbons from the oil wells [1]. Their existence is important as there is no suitable substitute for its usage. It is because pipelines are one of the most efficient ways to transport large amount of liquid. Table 1 exhibits the energy consumption of different methods of hydrocarbon transportation.

R. Silhavy et al. (Eds.): CoMeSySo 2018, AISC 859, pp. 236–254, 2019.
https://doi.org/10.1007/978-3-030-00211-4_22

Table 1. Energy consumption for different methods of hydrocarbon transportation [2]

Method of transportation	Energy consumption (BTU/Ton-mile)
Airplane	37,000
Truck	2,300
Railroad	680
Waterway	540
Pipeline	450

Based on Table 1, pipeline uses the least energy per ton-mile in comparison to other methods of hydrocarbon transportation. However, pipelines are prone to defects that could disturb oil and gas production. Inconsistent monitoring and maintenance also lead to tragedies and environmental catastrophe. Santa Barbara oil spill case is one of the tragedies that happened due to the corroded oil pipeline. Over 100,000 gallons of oil leaked along the Santa Barbara coast [3]. The leakage caused harm to the aquatic ecosystem and prompted legal cases towards the company responsible [4].

There are many causes to degradation of pipeline integrity but in this paper, we focus on CO_2 internal corrosion occurrence. Pipeline corrosion causes damage to the oil and gas pipelines [5]. Pipeline corrosion damages resulted in USD 26 billion cost to the USA oil and gas industry [2]. To mitigate the problem, oil and gas companies have invested millions of dollars for pipeline maintenance. However, the current methods are with inadequacies. Most of the industry solutions still depend on empirical model that are generated based on experiments conducted in lab scale environment. This could pose a problem as the empirical model does not possess intelligence in adapting to different environment. Further elaborate experiments and testing needed to be done in order to find a suitable model. Most of empirical models are also patented and copyrighted which make them highly expensive to be implemented.

The inadequacies of current mechanisms create a necessity to evaluate machine learning algorithms that will be able to predict the corrosion rate. Due to the recent advancement in this field, it creates an interest of selecting appropriate machine learning algorithm in providing predictions of corrosion occurrence. Hence, this paper will emphasize on evaluating machine learning algorithms in predicting CO_2 internal corrosion occurrence.

1.1 Scope of Paper

This paper will only focus on three types of machine learning algorithms, which are Artificial Neural Network, Random Forest and Support Vector Machine. The data that will be used in the project are synthetically generated. The synthetic data are generated based on experiments conducted by [2,6,7].

Based on the synthetic data generated, the corrosion rate is generated from de Waard-Milliams model. The model is widely used in determining the corrosion rate for CO_2 internal corrosion occurrence.

1.2 Structure of Paper

This paper is structured as follows: Sect. 2 presents the preliminaries or background information about CO_2 internal pipeline corrosion. Section 3 discusses the comparative analysis of the selected machine learning algorithms. Section 4 exhibits the experiment which includes the data preparation and experiment setup. Section 5 presents the results and also discussion and analysis of the results obtained. Section 6 concludes this paper and presents possible future work in this area.

2 Preliminaries

This section discusses the basic concept of CO_2 internal pipeline corrosion, factors leading to corrosion occurrence and the cost of pipeline corrosion maintenance. Lastly, existing empirical methods on predicting corrosion occurrence are explained.

2.1 Basic Concept of CO_2 Internal Pipeline Corrosion

Corrosion is defined as the deterioration of metal through chemical reactions induced by the environments [8]. Figure 1 depicts the effects of internal corrosion towards pipelines.

Fig. 1. Examples of internal pipeline corrosion

CO_2 internal pipeline corrosion occurs through an electrochemical reaction between the anodic dissolution of iron and the cathodic evolution of hydrogen [9]. The reaction is formulated as in (1):

$$Fe + CO_2 + H_2O \rightarrow FeCO_3 + H_2 \quad (1)$$

This reaction is coupled with the formulation of scales ($FECO_3$ or FE_3O_4) [9, 10]. The scales, under ideal situation, could create a protective layer that prevents further corrosion occurrence (formation of Fe). However, in other circumstances, the formation will deteriorate the condition of the pipeline.

Due to this being an aqueous CO_2 corrosion, the acid-base reaction is (2):

$$CO_2 + H_2O \rightarrow 2H_2CO_3 \tag{2}$$

H_2CO_3 or carbonic acid deprotonates (removing a proton) to form bicarbonate ion (3).

$$H_2CO_3 \rightarrow H^+ + HCO_3^- \tag{3}$$

Carbon dioxide will constantly converted into carbonic acid and carbonic acid constantly forms bicarbonate [11].

Based on the reaction, CO_2 dissolves into the boundary layer of pipeline. After it dissolves, it hydrates and form carbonic acid [12]. Next, it dissociates and form hydrogen ion, bicarbonate ion and carbonate ion. From this dissociation, carbonate ion is combined with ferrous ion to form $FeCO_3$. A graphical representation of the corrosion occurrence is as depicted in Fig. 2.

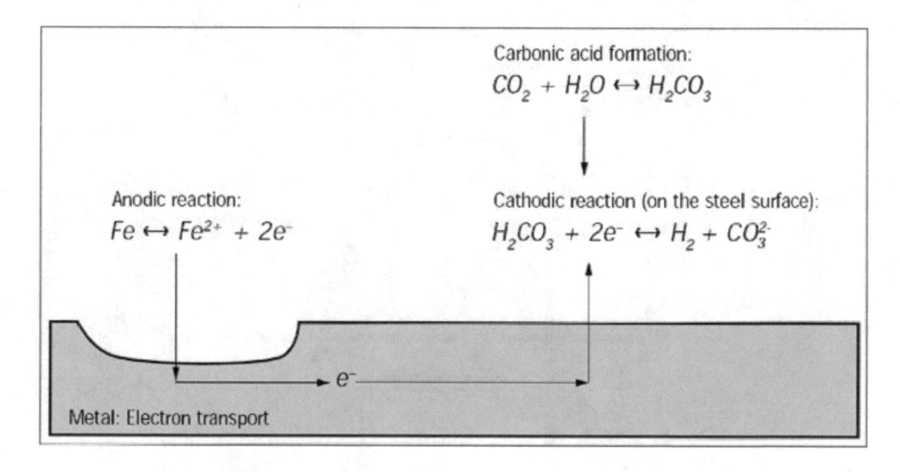

Fig. 2. Electrochemical reaction of pipeline corrosion [13]

There are several factors affecting the presence of corrosion occurrence, subsequently affecting the corrosion rate. This is discussed in the next subsection.

2.2 Factors Leading to Corrosion Occurrence

There are numerous factors affecting directly the occurrence of corrosion. In this paper, we will be limiting to two parameters which are partial pressure of carbon dioxide and temperature of the pipeline.

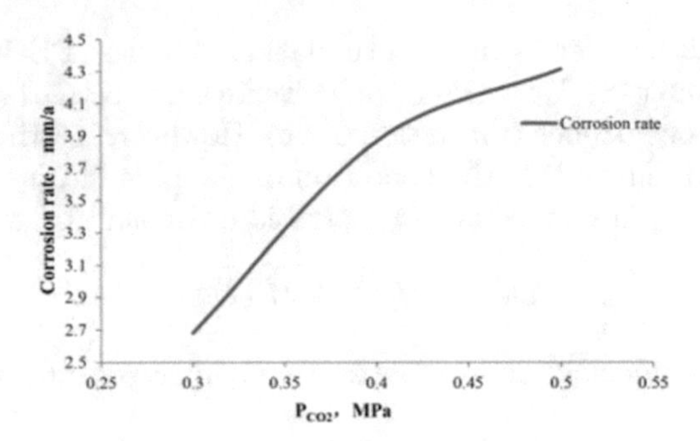

Fig. 3. Effects of partial pressure of carbon dioxide to corrosion rate [6]

Partial Pressure of Carbon Dioxide. The effect of partial pressure of carbon dioxide to the corrosion rate is exhibited in Fig. 3.

Based on the graph above, the corrosion rate increases when the partial pressure of carbon dioxide is between 0.3 MPa and 0.5 MPa.

Temperature. The effect of the temperature of the material to the corrosion rate is exhibited in Fig. 4.

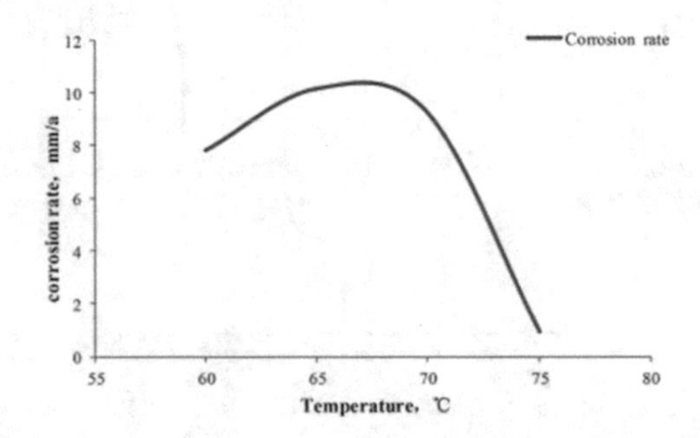

Fig. 4. Effects of temperature to corrosion rate [6]

Based on the graph above, the corrosion rate increases respectively with the increase of temperature until it reaches 65 °C. The next subsection discusses the cost of pipeline corrosion maintenance.

2.3 Cost of Pipeline Corrosion Maintenance

Corrosion poses a threat to the economic well-being of an oil and gas company. Table 2 depicts the annual corrosion cost in important sectors of USA oil and gas industry.

Table 2. Annual corrosion cost in US oil and gas sector [2]

Segment	Annual cost of corrosion in USA (in million US dollars)
Production	1372
Transmission - pipeline	6973
Transportation - tanker	2734
Storage	7000
Refining	3692
Distribution	5000
Total	26771

Annual corrosion costs for pipeline amounted to USD 6973 million, which is the second highest among all the sectors [2]. Papavinasam further emphasized that one of the major causes for transportation damage (both pipeline and tanker) is corrosion (both internal and external). 70% of the failures in production pipelines in Alberta, Canada were accounted by corrosion [2].

To address this problem, oil and gas companies have taken precautionary steps in preparing themselves before the disasters occur. It is estimated that to mitigate or control corrosion in the oil and gas industry, companies are investing close to USD 2.66 billion [2]. To put into perspective of operating expenditure, the oil and gas industry in the USA spends roughly 76% of their investment on handling corrosion issues [2]. This shows a significant amount of money is being utilized to either model, mitigate, monitor, maintain and manage corrosion in production. To model internal corrosion occurrence on pipeline, multiple empirical methods are being utilized by both the industry and academia. These empirical methods are formulated based on experiments conducted in laboratory environment [7,14,15]. The next subsection discusses the three existing methods used in predicting corrosion occurrence.

2.4 Existing Empirical Methods in Predicting Corrosion Occurrence

There are existing models that are used in predicting the corrosion rate at the oil and gas pipelines. Table 3 shows the prediction methods developed by the oil and gas industries.

Based on the table below, de Waard-Milliams model is selected due to it is foundation of corrosion prediction mechanism. The model is also recurrently referenced to predict pipeline corrosion [6].

de Waard-Milliams Model. It is one of the widely used model in predicting the corrosion rate of subsea pipeline. It was proposed by de Waard and Milliams of Shell. The model was developed empirically from experiments conducted in lab scale. The model is:

Table 3. Comparison between different empirical corrosion methods

Characteristics	de Waard-Milliams [16]	Cassandra [17]	Norsok M-506 [18]
Company	Shell	BP	Norsok Hydro, Saga Petroleum and Statoil
Weaknesses	Do not include protective corrosion films effect	Do not include oil wetting effects	Not accurate on very low temperature
Strengths	Extensively used and referenced in the industry	BP's representation of de Waard-Milliams model	Tested using a lot of data

$$logV_{corr} = 5.8 - \frac{1710}{T} + 0.68log(P_{co_2}) \tag{4}$$

Where:

V_{corr} is the corrosion rate in mm/y
T is the temperature in Kelvin (K)
Pco_2 is the partial pressure of Carbon Dioxide in Pascal.

de Waard-Milliams model is used as the empirical method chosen for determining the corrosion rate using synthetic data generated. This is explained in Sect. 4. Next section is the comparative analysis between machine learning methods for regression.

3 Comparative Analysis of Machine Learning Methods for Regression

In the field of Machine Learning and Artificial Intelligence, several algorithms have been formulated and designed for prediction or regression. Table above depicts the comparison between machine learning algorithms used for regression analysis.

3.1 Artificial Neural Network

Artificial Neural Network or ANN is an artificial intelligence algorithm that is able to perform prediction on continuous and discrete value [19]. It is mainly a supervised learning algorithm. Supervised learning algorithm requires correct responses or training data for it to perform its tasks. It was inspired by biological neural networks in human brains [19,20]. Basic structure of ANN is depicted in Fig. 5.

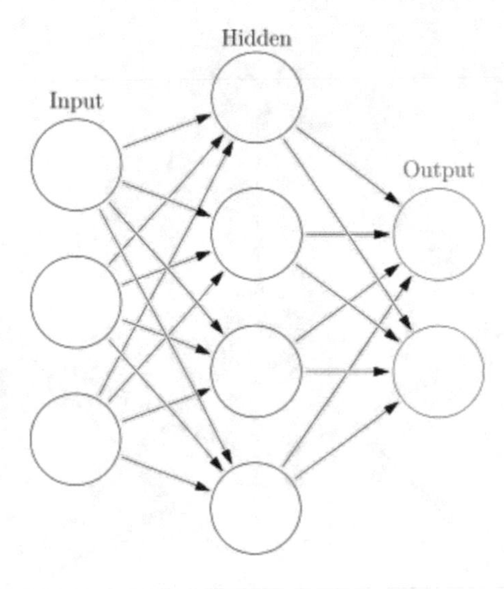

Fig. 5. Basic structure of artificial neural network [6]

The network consists of at least 3 layers; input, hidden and output layer. The structure could have multiple hidden layers, depending on the layers initialized [21]. The structure used in this evaluation is explained in detail in Sect. 3.

The network contains adopted learning rule, specific to each network. Learning is done through modification of the weights of the connections that link the input to the output. In this paper, we utilised the radial basis learning function or RBF. Radial function response monotonically with distance from the a central point [22]. A radial function can be formulated as:

$$f(x) = exp(-\frac{(x-c)^2}{r^2}) \tag{5}$$

Where:

c is the central point
r is the radius.

RBF in artificial neural network can be depicted as in Fig. 6.

Based on Fig. 6, for all n components of x are the input perceptron feed into function $h(x)$. The outputs for all m functions $h(x)$ are linearly combined with weights $\{w_j\}_{j=1}^{m}$. $f(x)$ is the output of the neural network.

One of the strengths of artificial neural network is that in both regression and classification problem, it provides a smooth transition from one observed value to another [23]. RBF also allows the functions to move or change size accordingly [22]. Detailed explanation about this algorithm is discussed in [24, 25].

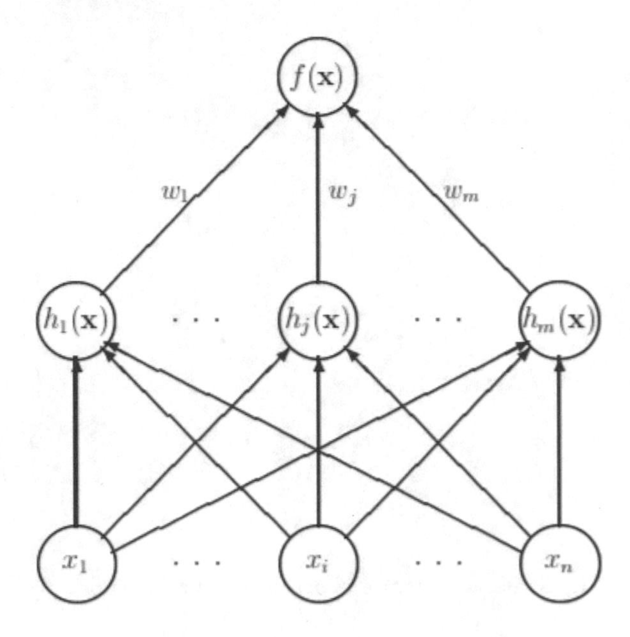

Fig. 6. RBF in artificial neural network [22]

3.2 Support Vector Machine

Support Vector Machine or SVM is a machine learning algorithm that is utilized for both prediction and classification problems [26–28]. It is also a supervised learning algorithm which needs training. An example of SVM performing its task is as depicted in Fig. 6.

SVM classifies the data using a hyperplane which calculates the biggest distance to the nearest point of data of a class [29]. As the scope of this paper focuses on the application of regression or prediction, support vector regression is used. The aim of the algorithm is to find a function $f(x)$ that has at most ϵ deviation from the real values, y_i. The linear function f can be defined as follows:

$$f(x) = \langle w, x \rangle + b \quad with \quad w \in X, b \in R \tag{6}$$

Where:

$\langle ., . \rangle$ is the dot product in X
X is the space of input patterns
w represents the flatness of the function.

Based on the linear function, the convex optimization problem that the model solves is as follows [30]:

$$minimize \quad \frac{1}{2}||w||^2 \tag{7}$$

$$subject\ to \quad \left\{ y_i - \langle w, x_i \rangle - bx \leq \epsilon \langle w, x_i \rangle + b - y_i \leq \epsilon \right. \tag{8}$$

where:

x_i is the training sample
y_i is the target value
$\langle$w, $x_i\rangle$ + b is the prediction based on the selected sample
$\in$ is the threshold. Threshold limits the range of true predictions.

As for regression problem (or prediction), the loss function is altered to take into consideration a distance measure [31]. Even though this modified SVM still uses all the main features in characterizing maximum margin algorithm, the linear learning is leaned by a non-linear function through mapping.

Hence, based on this modification, an error margin or “soft margin” loss function is introduced into the problem. This is done through introduction of slack variables ϵ_i and ϵ_i^* [32,33]. Slack variables help to relax the infeasible constraints to the earlier optimization problem. The optimization problem can now be reformulate as follows:

$$minimize \quad \frac{1}{2}||w||^2 + C\sum_{i=l}^{l}(\epsilon_i + \epsilon_i^*) \tag{9}$$

$$subject\ to \quad \{y_i - \langle w, x_i\rangle - b \leq \epsilon + \epsilon_i \langle w, x_i\rangle + b - y_i \leq \epsilon + \epsilon_i^* \epsilon_i, \epsilon_i^*, \geq 0 \tag{10}$$

In minimization equation, the constant C influences the trade-off between the linearity of f.

One of the strengths of SVM in solving regression problems is that the algorithm can avoid difficulties of implementing linear functions in large feature space. The introduction of error margin also give a larger representation of f function which give a representational advantage. This algorithm is explained in detail in [30,32,34].

3.3 Random Forest

Random Forest is an ensemble learning algorithm that is able to solve both prediction and classification problem [35,36]. It was proposed in improving bagging of classification trees. The algorithm adds another layer of randomness to bagging. An illustration of Random Forest is depicted in Fig. 7.

Random Forest splits each nodes based on the best subsets of predictors selected at that node [35]. Through this selection, it avoids correlation in bootstrapped sets [38]. The preferred subsets are denoted by having the majority of votes [39].

One of the strengths of Random Forest is it is able to avoid overfitting. This is done through the selection via majority vote which leads to a more accurate classification or forecasts [38]. This algorithm is explained in detail in [35,36,40].

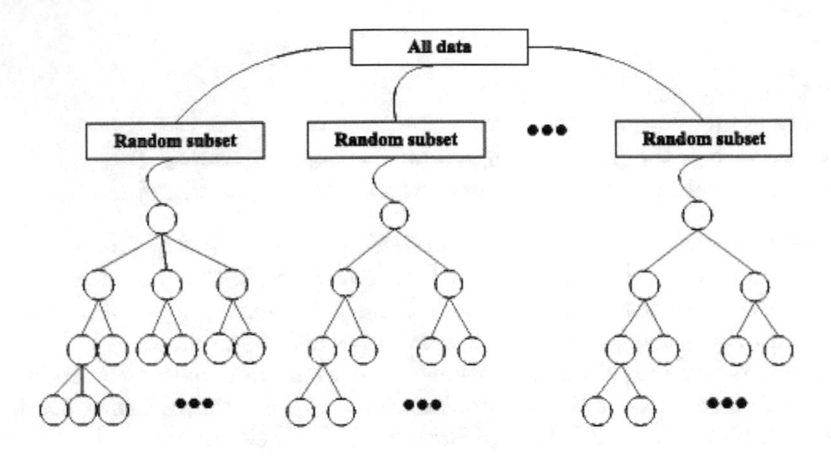

Fig. 7. Random forest [37]

4 Experiment

This section discusses the experiment conducted towards the machine learning algorithms selected. First, we describe the dataset prepared for the evaluation. Next, we outline how each of the experiments are conducted. This includes the settings used for each of the algorithms.

4.1 Data Preparation

For this paper, synthetic dataset is generated. The rules set in generating the synthetic dataset are based on experiments conducted by Nyborg and Zeng [6]. The parameters needed in this project are the pressure of carbon dioxide or pCO_2, the velocity of the corrosion medium and the temperature. Table 4 exhibits the parameters in the data preparation with corresponding units.

Table 4. Parameters in data preparation

Parameter	Unit
Pressure of carbon dioxide (pCO_2)	MPa (Megapascal)
Temperature	°C (celcius)

A program was created in Java to generate the synthetic dataset. The Gaussian function in the Java language was utilized in generating the random number. The ranges used in generating the values for the parameters that will be used in calculating the corrosion rates are as in Table 5.

These data are generated based on the experiment conducted by Peng and Zheng [6,41]. However, for the equation purposes, the unit or data was adapted

Table 5. Range of data generated

Parameter	Unit
Pressure of carbon dioxide (pCO_2)	0.03–0.65 (MPa)
Temperature	42.40–72.00 (°C)

into Pascal (Pa) and Kelvin (K) respectively. Based on the data generated, the corrosion rate is calculated using the chosen empirical method, de Waard-Milliams model.

4.2 Experiment Set-Up

Training and Testing Set Up. After the data has been generated, it is split into testing and training set. The split was done through train test split function in sklearn. It was chosen due to the random functionality which randomly pick and split the data generated. The percentage of split is 75% training and 25% testing. Each algorithm will go through learning process based on the training data.

Pre-processing. After the split between train and test dataset, feature scaling is performed. This is done to ensure that the data are normalized between the value of 0 and 1. The feature scaling is implemented using StandardScaler in sklearn package.

Artificial Neural Network. The settings used for Artificial Neural Network is as in Table 6.

Table 6. Settings used for artificial neural network

Features	Values
Activation function	Logistic
Training	Broyden-Fletcher-Goldfarb-Shanno algorithm
Hidden Layer	3
Alpha	0.0001
Epsilon	1e–08

Activation function chosen is the logistic sigmoid function. The function is defined as:

$$f(x) = 1/(1 + exp(-x)) \tag{11}$$

Broyden-Fletcher-Goldfarb-Shanno algorithm is selected as the solver for the weight optimization or learning algorithm. It belongs to the family of quasi-Newton methods.

Table 7. Settings used for random forest

Features	Values
Number of trees	10
Quality criterion	Mean squared error

Random Forest. The settings used for Random Forest is as in Table 7.

Support Vector Machine. The settings used for Support Vector Machine is as in Table 8.

Table 8. Settings used for support vector machine

Features	Values
Penalty parameter	100
Epsilon	0.01
Kernel	Radial basis function

For this experiment, the value of penalty parameter C of error term is increased to increase the fit to the training data. The higher penalized value is given if the model failed to fit to the training data.

5 Results and Discussion

In this section, the results of the evaluation of the machine learning algorithms is discussed. First, we conducted a data quality test on the synthetic dataset generated.

5.1 Data Quality Test

Data quality test is conducted to test whether the synthetic dataset is appropriate for the usage of the system. The two main characteristics tested in this test are the uniformity of the data and the predictability of the data. The scope of the testing will be limited to the synthetic data generated and the characteristics tested are the uniformity of the data and the repeating occurrence of the data generated.

Temperature. The synthetic data generated for temperature is as depicted in Fig. 8.

To view the dispersion of data generated, a boxplot has been generated based on the synthetic data. The boxplot is as depicted in Fig. 9.

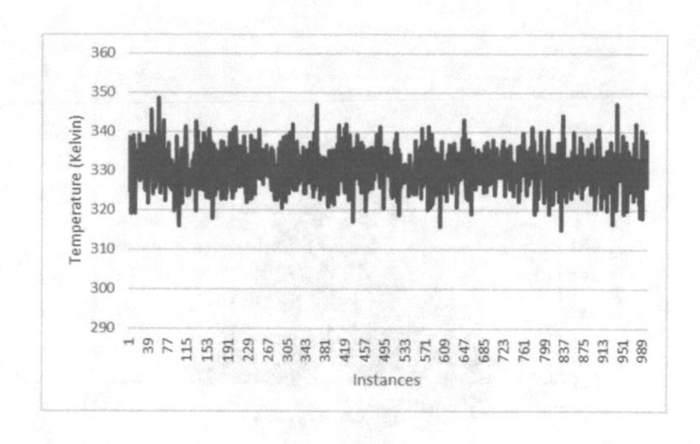

Fig. 8. Data of temperature generated

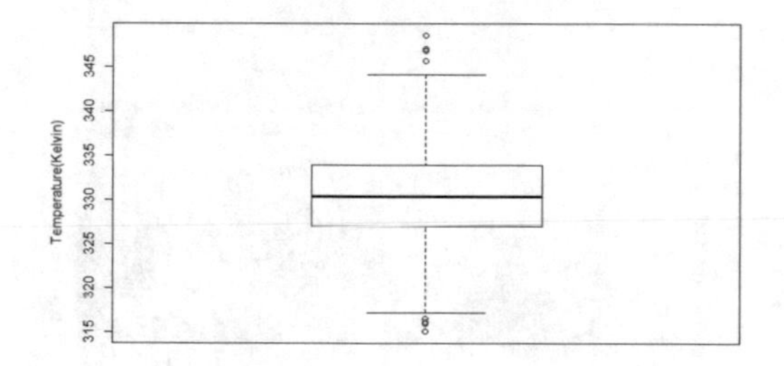

Fig. 9. Boxplot of temperature

Table 9. Characteristics of temperature

Characteristics	Values
Median	330.3
Mean	330.4

Based on the boxplot, the characteristics of the data can be extracted. The characteristics are summarized in Table 9.

As we generated the data based on normal distribution, a graph depicting the density probability of the samples is built. Figure 10 depicts the graph of the normal distribution of the data generated.

Partial Pressure of Carbon Dioxide. The synthetic data generated for Partial Pressure of CO_2 is as depicted in Fig. 11.

To view the dispersion of data generated, a boxplot has been generated based on the synthetic data. The boxplot is as depicted in Fig. 12.

Based on the boxplot, the characteristics of the data can be extracted. The characteristics are summarized in Table 10.

The density probability of the samples generated for partial pressure of CO_2 is also generated and depicted in Fig. 13.

Based on the data generated, the values of both median and mean from the synthetic dataset are close to each other. This depicts that the data has a

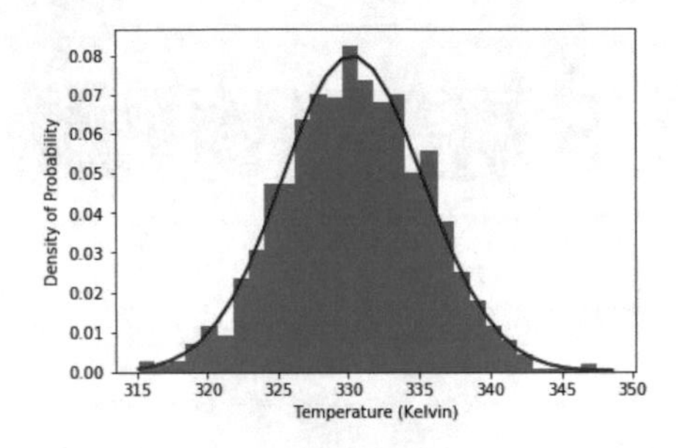

Fig. 10. Normal distribution of temperature generated

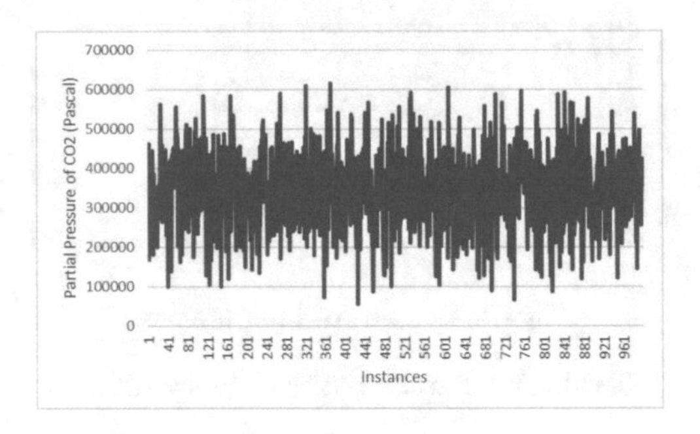

Fig. 11. Data of partial pressure of carbon dioxide generated

Table 10. Characteristics of partial pressure of carbon dioxide

Characteristics	Values
Median	345541
Mean	343354

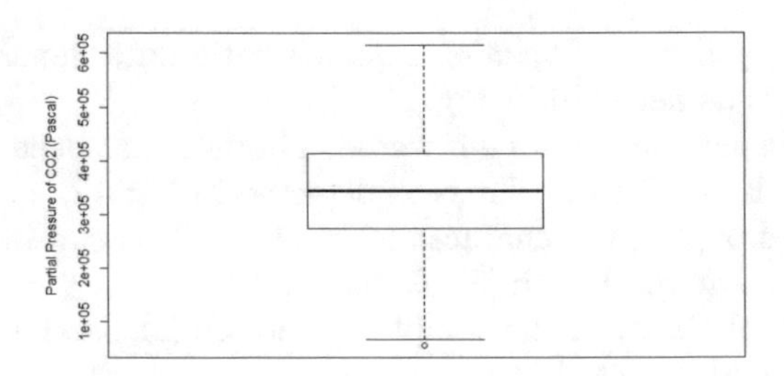

Fig. 12. Boxplot of partial pressure of carbon dioxide

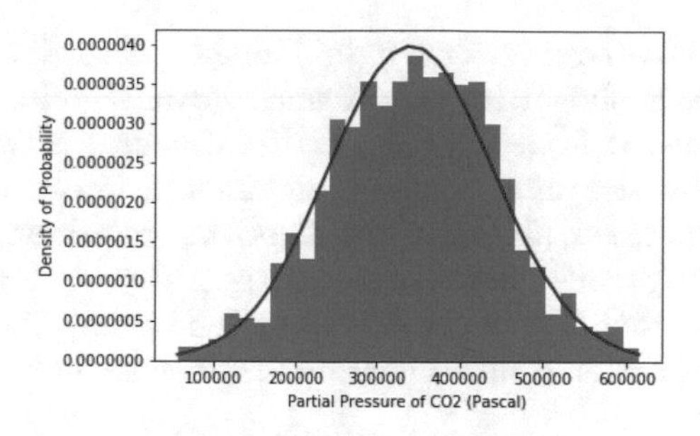

Fig. 13. Normal distribution of partial pressure of carbon dioxide generated

symmetrical distribution. Further analysis was conducted to view if there was any duplicate occurrence. This was to ensure that the data represents true randomness. Out of 1000 samples generated, there was no repeated occurrence of the temperature and partial pressure of Carbon Dioxide observed.

5.2 Accuracy Test

The aim of this test is to evaluate the accuracy of the machine learning algorithms in predicting the corrosion rate of internal corrosion occurrence. The evaluation methods used to evaluate individual algorithms are the mean-squared error (MSE) and also mean absolute error (MAE). The calculation for MSE and MAE are as follows:

$$MSE = \frac{1}{T}\sum_{t=1}^{T}(x_t - \bar{x}_t)^2 \tag{12}$$

$$MAE = \frac{1}{T}\sum_{t=1}^{T}(x_t - \bar{x}_t) \tag{13}$$

Where:

T is the size of testing set
x_t is the actual corrosion rate
$\bar{x}_t$ is the predicted corrosion rate.

Any outlier values are removed as it does not reflect the true environment of the internal pipeline [7,42].

Table 11. Accuracy test result

Algorithm	MSE	MAE
Artificial neural network	0.1108	2.171
Random forest	0.1113	2.213
Support vector Machine	0.2864	2.662

The values calculated are shown in Table 11. Based on Table 11, artificial neural network is shown to have a small advantage in comparison to the other two algorithms. It is because the artificial neural network can work better through extensive learning better than random forest and support vector machine. Besides that, the ability of the algorithm to adjust the weights of the outcome of the perceptrons offer a better output. However, as the advantage is small, it leaves more room for improvement on the tuning of artificial neural network in obtaining a better result. Future possible work in this area is discussed in the next section.

6 Conclusion and Future Work

This paper has presented its findings on the comparison between machine learning algorithms in predicting the corrosion rate occurrence. Based on the experiments conducted, artificial neural network shows a better result in comparison to other types of machine learning algorithms. Several suggestions could be made for further work in this area.

Using Real Industry Dataset for Training. This could significantly improve the ability of the machine learning algorithm to learn from real data instead of synthetic data which might not depict the real corrosion rate.

Further Refinement of the Parameters of the Machine Learning Algorithm. By tuning the parameters, a more accurate result could be obtained as the structure of the algorithm is altered to fit the data that it learns from. However, the tuning of the parameter is out of the scope of this study.

Test with a Wider Variety of Algorithms. This will also assist in determining the most suitable algorithm for predicting corrosion rates.

Evaluate the Models with Different Types of Test. For this paper, performance test was not conducted as it is out of the scope for the paper. Hence, an evaluation of performance can also assist in determining algorithm that is optimized in terms of accuracy and performance.

Acknowledgments. This work is supported by the Universiti Teknologi PETRONAS's Graduate Assistantship Scheme.

References

1. Kennedy, J.L.: Oil and Gas Pipeline Fundamentals. Pennwell Books, Houston (1993)
2. Papavinasam, S.: Corrosion Control in the Oil and Gas Industry. Elsevier, New York (2013)
3. Naylor, B.: Burst oil pipeline in California severely corroded, investigators say (2015)
4. Cart, J.P.: Ruptured oil pipeline was corroded, federal regulators say. Los Angeles Times (2015)

5. Papavinasam, S., Revie, R.W., Friesen, W.I., Doiron, A., Panneerselvan, T.: Review of models to predict internal pitting corrosion of oil and gaspipelines. Corros. Rev. **24**(3–4), 173–230 (2006)
6. Peng, S., Zeng, Z.: An experimental study on the internal corrosion of a subsea multiphase pipeline. Petroleum (2015)
7. Nyborg, R., et al.: Overview of CO_2 corrosion models for wells and pipelines. In: CORROSION 2002, NACE International (2002)
8. NP Laboratory: A short introduction to corrosion and its control (2015)
9. Nešić, S.: Key issues related to modelling of internal corrosion of oil and gas pipelines - a review. Corros. Sci. **49**(12), 4308–4338 (2007)
10. Nyborg, R.: Guidelines for prediction of CO_2 corrosion in oil and gas production systems. Institute for Energy Technology (2009)
11. Jenkins, H.: Le Chatelier's Principle. Chemical Thermodynamics at a Glance, pp. 160–163 (2008)
12. Song, F.: A comprehensive model for predicting CO_2 corrosion rate in oil and gas production and transportation systems. Electrochimica Acta **55**(3), 689–700 (2010)
13. Videm, K., Dugstad, A., Lunde, L.: Parametric study of CO_2 corrosion of carbon steel. CORROSION/94, paper (14) (1994)
14. Nesić, S., Postlethwaite, J., Vrhovac, M.: CO_2 corrosion of carbon steel-from mechanistic to empirical modelling (1997)
15. Zubir, W.M.A.M., Aziz, I.A., Haron, N.S., Jaafar, J., Mehat, M.: CO_2 corrosion rate determination mechanism implementing de Waard-Milliams model for oil gas pipeline. In: 2016 3rd International Conference on Computer and Information Sciences (ICCOINS), pp. 298–303, August 2016
16. Pots, B.F., Kapusta, S.D., John, R.C., Thomas, M., Rippon, I.J., Whitham, T., Girgis, M.: Improvements on de Waard-Milliams corrosion prediction and applications to corrosion management. In: CORROSION 2002, NACE International (2002)
17. Moghissi, O., Burwell, D., Eckert, R., Vera, J., Sridhar, N., Perry, L., Matocha, G., Adams, D.: Internal corrosion direct assessment for pipelines carrying Wetgas-methodology. In: 2004 International Pipeline Conference, American Society of Mechanical Engineers, pp. 1111–1119 (2004)
18. Olsen, S.: CO_2 corrosion prediction by use of the NORSOK M-506 model-guidelines and limitations. In: CORROSION 2003, NACE International (2003)
19. Wang, S.C.: Artificial Neural Network, pp. 81–100. Springer, Heidelberg (2003)
20. Zhang, Z.: Artificial Neural Network, pp. 1–35. Springer, Heidelberg (2018)
21. Hill, T., Marquez, L., O'Connor, M., Remus, W.: Artificial neural network models for forecasting and decision making. Int. J. Forecast. **10**(1), 5–15 (1994)
22. Orr, M.J., et al.: Introduction to radial basis function networks (1996)
23. Specht, D.F.: A general regression neural network. IEEE Trans. Neural Netw. **2**(6), 568–576 (1991)
24. Kröse, B., Krose, B., van der Smagt, P., Smagt, P.: An introduction to neural networks (1993)
25. Lawrence, J.: Introduction to Neural Networks: Design, Theory, and Applications. California Scientific Software, Nevada City (1994)
26. Cortes, C., Vapnik, V.: Support vector machine. Mach. Learn. **20**(3), 273–297 (1995)
27. Burges, C.J.: A tutorial on support vector machines for pattern recognition. Data Min. Knowl. Discov. **2**(2), 121–167 (1998)

28. Vapnik, V.N., Vapnik, V.: Statistical Learning Theory, vol. 1. Wiley, New York (1998)
29. Hua, S., Sun, Z.: Support vector machine approach for protein subcellular localization prediction. Bioinformatics **17**(8), 721–728 (2001)
30. Basak, D., Pal, S., Patranabis, D.C.: Support vector regression. Neural Inf. Process.-Lett. Rev. **11**(10), 203–224 (2007)
31. Smola, A.J., et al.: Regression estimation with support vector learning machines. Ph.D. thesis, Master's thesis, Technische Universität München (1996)
32. Smola, A.J., Schölkopf, B.: A tutorial on support vector regression. Stat. Comput. **14**(3), 199–222 (2004)
33. Bennett, K.P., Mangasarian, O.L.: Robust linear programming discrimination of two linearly inseparable sets. Optim. Methods Softw. **1**(1), 23–34 (1992)
34. Welling, M.: Support vector regression. Department of Computer Science, University of Toronto, Toronto (Kanada) (2004)
35. Liaw, A., Wiener, M.: Classification and regression by randomforest. R News **2**(3), 18–22 (2002)
36. Kam, H.T.: Random decision forest. In: Proceedings of the 3rd International Conference on Document Analysis and Recognition, Montreal, Canada, August, pp. 14–18 (1995)
37. Ma, L., Fan, S.: Cure-smote algorithm and hybrid algorithm for feature selection and parameter optimization based on random forests. BMC Bioinform. **18**(1), 169 (2017)
38. Barboza, F., Kimura, H., Altman, E.: Machine learning models and bankruptcy prediction. Expert Syst. Appl. **83**, 405–417 (2017)
39. Breiman, L.: Random forests. Mach. Learn. **45**(1), 5–32 (2001)
40. Svetnik, V., Liaw, A., Tong, C., Culberson, J.C., Sheridan, R.P., Feuston, B.P.: Random forest: a classification and regression tool for compound classification and qsar modeling. J. Chem. Inf. Comput. Sci. **43**(6), 1947–1958 (2003)
41. Zhang, Y., Pang, X., Qu, S., Li, X., Gao, K.: Discussion of the CO_2 corrosion mechanism between low partial pressure and supercritical condition. Corros. Sci. **59**, 186–197 (2012)
42. Rousseeuw, P.J., Leroy, A.M.: Robust Regression and Outlier Detection, vol. 589. Wiley, Hoboken (2005)

An Approach of Filtering to Select IMFs of EEMD in Acoustic Emission [AE] Sensors for Oxidized Carbon Steel

Nur Syakirah Mohd Jaafar(✉), Izzatdin Abdul Aziz, Jafreezal Jaafar, and Ahmad Kamil Mahmood

Centre for Research in Data Science, Universiti Teknologi PETRONAS, Seri Iskandar, Malaysia
{nur_16001470, izzatdin, jafreez, kamilmh}@utp.edu.my

Abstract. Number of existing signal processing methods can be used for extracting useful information. However, the problem of signal processing method, essential to highlight the wanted information and attenuate the undesired signal is trivial. Several signal processing methods have been implemented to solve this issue. Research using Empirical Mode Decomposition (EMD) algorithm shows promising results in comparison to other signal processing methods, especially in the accuracy showing the relationship between signal energy and time – frequency distribution by represents series of the stationary signals with different amplitudes and frequency bands. However, this EMD algorithm will still have noise contamination that may compromise the accuracy of the signal processing to highlight the wanted information. It is because the mode mixing phenomenon in the Intrinsic Mode Function's (IMF) due to the undesirable signal with the mix of additional noise. There is still room for the improvement in the selective accuracy of the sensitive IMF after decomposition that can influence the correctness of feature extraction of the oxidized carbon steel. Using four datasets, analysis parameters of the Ensemble Empirical Mode Decomposition (EEMD) algorithm has been conducted.

Keywords: Signal processing method · Empirical Mode Decomposition Ensemble Empirical Mode Decomposition · Intrinsic Mode Function's Noise contamination · Mode mixing · Selective accuracy of IMF

1 Introduction

Improved signal denoising method is defined by reducing the unwanted signals from the raw signal which are collected from the sensors that installed outside of the pipeline network [1]. The ultrasonic inspection and magnetic flux leakage inspection technique are the common methods for evaluating and monitoring the state of pipelines through the collected signals from the interior of the offshore pipeline [2]. The signal characteristics will have derived the anomalies, signal amplitudes, signal phases and signal frequency to ascertain actions to be taken ahead.

Lacking techniques for evaluating and monitoring the state of pipelines through the collected signals from the interior of the pipeline leads to inaccurate corrosion

R. Silhavy et al. (Eds.): CoMeSySo 2018, AISC 859, pp. 255–273, 2019.
https://doi.org/10.1007/978-3-030-00211-4_23

detection. This happens due to high background noise and noise condition which will affect the actual signal. For example, the signals which can reveal the anomalies can be overlapped. Thus, the recovery of a signal from observed noisy data is important. The analysis of techniques of signal processing is essential to highlight the wanted information and attenuate the undesired one [3]. Butterworth low pass filter, wavelet-based thresholding filter (Wavelet Transform) and Hilbert Huang Transform (HHT) are few examples of complex decomposition signal breaking into finite components [4]. These existed denoising methods in signal processing still have limitations in the removal of noise reduction to achieve the accurate reading of signal signatures. In general, most often, signal processing methods are applied to solve the hidden results from the raw signal including removing noisy signals from the data collected through sensors which are installed outside of the pipeline network.

Empirical Mode Decomposition (EMD) algorithm in Hilbert Huang Transform (HHT) contains undesirable signal signatures. Undesirable signal signature contains anomalies such as noises interference, which leads to signal misinterpretation. Signal is decomposed into many IMFs by EMD in HHT. Based on the analysis on EMD algorithm, the mode mixing problems are encountered when signal contains intermittency. The mode mixing will affect the decomposition of signal into different IMFs on the time series as similar scale is residing in different IMF. Thus, this drawback creates a necessity for a proposal and development of an improved decomposition of signal using Ensemble Empirical Mode Decomposition (EEMD).

The problem found creates a necessity for a proposal and development of an improved method to decompose signals using Ensemble Empirical Mode Decomposition (EEMD). This is to improve the reconstruction of noise removed signal when the selection of the Intrinsic Mode Functions (IMFs) is relevant to the corresponding of the most important structures of the signal. The following research questions are extracted from the problem statement.

[RQ1]: What are the correct values for each parameter that are relevant with signal processing method using Ensemble Empirical Mode Decomposition (EEMD)?

[RQ2]: To what extent does Ensemble Empirical Mode Decomposition (EEMD) that takes advantage of Intrinsic Mode Functions (IMF), able to analyses filtering in frequency-time domain?

[RQ3]: To what extent does the selection of the Intrinsic Mode Functions (IMFs), able to reconstruct the noise removed signal after the signal – filtering method been proposed?

The research aims are to improve the existing Empirical Mode Decomposition (EMD) algorithm using Hilbert Huang Transform (HHT). Through the preliminary research conducted, several research questions had been brought forward. To address the research questions arose; three research objectives have been devised:

[RO1]: To evaluate the correct values for each parameter that is relevant to Hilbert Huang Transform (HHT) using Ensemble Empirical Mode Decomposition (EEMD).

[RO2]: To compare and analyze filtering used in the frequency -time domain analysis.
[RO3]: To construct signal signature based on the selected principal Intrinsic Mode Functions (IMF) components to indicate the noise removed signal using data from oxidized carbon steel.

The study focuses on the existing EMD algorithm using HHT to recover a signal from the raw signals. The improved signal processing methods, which the extent of the EMD algorithm; are Ensemble Empirical Mode Decomposition (EEMD) algorithm are tested against the dataset obtained through field testing. The parameters tested for this research work are limited to the analysis of the EEMD algorithm which is many ensemble members, fixed sifting numbers and amplitude of the added white noise.

The algorithm used in this research is EEMD algorithm and for the representations in the frequency-time domain, HHT is used. The signal processing methods, EEMD will be improved to achieve higher accuracy based on the selection of the Intrinsic Mode Function (IMFs) leading to the noise removal signal for the oxidized carbon steel. High accuracy is defined as the ability of the signal processing algorithm, EEMD to the selection process of IMFs adaptive, and this selection IMFs achieves high Signal to Noise Ratio (SNR) while the Percentage of RMS Difference (PRD) and Max Error values are low.

However, for proof of concept, the existing algorithm without any improvements is used. The existing implementation of the EEMD algorithm is utilized to perform signal decomposition to enhance the thoroughness of denoising by fixing the parameters analysis of the EEMD. The algorithm is tested against four data sets obtained from the field testing. The study is limited to the signals that were corrupted by additive white Gaussian noise and is conducted based on extended numerical experiments. The algorithm is tested again four data sets obtained from the field testing by using the acoustic emission sensors SR10b low frequency AE sensor. The four data sets are data sets for test cases healthy, leak, no-leak and eroded test cases.

2 Literature Review

2.1 Selection of Signal Processing Method

Table 1 describes the existing method in signal processing methods. For this research, EEMD by HHT is selected. The main reason of EEMD by HHT is selected due to the feature that it is data driven basis and suitable for analysis on the signal interpretation algorithm. It is important that the signal processing method can decomposed the signal into mono component functions. EEMD by HHT works through performing a time-adaptive decomposition operation.

By having the decomposition signal without requires no prior knowledge of the target signal, this will be more efficient in filtering accuracy. Even though EEMD can filter between noise and denoised signal, it still lacks in accuracy to filter between selections of the IMFs. As EEMD, the method still encounters the problem of mode mixing when the signal contains intermittency when the parameters of the EEMD is not set up. The parameter of the EEMD is based on the knowledge of the researchers [11–13].

Table 1. Comparison between signal processing methods

Name	Advantages	Disadvantages
Fourier transform by (Fast Fourier Transform) [5]	Fast Fourier Transform has accuracy in analytical form of smoothness signal of the underlying function [5]	Limitations for Fast Fourier Transform for pipeline having no localization of signal discontinuity [6]
Wavelet based thresholding filter by (Wavelet Transform) [7]	Wavelets have allowed filters to be constructed for signal localization for both time and frequency domain [7]	Limitations for the wavelet transform in signal generated for length of the basic wavelet function, wavelet base, scale, threshold function and optimal threshold values are fixed [8, 9]
Empirical Mode Decomposition by Hilbert Huang Transform (HHT) [10]	No pre-determine filter or wavelet function for denoising the signal [11]	Mode mixing occur in the Empirical Mode Decomposition (EMD) [11]
Ensemble Empirical Mode Decomposition by Hilbert Huang Transform (HHT) [10]	No pre-determine filter or wavelet function for denoising the signal [12]	Parameters analysis of the ensemble Empirical Mode Decomposition (EEMD) is not fixed for the decomposition of the signal leading to the mode mixing [11]

Wavelet Transform method are more effective in performance compared with the EMD. Still this method involves major drawbacks when it comes to the selection of the wavelet based [12, 13]. The next subsection discusses in brief on EEMD algorithm.

2.2 Ensembles Empirical Mode Decomposition (EEMD)

EEMD, means from the numbers of IMF's been extracted from the adapted of EMD to the original signal. The ensemble mean is calculated with a different addition of white noise.

Figure 1 shows the process of EEMD algorithm. Given the signal, EEMD can be summarized as follows. By having many amplitudes of added white noise to the original signal resulted as noise- added signal. Next, the EMD operation will be implemented with the noise-added signal. Lastly, the ensemble means of each IMF component will be calculated.

Figure 2 shows an example of an IMF (green curve) with corresponding envelopes. A signal consists of higher-frequencies waves riding upon lower-frequencies carrying waves. The characteristic of time scales is used to find the intrinsic modes.

The main idea of having the sifting process which is the IMF is to separate the data into a slowly varying local mean part and a fast-varying symmetric oscillatory part. The latter part becoming the IMF represented by the green envelope and the local mean represented by the black line defining a residue [14]. This residue serves as input for

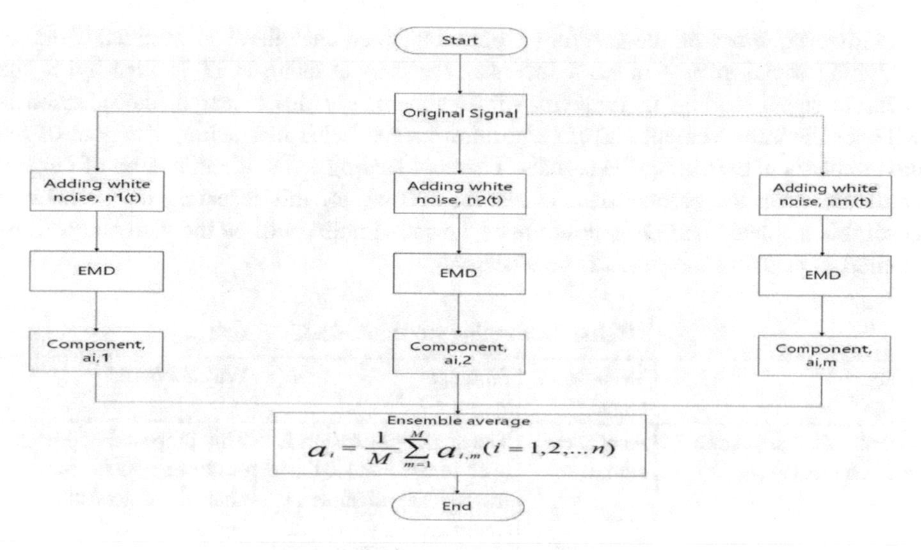

Fig. 1. Process of EEMD algorithm

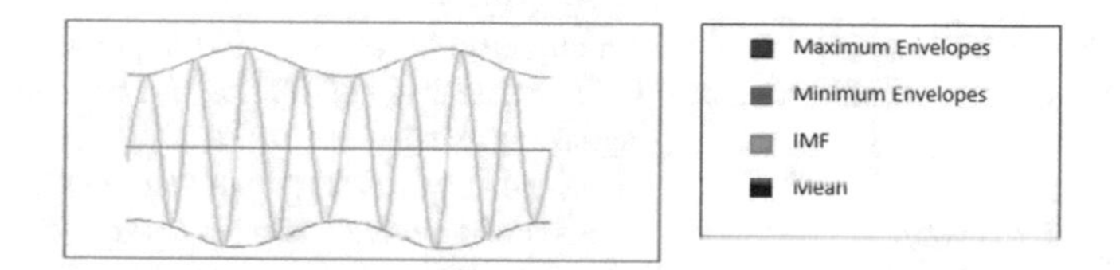

Fig. 2. Example of an Intrinsic Mode Function (IMF)

further decomposition, with the process being repeated until no more oscillations can be obtained [14].

To overcome partially on the mode mixing phenomenon within the IMF, EEMD is proposed [15, 16]. This EEMD defines the true IMF components as the mean of an ensemble of trials, whereby each of the signals consist of the added white noise of finite amplitude. With this EEMD approach, the mode mixing scales in the IMF could be partially resolves based on the recent studies of the statistical properties of white noise [17]. The next section discusses existing works that utilize EEMD.

2.3 Existing Work for Ensembles Empirical Mode Decomposition (EEMD)

The content in the Table 2 explains the comparison between works implementing EEMD in decomposing the signal. Based on the existing works, all of them proved that EEMD could perform decomposition of signal. Either it is decomposition of signal in acoustic, ultrasonic guided waves, or any other types of signals, EEMD can decompose the signal into mono components to be more meaningful to be interpreted.

However, most of the existing works explained that there is minimal work in analyzing signal energy of each IMF's in the EEMD using HHT in identifying the accurate signal reading in frequency-time domain. For this research, the question is whether the improvement EEMD algorithm indeed helps in reaching the goal of the data analysis in extracting the signals. Thereby, leading to the identification of t signal signature when the carbon steel is oxidized. Thus, for this research, the number of ensemble numbers, and the amplitude of the added noise, will be the parameters to be studied in reaching the goal of the research.

Table 2. Existing works of EEMD

Name	Type of signal	Findings	Way forward
Adnan, Ghazali, Amin and Hamat [8]	Acoustic signal	The signal processing is used to decompose the raw signal and show in time-frequency. Acoustic signal and HHT is the best method to detect leak in gas pipelines	This proposed approach could be applied for analysis of acoustic signal for the Acoustic Emission [AE] sensors in pipeline
Camarena-Martinez, Valtierra-Rodriguez, Perez- Ramirez, Amezquita-Sanchez, de Jesus Romero-Troncoso, and Garcia-Perez [18]	PQ disturbances	An approach based on EMD for analysis of PQ signals has been presented. The method consists of iterative down sampling stage that helps to extract the fundamental component as the first IMF	This proposed approach could be applied for analysis of acoustic signal for the Acoustic Emission [AE] sensors in pipeline
Xu, Wang, Tan, Si and Liu [19]	Acoustic signal	An adaptive EMD threshold denoising method for acoustic signal has been presented. The method is introduced into the threshold determination process of the denoising approach	This proposed approach could be applied for analysis of acoustic signal for the Acoustic Emission [AE] sensors in pipeline
Siracusano, Lamonaca, Tomasello, Garescì, La Corte, Carpentieri, Grimaldi and Finocchio [20]	Acoustic emission signal	An approach on a framework based on the Hilbert Huang Transform for the analysis of the acoustic emission data. By having the EMD features, the acoustic signal could isolate and extract the signal to be interpreted such as the amplitude to be more meaningful	This proposed approach could be applied for analysis of acoustic signal for the Acoustic Emission [AE] sensors in pipeline

3 Experiment Setup

3.1 Test Rig and Sensor Positioning Layout

Experimental tests were carried-out to compare the healthy, leak, eroded and no-leak conditions pipe operations using four sensors at two positions in Fig. 3. The acoustic emission sensors are SR10b low frequency AE sensors that have pre-amplify between 1 kHz and 1.5 MHz. The number of sensors is eight and the sensors are placed at N-S-W-E directions on two locations of the pipe. The sampling frequency is 51.2 kHz.

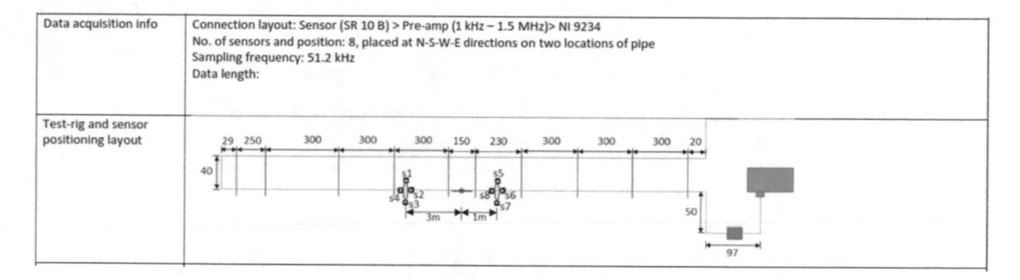

Fig. 3. Scheme of the experimental setup

Elements/Tests	*Conditions*
Test 1	Healthy
Test 2	Leak
Test 3	Eroded
Test 4	No Leak [use the leak test pipe section with leak closed]
Pre -amplifier gain	20 db [only for the leak and no - leak] 40 db [for all tests]
Liquid pressure	2 bar pressures

Fig. 4. Test conditions

Experiments were conducted corresponding to Fig. 4. The algorithm is tested against four data sets obtained from the field testing by using the acoustic emission sensors SR10b low frequency AE sensor. The four data sets are data sets for test cases healthy, leak, no-leak and eroded test cases. For this research, currently four data sets that have been tested which is test case for healthy 2 bar pressure, test case for leak with 20 db, test case for eroded 2 bar pressure and test case for no leak with 20 db.

4 Results and Discussions

4.1 Low Pass Filter and Discussion

A. Test case for Healthy

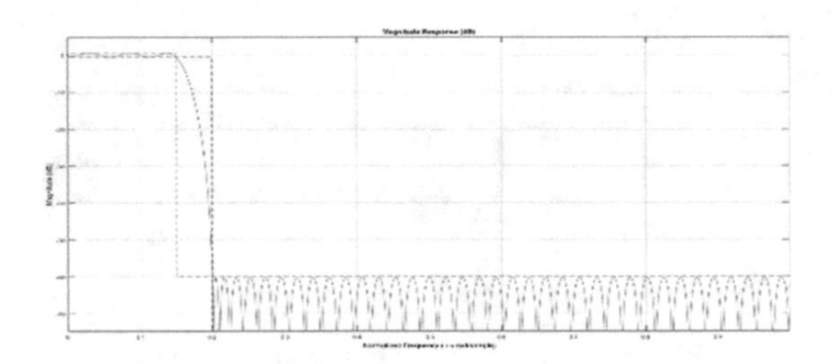

Fig. 5. Low pass filtering for healthy

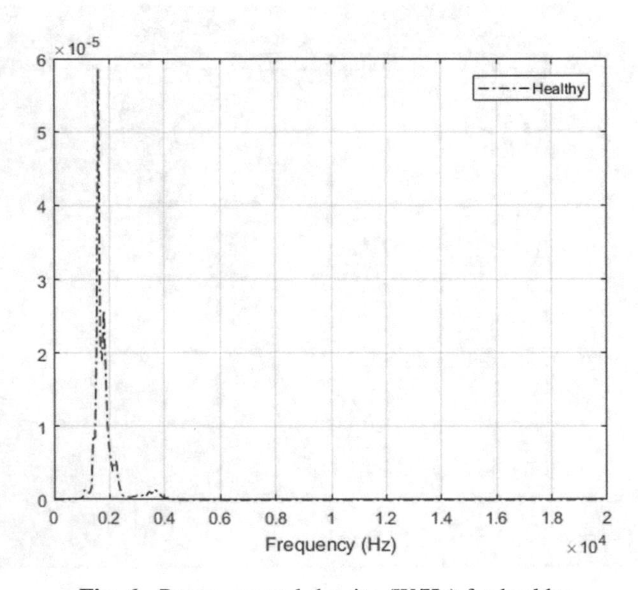

Fig. 6. Power spectral density (W/Hz) for healthy

B. Test case for Leak

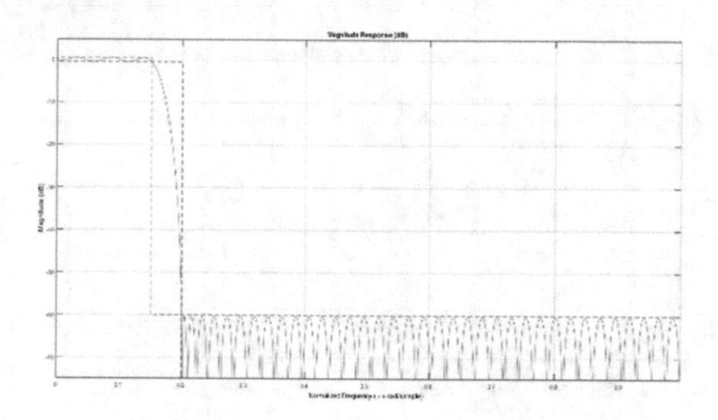

Fig. 7. Low pass filtering for leak

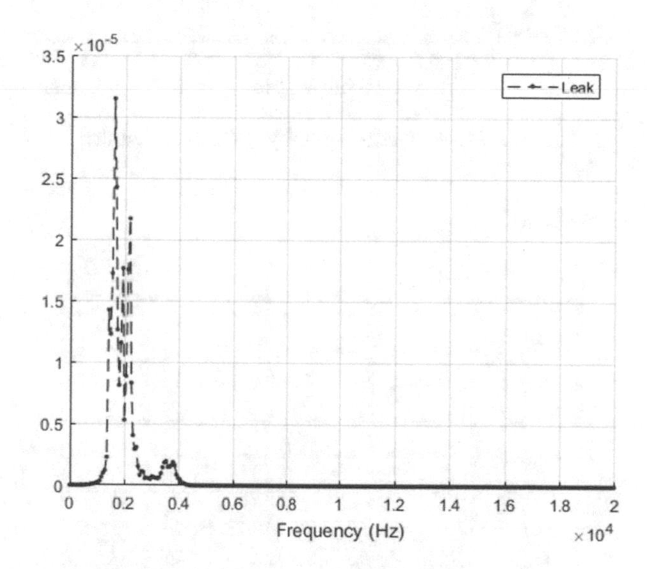

Fig. 8. Power spectral density (W/Hz) for leak

C. Test case for Eroded

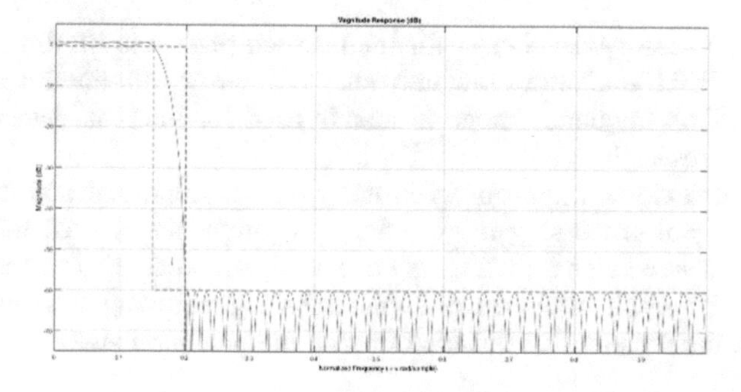

Fig. 9. Low pass filtering for eroded

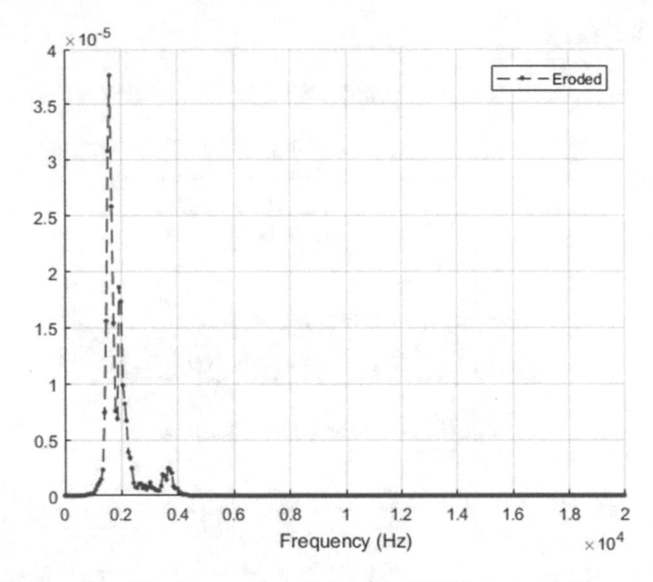

Fig. 10. Power spectral density (W/Hz) for eroded

D. Test case for No-Leak

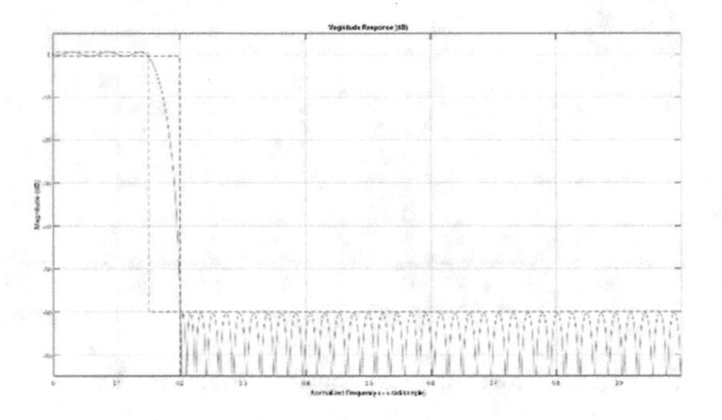

Fig. 11. Low pass filtering for no-leak

The acoustic sensor was chosen for the research project is SR10b low frequency AE sensor. This SR10b is ultra - low frequency AE sensor with the frequency range of 1 Hz–15 kHz. This frequency range is mainly used for the leak detection thus it is suitable for this research.

Recovery of a signal from observed noisy data is important and the analysis of techniques of signal processing is essential to highlight the wanted information and attenuate the undesired signal [3, 21]. In general, signal processing methods were used to solve the hidden result from the raw signal including removing noisy signal that been collected from the sensors that installed outside the pipeline network.

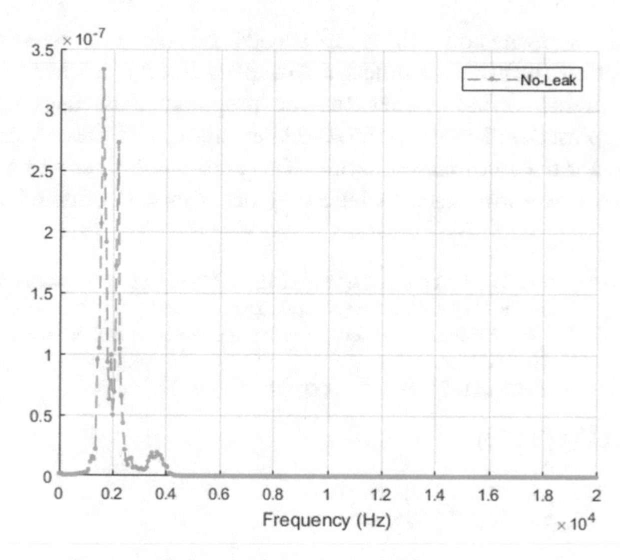

Fig. 12. Power spectral density (W/Hz) for no-leak

A filter is a process to remove the unwanted components or features from a signal. Filtering is also a class of signal processing where the unwanted components or features from a signal means removing frequency bands. A low-pass filter is a filter that allows signals below a cutoff frequency (known as the pass band) and attenuates signals above the cutoff frequency (known as the stop band).

As shown from the Figs. 6, 7, 8, 9, 10, 11 and 12, the low pass filter has been used for this research. The cut off frequency (known as the pass band) is 01.5 Hz and attenuates signals above the cutoff frequency (known as the stop band) which is 0.20 Hz. Low pass filter has been used for this research due to this filter is suitable to validate the range for the SR10b low frequency AE sensor.

For this research, there are four data sets that been tested with the low pass filter. The four data sets are test case for healthy condition, test case for leak condition, test case for eroded condition and test case for no leak condition. Figure 5 shown the low pass filter for the healthy condition. Power spectral density (PSD) is shown in Fig. 6 for healthy condition. Figure 7 shown the low pass filter for the leak condition. Power spectral density (PSD) is shown in Fig. 8 for leak condition. The power spectral density is higher in healthy condition compared with the leak condition.

Figure 9 shown the low pass filter for the eroded condition. Power spectral density (PSD) is shown in Fig. 10 for eroded condition. Figure 11 shown the low pass filter for the no leak condition. Power spectral density (PSD) is shown in Fig. 12 for no leak condition. The power spectral density is higher in no leak condition compared with the eroded condition.

The difference between power spectral density for the all the test cases are due to change to pipe spool (of leak, healthy, eroded and no-leak) in the test-rig, changed the system stiffness, and sensor sensitivity. Moreover, each of the mechanism is different, so these different test cases supposed to transfer signal energy to a different band.

Figure 13 is showing on the snippet code on the low pass filtering. This **MATLAB** function [**bpFilt**] constructs a **low pass filter** specification for test case healthy[**xh**], applying default values for the properties PassbandFrequency, StopbandFrequency, PassBandRipple, StopBandAttenuation, and Design Method. **Filtfilt** function is used for the zero-phase filtering. This zero-phase filtering preserve features in a filtered time waveform exactly where they occur in the unfiltered signal.

```
bpFilt= designfilt('lowpassfir','PassbandFrequency',
        0.15,'StopbandFrequency',0.2,. . .
        'PassbandRipple',1,'StopbandAttenuation',
        60, . . .
        'DesignMethod','equiripple');

fvtool(bpFilt)
dataln=xh;
xhf=filtfilt(bpFilt,xh);
xhf1=filter(bpFilt,xh);
```

Fig. 13. Snippet code on the low pass filtering for test case healthy

The frequency response describes which frequency bands the filter passes 0.15 kHz (the pass band) and which it rejects 0.20 kHz (the stop band). Thus, for this low pass filtering, low frequencies are passed which is 0.15 kHz for all the test cases, and high frequencies which are 0.20 kHz are attenuated.

4.2 Band Pass Filter and Discussion

A. Test case for Healthy

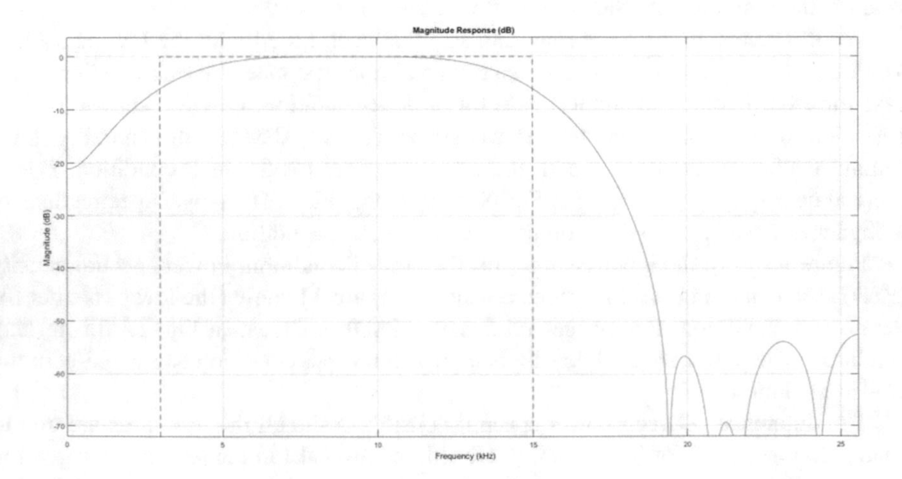

Fig. 14. Band pass filtering for healthy

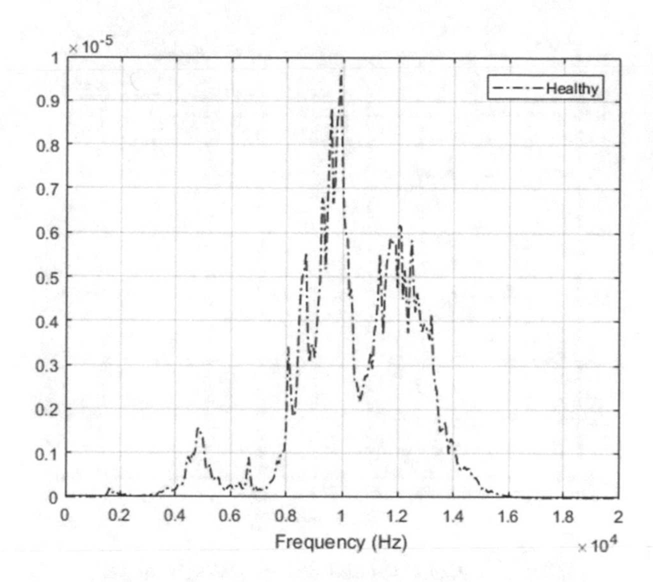

Fig. 15. Power spectral density (W/Hz) for healthy

B. Test case for Leak

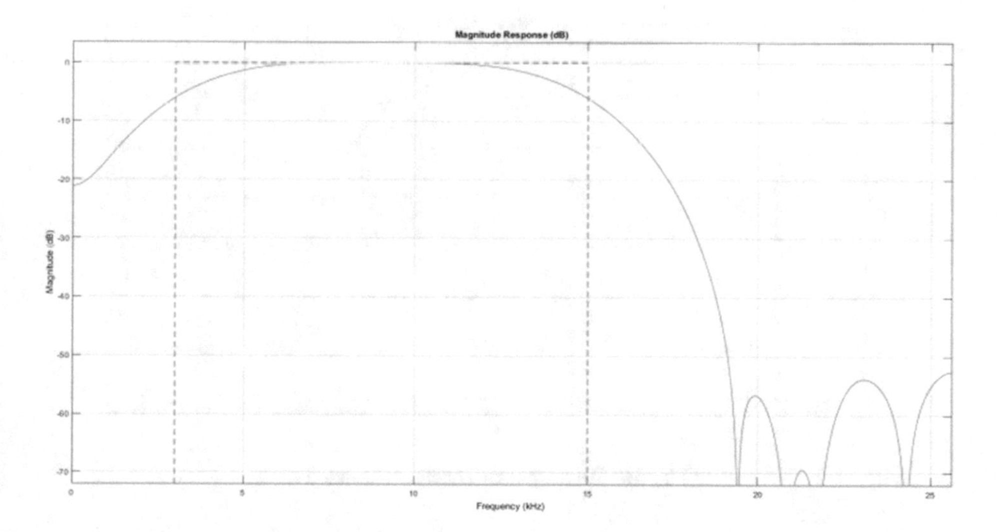

Fig. 16. Band pass filtering for leak

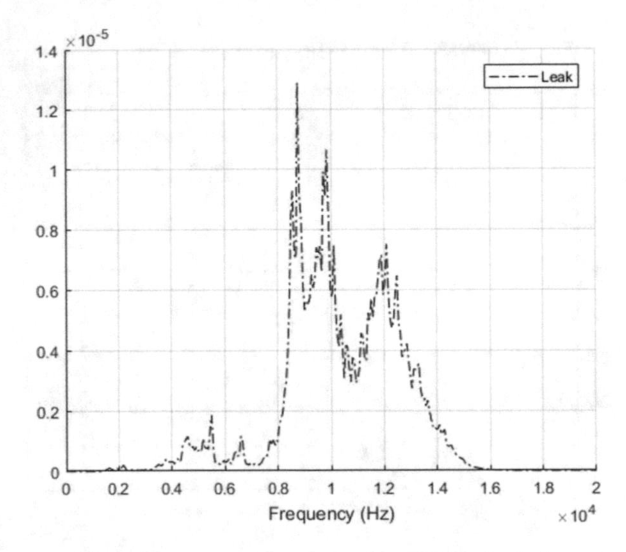

Fig. 17. Power spectral density (W/Hz) for leak

C. Test case for Eroded

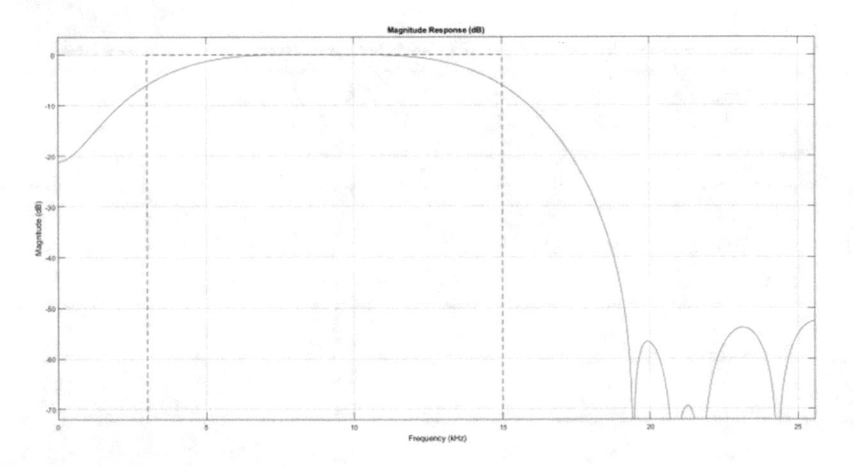

Fig. 18. Band pass filtering for eroded

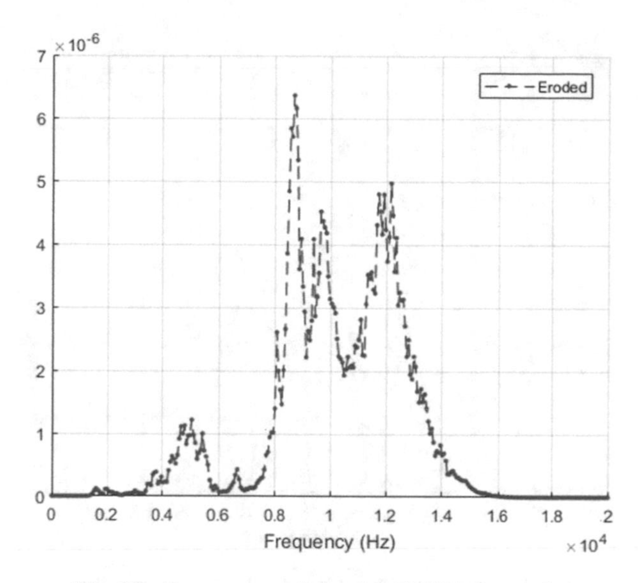

Fig. 19. Power spectral density (W/Hz) for eroded

D. Test case for No-Leak

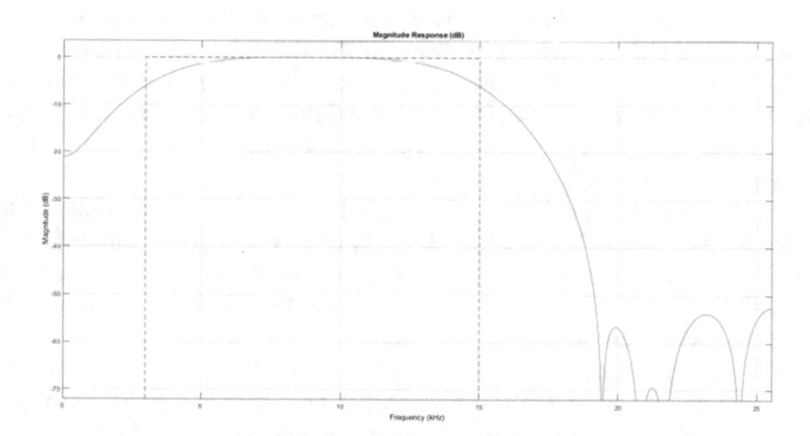

Fig. 20. Band pass filtering for no-leak

The acoustic sensor was chosen for the research project is SR10b low frequency AE sensor. This SR10b is ultra - low frequency AE sensor with the frequency range of 1 Hz–15 kHz. This frequency range is mainly used for the leak detection thus it is suitable for this research.

Recovery of a signal from observed noisy data is important and the analysis of techniques of signal processing is essential to highlight the wanted information and attenuate the undesired signal [3]. In general, signal processing methods were used to

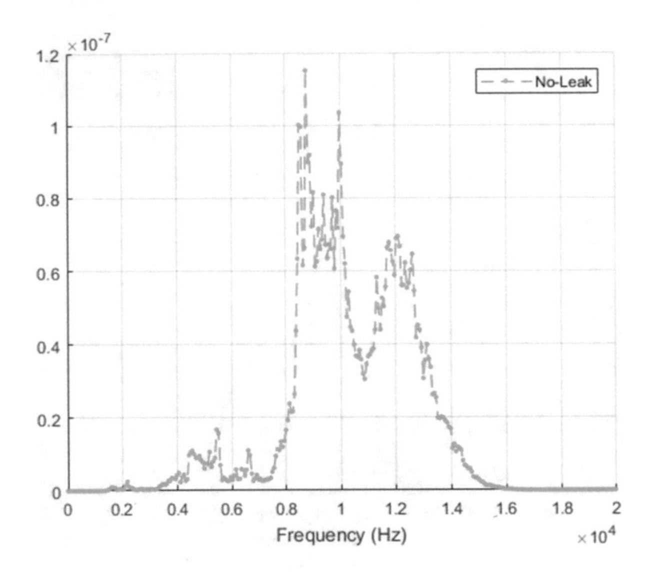

Fig. 21. Power spectral density (W/Hz) for no-leak

solve the hidden result from the raw signal including removing noisy signal that been collected from the sensors that installed outside the pipeline network.

A filter is a process to remove the unwanted components or features from a signal. Filtering is also a class of signal processing where the unwanted components or features from a signal means by removing frequency bands. Band Pass Filters passes signals within a certain band of frequencies without distorting the input signal or introducing extra noise. Pass band bandwidth is the difference between the upper [f_cut 2] and lower [f_cut1] cutoff frequencies.

As shown from the Figs. 14, 15, 16, 17, 18, 19, 20 and 21, the band pass filter has been used for this research. The upper [f_cut 2] and lower [f_cut1] cutoff frequencies are only frequencies in a frequency band are passed. Band pass filter has been used for this research due to this filter is suitable to remove the electrical components for the range for the SR10b low frequency AE sensor.

For this research, there are four data sets that been tested with the band pass filter. The four data sets are test case for healthy condition, test case for leak condition, test case for eroded condition and test case for no leak condition. Figure 14 shown the band pass filter for the healthy condition. Power spectral density (PSD) is shown in Fig. 15 for healthy condition. Figure 16 shown the band pass filter for the leak condition. Power spectral density (PSD) is shown in Fig. 17 for leak condition. The power spectral density is higher in healthy condition compared with the leak condition. Figure 18 shown the band pass filter for the eroded condition. Power spectral density (PSD) is shown in Fig. 19 for eroded condition. Figure 20 shown the low pass filter for the no leak condition. Power spectral density (PSD) is shown in Fig. 21 for no leak condition. The power spectral density is higher in no leak condition compared with the eroded condition.

The difference between power spectral density for the all the test cases are due to change to pipe spool (of leak, healthy, eroded and no-leak) in the test-rig, changed the system stiffness, and sensor sensitivity. Moreover, each of the mechanism is different, so these different test cases supposed to transfer signal energy to a different band. Figure 22 is showing on the snippet code on the band pass filtering.

This **MATLAB** function [**bpFilt**] constructs a **band pass filter** specification for test case healthy [**xh**], applying default values for the properties Filter Order, CutoffFrequency1, CutoffFrequency2, and Sample Rate. **Filtfilt** function is used for the band pass filtering. This band pass filtering preserve only frequencies in a frequency band are passed which are f_cut1: 3e3; and f_cut2: 15e3.

```
f_cut1= 3e3;
f_cut2= 15e3;
bpFilt= designfilt('bandpassfir','Filterorder',20,. . .
        'CutoffFrequency1', f_cut1, 'CutoffFrequency2',
        f_cut2, . . .
        'SampleRate',Fs);

fvtool(bpFilt)
dataln=xh;
xhf=filtfilt(bpFilt,xh);
xhf1=filter(bpFilt,xh);
```

Fig. 22. Snippet code for band pass filtering for test case healthy

The frequency response describes which frequency bands the filter which are f_cut1: 3e3; and f_cut2: 15e3; and which it rejects 0.20 kHz (outside of the frequency band) in Fig. 22. Thus, for this band pass filtering, which are f_cut1: 3e3; and f_cut2: 15e; are the acceptable frequency bands for all the test cases. Cut off frequency refers to f_cut1: 3e3; and f_cut2: 15e; which the frequency beyond when the filter will not pass signals. It is usually measured at a specific attenuation such as 3 dB. If the cut off frequency 2 dB, roll off will happening.

5 Conclusion

A filter is a process to remove the unwanted components or features from a signal. Filtering is also a class of signal processing where the unwanted components or features from a signal means by removing frequency bands. The cut off frequency (known as the pass band) is 01.5 Hz and attenuates signals above the cutoff frequency (known as the stop band) which is 0.20 Hz. Low pass filter has been used for this research due to this filter is suitable to validate the range for the SR10b low frequency AE sensor. Band pass filter which are f_cut1: 3e3; and f_cut2: 15e; has been used for this research due to this filter is suitable to remove the electrical components for the range for the SR10b low frequency AE sensor.

The difference between power spectral density for the all the test cases are due to change to pipe spool (of leak, healthy, eroded and no-leak) in the test-rig, changed the system stiffness, and sensor sensitivity. Moreover, each of the mechanism is different, so different test cases supposed to transfer signal energy to a different band. Thus, further analysis needed to be conducted for more analysis on healthy, leak, eroded, and no leak events to identify the signal signature.

Different test cases supposed to transfer signal energy to a different band for a narrow frequency band (SR10B operating range). Once the signature of each case been identified, the IMF's for the different test cases could be analyzed and reconstruction.

Acknowledgments. This work was supported by Development of Intelligent Pipeline Integrity Management System (I-PIMS) Grant Scheme from Universiti Teknologi PETRONAS.

References

1. Underground pipeline corrosion
2. Shi, Y., Zhang, C., Li, R., Cai, M., Jia, G.: Theory and application of magnetic flux leakage pipeline detection. Sensors **15**(12), 31036–31055 (2015)
3. Gaci, S.: A new ensemble empirical mode decomposition (EEMD) denoising method for seismic signals. Energy Proc. **97**, 84–91 (2016)
4. Agarwal, M., Jain, R.: Ensemble empirical mode decomposition: an adaptive method for noise reduction. IOSR J. Electron. Commun. Eng. **5**, 60–65 (2013)
5. Karkulali, P., Mishra, H., Ukil, A., Dauwels, J.: Leak detection in gas distribution pipelines using acoustic impact monitoring. In: IECON 2016-42nd Annual Conference of the IEEE Industrial Electronics Society. IEEE (2016)
6. Datta, S., Sarkar, S.: A review on different pipeline fault detection methods. J. Loss Prev. Process Ind. **41**, 97–106 (2016)
7. Jiao, Y.-L., Shi, H., Wang, X.-H.: Lifting wavelet denoising algorithm for acoustic emission signal. In: 2016 International Conference on Robots and Intelligent System (ICRIS). IEEE (2016)
8. Adnan, N.F., Ghazali, M.F., Amin, M.M., Hamat, A.M.A.: Leak detection in gas pipeline by acoustic and signal processing: a review. In: IOP Conference Series: Materials Science and Engineering. IOP Publishing (2015)
9. Fang, Y.-M., Feng, H.-L., Li, J., Li, G.-H.: Stress wave signal denoising using ensemble empirical mode decomposition and instantaneous half period model. Sensors **11**(8), 7554–7567 (2011)
10. Yang, J., Wang, X., Feng, Z., Huang, G.: Research on pattern recognition method of blockage signal in pipeline based on LMD information entropy and ELM. Math. Probl. Eng. **2017** (2017)
11. Kevric, J., Subasi, A.: Comparison of signal decomposition methods in classification of EEG signals for motor-imagery BCI system. Biomed. Signal Process. Control **31**, 398–406 (2017)
12. Rostami, J., Chen, J., Tse, P.W.: A signal processing approach with a smooth empirical mode decomposition to reveal hidden trace of corrosion in highly contaminated guided wave signals for concrete-covered pipes. Sensors **17**(2), 302 (2017)
13. Samadi, S., Shamsollahi, M.B.: ECG noise reduction using empirical mode decomposition based combination of instantaneous half period and soft-thresholding. In: 2014 Middle East Conference on Biomedical Engineering (MECBME). IEEE (2014)

14. Saeed, B.S.: De-noising seismic data by Empirical Mode Decomposition (2011)
15. Huang, Y., Wang, K., Zhou, Z., Zhou, X., Fang, J.: Stability evaluation of short-circuiting gas metal arc welding based on ensemble empirical mode decomposition. Meas. Sci. Technol. **28**(3), 035006 (2017)
16. Potty, G.R., Miller, J.H.: Acoustic and seismic time series analysis using ensemble empirical mode decomposition. J. Acoust. Soc. Am. **140**(4), 3423–3424 (2016)
17. Honório, B.C.Z., de Matos, M.C., Vidal, A.C.: Progress on empirical mode decomposition-based techniques and its impacts on seismic attribute analysis. Interpretation **5**(1), SC17–SC28 (2017)
18. Camarena-Martinez, D., et al.: Novel down sampling empirical mode decomposition approach for power quality analysis. IEEE Trans. Ind. Electron. **63**(4), 2369–2378 (2016)
19. Xu, J., Wang, Z., Tan, C., Si, L., Liu, X.: A novel denoising method for an acoustic-based system through empirical mode decomposition and an improved fruit fly optimization algorithm. Appl. Sci. **7**(3), 215 (2017)
20. Siracusano, G., Lamonaca, F., Tomasello, R., Garescì, F., La Corte, A., Carnì, D.L., Carpentieri, M., Grimaldi, D., Finocchio, G.: A framework for the damage evaluation of acoustic emission signals through Hilbert-Huang transform. Mech. Syst. Signal Process. **75**, 109–122 (2016)
21. Wu, Z., Huang, N.E.: Ensemble empirical mode decomposition: a noise-assisted data analysis method. Adv. Adapt. Data Anal. **1**(01), 1–41 (2009)

Qualitative Study of Security Resiliency Towards Threats in Future Internet Architecture

M. S. Vidya[1(✉)] and Mala C. Patil[2]

[1] Department of Computer Science and Engineering, BMSIT and M, Bengaluru, India
rvidyapai@bmsit.in
[2] Department of Computer Science, COHB, University of Horticultural Sciences, Bagalkot, India
Malapatil2002@yahoo.co.in

Abstract. With the consistent evolution of distributed networks and cloud computing, the communication process of existing Internet architecture are not at par to offer comprehensive security; thereby fails to cater up the dynamic web security demands of online users. This leads to the evolution of Future Internet Architecture (FIA) that claims of enhanced security system. However, a closer look into both existing web security and security of existing FIA projects shows that there are enough security loopholes in both that demands a novel security solution. Therefore, this paper contributes to investigate the effectiveness of existing web security approaches as well as security approaches of FIA in order to explore the open research issues. The finding of the study shows that there is a significant research gap that demands extreme improvement of existing web security in order to fit for secure communication in FIA.

Keywords: Web security · Internet · Future internet architecture Attacks · Vulnerability · Denial of service · Distributed security IoT

1 Introduction

The existing version of Internet architecture has been serving since 40 years for all communication needs and has also contributed many successful implementation of communication protocols [1]. However, with the rising demands of distributed network system, they are no more potential especially with respect to security factor. There are various reasons behind this viz. (i) they use IP network that is host centric and cannot cater up any form of distributed communication system, which is the need of present age, (ii) the policy of IP address is encountering exhaustion with the exponential rise of users, (iii) no inclusion of security attributes (except IPv6 that offers distinct and unique IP addresses to promote security), (iv) highly non-flexible posing difficult to incorporate new operation [2]. Therefore, web security has become one of the essential concerns among the network communities [3–5] and this leads to the formation of Future Internet Architecture (FIA) [6]. One of the essential part of FIA is its security

R. Silhavy et al. (Eds.): CoMeSySo 2018, AISC 859, pp. 274–284, 2019.
https://doi.org/10.1007/978-3-030-00211-4_24

inclusion. Funded by National Science Foundation (NSF) in United States, FIA project has been initiated that has transformed the existing host centric web design to content centric architecture [7] with more capability to perform an effective traffic management. Some of the reputed projects of FIA are Mobility First [8], XIA [9], SCION [10], COAST [11], etc. However, work carried out by Ding et al. [12] has reported significant level of problems associated with almost all the evolved projects of NSF. This leads to a conclusion that there is a need to investigate the security demands of FIA right from gauging strength of existing web security approaches. This paper discusses about the some of the recent work being carried out toward security existing web and FIA. Section 2 discusses about the taxonomies of attacks in internet followed by studies on existing web security approaches in Sect. 3. Section 4 discusses about studies towards FIA security while open research issues are highlighted in Sect. 5, while Sect. 6 discusses about primary findings of proposed review work.

2 Taxonomies of Attack

There are various types of security vulnerability in existing internet architecture that is yet not met with 100% full proof solution. It is also likely that some of such attacks will also continue to lead in its enhanced version of internet architecture in future too. The most frequently found internet-based attacks are malware, phishing, SQL injection attack, cross-site scripting, Denial-of-Service, Session Hijacking, Man-in-Middle attack, Credential reuse, etc. In existing system, there are two forms of tools where one is used for attack and another is used for defender. The tools used by attackers are mainly used for launching malicious programs e.g. Trojans, Denial-of-Service, packet forging attack, fingerprinting attack, application layer attack, user attack. The initiation of Denial-of-Service is carried out by degree of automation, by exploited vulnerability, by attack network used, by attack rate dynamics, by victim type, by impact, and by agent. The application layer attack is again classified into types i.e. browser attack and server attack. The browser attack is again initiated using XSS request while server attack is initiated by protocols, SQL injection, code injection, buffer error, URL misinterpretation. The tools used by defender mainly user network monitoring tools e.g. visualization and analysis. Interestingly, there is a common tool usage found between attacker and defender i.e. *information gathering tool*. This tool is used for sniffing and network mapping operation where the intention of attacker is to look for suitable environment of launching attack while intention of defender is to capture the presence of abnormal or vulnerable network condition to report a problem. There is no doubt maximum information is circulated through the web and it becomes the hub of all the attackers too. Unfortunately, existing internet architectures lacks robustness as well as capabilities to identify and prevent all the lethal forms of attacks irrespective of usage of expensive firewall system. There are various researchers e.g. [13] who has already stated the challenges associated with existing cyber security that possible immense threat to existing web users. This leads to evolution of FIA, however, it is still unknown how much existing web security is prepared to be customized for shaping itself to offer maximum security when used with FIA. The next section discusses about existing web security approaches.

3 Studies on Web Security

Web services as well as its corresponding application is frequently used and accessed by millions of users worldwide. With increasing incorporation of securing the web, it is equally important to enhance the user's experience too. According to Stritter et al. [14], Web 2.0 is highly prone to different forms of intrusion over its cross-site scripting while adversely affecting Domain Name System. Usage of public key infrastructures for securing Web 2.0 is also claimed to be unjustified. Study towards identification of attacks on web-application is carried out by Kozik et al. [15] by introducing a joint implementation of machine learning and clustering approaches. A simple classification approach was stated by authors as,

$$c \rightarrow \sum_{i=0}^{n} \frac{(O_i - E_i)^2}{E_i} \tag{1}$$

The variables in expression (1) are O_i (sample of data that is required to be classified) and E_i (character histogram) while the outcome was assessed using accuracy factors. Although, the method offers simplified approach but it doesn't deal with any form of assessment of algorithm complexity. It is essential to assess algorithm complexity without which the practicality of the algorithm cannot be defined. According to Huang et al. [4] and Srinivasan [16], none of the existing web security solutions adheres to penetration test that lacks the capability to resist false positive very often. A very unique prototype has been introduced by Tsalaportas et al. [17] that uses cross-platform environment. The complete prototype is designed by making it independent from any server side and is purely self-contained. This is a lightweight concept of security but lacks consideration of identification of any specific security indicators. Study towards this direction has been carried out by Amrutkar et al. [18]. The study also offers browser security considering mobility factor and all sorts of W3C guidelines. The study is specific to mobile devices only that are usually biased as browsers for web and mobile has only slightest difference, which is privacy factor. According to Grandison and Koved [19], the privacy factors are highly essential for web security. The work carried out by Tatli and Urgun [20] have evolved up with a benchmarked tool with capability of investigating intrusion while searching the web using Java scripts. Apart from crawling-based intrusion, web attacks are mainly caused by phishing. A robust tool to capture phishing event was discussed by Mao et al. [21]. The authors have used certain feature that are not possible to be corrupted by attackers and has emphasized more on script security. The features were formed from whitelisted database, target page, and suspicious page which when subjected to similarity comparison detects phishing. Study towards resisting phishing has been carried out by Marchal et al. [22] where the approach is to identify different constraint encountered by attacker to construct forged page. Literature has also witnessed certain hybridized technique for securing web application. Goldsteen et al. [23] have hybridized network layer and presentation layer to introduce a secure masking policy to safeguard sensitive information. However, the loophole of this work is that it doesn't support securing any contents generated by multiple parties on web. This problem is reported to be addressed by Phung et al. [24] where security of JavaScript and mixed flash contents are subjected to security. However, the work doesn't safeguard many other forms of attacks e.g. denial-of-Service (DoS) attack. Therefore, it is essential to study the internal knowledge about the attack strategy and mining approach is most suitable for this task. Adoption of mining approach was seen in work of Medieros et al. [25] where policy for attack detection was formulated and then it was used to capture diverse attacks. However, owing to inclusion of iterative rules, it could offer computational complexity that has not been assessed. Similar direction of research was also carried out by Antunes and Viera [26] considering benchmarking approach for SQL injection and work of Gillman et al. [27] who discussed about delivering contents securely.

Table 1. Summary of existing research solutions towards web security issues

Authors	Problems	Techniques	Pros	Cons
Kozik [15]	Identification of attacks	Machine learning, clustering	Independent of prior protocol info	Response time w.r.t. large traffic not assessed
Tsalaportas et al. [17]	Web security	Prototyping, XML security	Self contained	Not benchmarked
Amrutkar et al. [18]	Mobile Browser security	Usage of TLS indicators	Comprehensice analysis	Specific to mobile device only
Tatli and Urgun [20]	Intrusion by web crawler	Benchmarked tool	Wide spread web security	No comparative analysis
Mao et al. [21], Marchal et al. [22]	Phishing	Prototyping, feature extraction	Good accuracy	Specific to phishing attacks
Goldsteen et al. [23]	Data security over web	Network layer, presentation layer, masking	Supports visual rule authoring	Not completely effective against all client-side application
Phung et al. [24]	Data security over web	Security by Flash, JavaScript	Transparent policy	Doesnt identify DoS attack
Medieros et al. [25]	Attack identification	mining	More information about attacks	Computationally complex

Table 1 highlights the summary of the essential work carried out on web security, where it can be seen that majority of the existing system towards web security has associated advantages as well as limitations. However, one thing is very clear that there are few research implementations towards web security considering FIA in mind. Therefore, the next section discusses about the research contribution towards securing FIA.

4 Studies Towards Securing Future Internet

Basically, Future Internet Architecture (FIA) is meant for overcoming the pitfalls of conventional architecture of current version of internet. The discussion of Pan et al. [28] has elaborated about different scale of contributions towards future internet architecture. However, projects carried out in FIA still have open scope of security incorporation. The manuscript drafted by Ambrosin et al. [6] have discussed about elementary requirements required for securing FIA and to boost up its privacy factor. After reviewing his work, it was found that none of the projects of future internet architecture offer full fledge security especially with respect to privacy. Figure 1 highlights all the essential factors that are required to be incorporated to ensure optimal security.

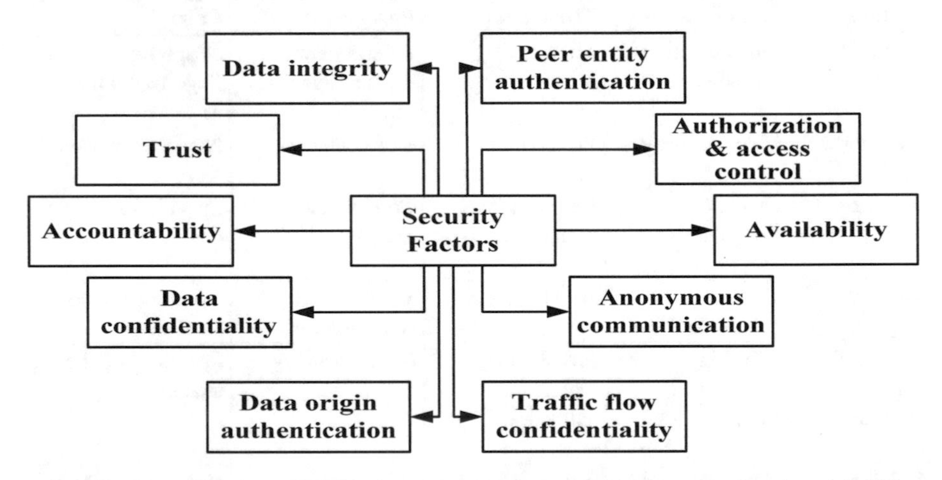

Fig. 1. Dependable security factors of future internet architecture (Ambrosin et al. [6])

A good initiative towards investigating security over FIA (called as CoLoR) was carried out by Chen et al. [29] where the authors have carried out comparative strength analysis of various threat levels. According to authors, designing FIA considering service location and inter-domain routing could offer better resource management as well as its packet forwarding strategy is highly resistive against reflection attacks, poisoning of DNS cache, prefix hijacking, bandwidth depletion, falsification of routing path, etc. However, the study doesn't consider any form of enhancing task for network security. This problem was further investigated by Chen et al. [30] where it was discussed that path identifiers plays a critical role in securing communication in FIA. It was discussed that if the path identifiers were changed dynamically that it could significantly escalate network security. Literatures have also witnessed that upcoming and prominent networking concept like *Internet-of-Things* (IoT) is the primary role player in enabling FIA. However, various researchers still debate that IoT has not achieved ultimate security as it has to be in order to support hosting FIA. The findings of Malyuk and Miloslavskaya [31] shows that a highly structured form of data security is demanded in FIA. Just like IoT, *Reconfigurable Computing* is also claimed to be used extensively in FIA. The discussion of Mesquita and Rosa [32] shows that there is an ongoing research towards securing incorporation of such concepts in FIA. As cloud and virtualization environment is highly required for secure hosting of applications on FIA so *Network virtualization* becomes highly essential to be considered. Chances to intrude network virtualization is more owing to vulnerability of virtual machines irrespective of various existing studies towards securing virtual machines over cloud [33, 34]. There are certain desperate research attempts towards securing FIA. A unique test-bed of security has been formulated by Ozcelik et al. [35] that deals with data security mainly. The work carried out by Rebahi et al. [36] has investigated about the security strength of network technologies of autonomic type. The authors have discussed about Generic Autonomic Network Architecture (GANA) as a reference model and discusses various hypothetical means to elevate the security strength of GANA. Samad et al. [37] have discussed about different strategies that could protect IoT from different risk factors. The authors have developed a technique that performs assessment of risk using contextual attributes.

There are also work being carried out towards resisting specific forms of attacks in cloud environment as well as cyber-physical system owing to increase of complexity in design. Work of Ge et al. [38] have presented architecture with double loop system for leveraging the optimal security system. The problem of pollution attack was addressed by Guo et al. [39] using simulation-based approach that uses diversity factor of path in point-of-presence network. The authors have used bloom filter for incorporating security. IoT and adhoc networks are some of the prominent backbone of FIA which requires higher degree of security. A different form of study was introduced by Jhaveri et al. [40] by presenting a trust-based technique to investigate an explicit pattern of attack in adhoc network and IoT. A communication policy has been introduced based on trust factor that consistently monitor patterns of dropping packets during path exploration phase. The authors have selected mobility environment to ensure that communication is highly secured. Similar line of security towards mobility of nodes has been introduced by Kang et al. [41] that deals with mainly resisting attacks related to location information. The authors have addressed the security problems associated with virtual migration and offer a pseudonym based policy to secure the sensitive data of location for mobile nodes. Solution towards collusion attacks has been addressed by Li et al. [42] where the authors have used fog computing along with the combination of cryptographic measures for access control. The problems of suppression attack in IoT has been addressed by Perazzo et al. [43]. An experimental analysis shows that such attacks adversely affect the energy state of the IoT nodes. The most hyped blockchain process was also reported to be adopted in existing system. The work of Sharma et al. [44] has reported to use blockchain based approach in order to secure IoT and thereby protect Software Defined Network (SDN). Security of SDN has been emphasized by various researchers as it forms one of the main building blocks in FIA. Study towards SDN security has been carried out by Xu et al. [45]. The authors have constructed a technique of extracting explicit attack based feature that assists in capturing the real-time information about the adversaries. The study outcome has shown that it is effective against identifying abnormal and malicious behaviour of a node and hence could contribute in safeguarding FIA. Table 2 highlights the summarized version of the existing contribution where existing security approaches towards protecting FIA is found to be resilient against certain set of lethal attacks. However, problems continue to exist as these solutions have their own scope and own limitations that couldn't encapsulate all the security problems associated with FIA.

Table 2. Summary of existing security of future internet

Authors	Problems	Techniques	Pros	Cons
Chen et al. [29]	Comparative study of FIA	Quantitative	CoLoR is resistive against many adversaries	No analysis of computational complexity, No benchmarking
Chen et al. [30]	Network security	Simulation, Path identifiers	Shorter timeout period	Leads overhead
Malyuk and Miloslavskaya [31]	Data security	Conceptualized model	Performs classification	No extensive analysis
Ozcelik et al. [35]	Data security	Developed a testbed	Collaborative network	No extensive analysis
Samad et al. [37]	Resisting risk	Conceptualized model of risk mitigation	Effective risk planning	No extensive analysis
Ge et al. [38]	Cyber attacks	Mathematical modelling, Architecure of double loop	Capable of identifying and filtering different attacks	No benchmarking
Guo et al. [39]	Pollution attack	Path diversity, bloom filter	Practical approach	Network specific solution
Jhaveri et al. [40]	Secure routing	Simulation, Discovery of attack patterns	Low overhead	Specific to some sets of attacks only
Kang et al. [41]	Location privacy	Analytical, Mapping-based	Increased observation of privacy	Increased overheads
Li et al. [42]	Collusion attack	Fog computing, cryptography	Energy efficient	Will lead to increased response time
Perazzo et al. [43]	Suppression attack	Experimental, cooja, contiki	Simpler implementation	No extensive analysis
Sharma et al. [44]	SDN security	Blockchain	Scalable	No extensive analysis
Xu et al. [45]	Table overflow attack	Analytical modelling, OpenFlow	Precise identification	Affects throughput

5 Open Research Issues

From the prior section, it is evident that there is various research works being carried out towards web security as well as towards securing FIA. These section further briefs of the open research issues explored as follows:

- **Gap between Web Security and FIA security**: A closer look into existing literatures shows that there are quantitatively more research papers for web security as compared to FIA security. However, neither research towards web security has no consideration of FIA standards and vice-versa. Hence, applicability of both on each other remains undisclosed at present. Moreover, very few research work is found to be benchmarked as the concept of FIA security is quite novel.
- **Attack specific solution**: Majority of the existing security solutions towards FIA is developed on the basis of specific attacks e.g. pollution attacks, table overflow attack, suppression attack, collusion attack, etc. There are few studies to emphasize the effectiveness of such attack-specific solution to be functional if the adversary changes.
- **More focus on architecture less on component**: FIA is considered as a wholesome architecture of communication that is constructed on multiple components. Since, last decade, there have been various changes in every projects of FIA with evolution of new characteristics. However, none of the projects on FIA (e.g. NEBULA, Mobility First, etc.) have been found to emphasize on the changing trend of robust components for developing potential FIA e.g. Content Centric Network (CCN), Internet-of-Things, and Software Defined Network. There is no doubt there exists massive literature talking about problems associated with above mentioned components but none of them are actually correlated with FIA or there was no visualization about its usage in FIA. This leads to avenue of more set of problems in these component inclusion with respect to security as:
  - **Security Threats in CCN**: There are various studies that has claimed that content centric network is one of the most effective and secure data delivery system for FIA owing to verifiable integrity, absence of address, resistance against DoS, and name resolution. However, they also present some of the potential challenges towards privacy e.g. conversation cloning, timing attack, etc.
  - **Security Threat in IoT**: The concept of IoT in FIA is very vast and hence is also vulnerable to attacks on different layers of protocol stack. Although, there exists some novel security solution in IoT, but it is yet unknown that which one of them could be possible used for a given scenario of resisting threats. Rather than resisting threat, IoT requires a universal framework that could offer true picture of threat specifically.
  - **Security Threat in SDN**: SDN offers a suitable network management for IoT operations in FIA but is vulnerable to various forms of attacks. The existing interface design of SDN is never meant for controlling its controller system completely and securely. This rests in poor access control system when existing SDN system is used in FIA.

6 Conclusion

After reviewing this review work, it is realized now that there is a long way to go for customizing the existing web security approaches. The essential findings of the proposed review work are: (i) Existing Internet architecture cannot be modified or changed as it is not flexible and hence a completely novel approach is required for enhanced web security, (ii) existing approaches of web security as well as some of the novel research attempts towards securing FIA are quite experimental and less practical as they were never being jointly studied. There could be fair possibilities that some of the existing web security approaches could be designed considering FIA in mind, but there was no such research attempt and (iii) even the security solutions considering FIA has not considered some of the essential components e.g. IoT, SDN, CCN, and many more. Without such consideration, the claimed system may offer security but they may be less functional as compared to the claimed enhanced functionality of FIA. Hence, a serious research gap existed between existing approaches and what is actually demanded. Therefore, our future work will be in direction of addressing such research gap. We will like to investigate by inclusion of SDN, CCN, IoT very specifically using mathematical modeling to evolve up with more secure solution in FIA with more practical assumption.

References

1. van Schewick, B.: Internet Architecture and Innovation. MIT Press, Cambridge (2012)
2. Xu, K., Zhu, M., Wu, G.: Towards evolvable Internet architecture-design constraints and models analysis. Sci. Chin. Inf. Sci. **57**(11), 1–24 (2014)
3. Zhou, B., Shi, Q., Yang, P.: A survey on quantitative evaluation of web service security. In: 2016 IEEE Trustcom/BigDataSE/ISPA, pp. 715–721. Tianjin (2016)
4. Huang, H.C., Zhang, Z.K., Cheng, H.W., Shieh, S.W.: Web application security: threats, countermeasures, and pitfalls. Computer **50**(6), 81–85 (2017)
5. Buchanan, W.J., Helme, S., Woodward, A.: Analysis of the adoption of security headers in HTTP. IET Inf. Secur. **12**(2), 118–126 (2018)
6. Ambrosin, M., Compagno, A., Conti, M., Ghali, C., Tsudik, G.: Security and privacy analysis of national science foundation future internet architectures. IEEE Commun. Surv. Tutor. **20**, 1418–1442 (2018). https://doi.org/10.1109/comst.2018.2798280
7. Su, Z., Hui, Y., Yang, Q.: The next generation vehicular networks: a content-centric framework. IEEE Wirel. Commun. **24**(1), 60–66 (2017)
8. Venkataramani, A., Kurose, J., Raychaudhuri, D., Nagaraja, K., Mao, M., Banerjee, S.: Mobilityfirst: a mobility-centric and trustworthy internet architecture. ACM SIGCOMM Comput. Commun. Rev. **44**(3), 74–80 (2014)
9. Naylor, D., et al.: XIA: architecting a more trustworthy and evolvable internet. ACM SIGMOBILE Comput. Commun. Rev. **44**(3), 50–57 (2014)
10. Barrera, D., Reischuk, R.M., Szalachowski, P., Perrig, A.: SCION five years later: revisiting scalability, control, and isolation on next-generation networks. http://arxiv.org/abs/1508.01651 (2015)
11. COAST: Content Aware Searching Retrieval and Streaming. http://www.synelixis.com/coast/. Accessed 9 Nov 2015
12. Ding, W., Yan, Z., Deng, R.H.: A survey on future internet security architectures. IEEE Access **4**, 4374–4393 (2016)
13. Kotut, L., Wahsheh, L.A.: Survey of cyber security challenges and solutions in smart grids. In: 2016 Cybersecurity Symposium (CYBERSEC), pp. 32–37. Coeur d'Alene, ID (2016)
14. Stritter, B., et al.: Cleaning up Web 2.0's security mess-at least partly. IEEE Secur. Priv. **14**(2), 48–57 (2016)

15. Kozik, R., Choraś, M., Hołubowicz, W.: Packets tokenization methods for web layer cyber security. Log. J. IGPL **25**(1), 103–113 (2017)
16. Srinivasan, S.M., Sangwan, R.S.: Web App security: a comparison and categorization of testing frameworks. IEEE Softw. **34**(1), 99–102 (2017)
17. Tsalaportas, P.G., Kapinas, V.M., Karagiannidis, G.K.: Solar lab notebook (SLN): an ultra-portable web-based system for heliophysics and high-security labs. IEEE J. Select. Top. Appl. Earth Obs. Remote Sens. **8**(8), 4141–4150 (2015)
18. Amrutkar, C., Traynor, P., van Oorschot, P.C.: An empirical evaluation of security indicators in mobile web browsers. IEEE Trans. Mob. Comput. **14**(5), 889–903 (2015)
19. Grandison, T., Koved, L.: Security and privacy on the web [Guest editors' introduction]. IEEE Softw. **32**(4), 36–39 (2015)
20. Tatli, Eİ., Urgun, B.: WIVET—benchmarking coverage qualities of web crawlers. Comput. J. **60**(4), 555–572 (2017)
21. Mao, J., Tian, W., Li, P., Wei, T., Liang, Z.: Phishing-alarm: robust and efficient phishing detection via page component similarity. IEEE Access **5**, 17020–17030 (2017)
22. Marchal, S., Armano, G., Gröndahl, T., Saari, K., Singh, N., Asokan, N.: Off-the-hook: an efficient and usable client-side phishing prevention application. IEEE Trans. Comput. **66** (10), 1717–1733 (2017)
23. Goldsteen, A., Kveler, K., Domany, T., Gokhman, I., Rozenberg, B., Farkash, A.: Application-screen masking: a hybrid approach. IEEE Softw. **32**(4), 40–45 (2015)
24. Phung, P.H., Monshizadeh, M., Sridhar, M., Hamlen, K.W., Venkatakrishnan, V.N.: Between worlds: securing mixed javascript/actionscript multi-party web content. IEEE Trans. Dependable Secure Comput. **12**(4), 443–457 (2015)
25. Medeiros, I., Neves, N., Correia, M.: Detecting and removing web application vulnerabilities with static analysis and data mining. IEEE Trans. Reliab. **65**(1), 54–69 (2016)
26. Antunes, N., Vieira, M.: Assessing and comparing vulnerability detection tools for web services: benchmarking approach and examples. IEEE Trans. Serv. Comput. **8**(2), 269–283 (2015)
27. Gillman, D., Lin, Y., Maggs, B., Sitaraman, R.K.: Protecting websites from attack with secure delivery networks. Computer **48**(4), 26–34 (2015)
28. Pan, J., Paul, S., Jain, R.: A survey of the research on future internet architectures. IEEE Commun. Mag. **49**(7), 26–36 (2011)
29. Chen, Z., Luo, H., Cui, J., Jin, M.: Security analysis of a future internet architecture. In: 2013 21st IEEE International Conference on Network Protocols (ICNP), pp. 1–6. Goettingen (2013)
30. Chen, Z., Luo, H., Zhang, M., Li, J.: Improving network security by dynamically changing path identifiers in future internet. In: 2015 IEEE Global Communications Conference (GLOBECOM), pp. 1–7. San Diego, CA (2015)
31. Malyuk, A., Miloslavskaya, N.: Information security theory for the future internet. In: 2015 3rd International Conference on Future Internet of Things and Cloud, pp. 150–157. Rome (2015)
32. Mesquita, D.G., Rosa, P.F.: Reconfigurable computing and future internet: considerations on flexibility and security. IEEE Latin Am. Trans. **15**(7), 1326–1334 (2017)
33. Aslam, M., Gehrmann, C., Björkman, M.: Security and trust preserving VM migrations in public clouds. In: 2012 IEEE 11th International Conference on Trust, Security and Privacy in Computing and Communications, pp. 869–876. Liverpool (2012)
34. Mao, Y., Chen, X., Luo, Y.: HVSM: an in-out-VM security monitoring architecture in IAAS cloud. In: International Conference on 2014–2014 Information and Network Security ICINS, pp. 185–192. Beijing (2014)

35. Ozcelik, I., Ozcelik, I., Akleylek, S.: TRCyberLab: an infrastructure for future internet and security studies. In: 2018 6th International Symposium on Digital Forensic and Security (ISDFS), pp. 1–5. Antalya (2018)
36. Rebahi, Y., Tcholtchev, N., Chaparadza, R., Merekoulias, V.N.: Addressing security issues in the autonomic future internet. In: 2011 IEEE Consumer Communications and Networking Conference (CCNC), pp. 517–518. Las Vegas, NV (2011)
37. Samad, J., Reed, K., Loke, S.W.: A risk aware development and deployment methodology for cloud enabled internet-of-things. In: 2018 IEEE 4th World Forum on Internet of Things (WF-IoT), pp. 433–438. Singapore (2018)
38. Ge, H., Zhao, Z.: Security analysis of energy internet with robust control approaches and defense design. IEEE Access **6**, 11203–11214 (2018)
39. Guo, H., Wang, X., Chang, K., Tian, Y.: Exploiting path diversity for thwarting pollution attacks in named data networking. IEEE Trans. Inf. Forensics Secur. **11**(9), 2077–2090 (2016)
40. Jhaveri, R.H., Patel, N.M., Zhong, Y., Sangaiah, A.K.: Sensitivity analysis of an attack-pattern discovery based trusted routing scheme for mobile ad-hoc networks in industrial IoT. IEEE Access **6**, 20085–20103 (2018)
41. Kang, J., et al.: Location privacy attacks and defenses in cloud-enabled internet of vehicles. IEEE Wirel. Commun. **23**(5), 52–59 (2016)
42. Li, G., Wu, J., Li, J., Guan, Z., Guo, L.: Fog computing-enabled secure demand response for internet of energy against collusion attacks using consensus and ACE. IEEE Access **6**, 11278–11288 (2018)
43. Perazzo, P., Vallati, C., Anastasi, G., Dini, G.: DIO suppression attack against routing in the internet of things. IEEE Commun. Lett. **21**(11), 2524–2527 (2017)
44. Sharma, P.K., Singh, S., Jeong, Y.S., Park, J.H.: DistBlockNet: a distributed blockchains-based secure SDN architecture for IoT networks. IEEE Commun. Mag. **55**(9), 78–85 (2017)
45. Xu, T., et al.: Mitigating the table-overflow attack in software-defined networking. IEEE Trans. Netw. Serv. Manag. **14**(4), 1086–1097 (2017)

A Binary Percentile Sin-Cosine Optimisation Algorithm Applied to the Set Covering Problem

Andrés Fernández, Alvaro Peña(✉), Matías Valenzuela, and Hernan Pinto

School of Engineering, Pontificia Universidad Católica de Valparaíso, Valparaíso, Chile
{andres.fernandez,alvaro.pena,matias.valenzuela,hernan.pinto}@pucv.cl

Abstract. Today there is a line of research-oriented to the design of algorithms inspired by nature. Many of these algorithms work in continuous spaces. On the other hand, there is a great amount of combinatorial optimization problems (COP) which have application in the industry. The adaptation of these continuous algorithms to resolve COP is of great interest in the area of computer science. In this article we apply the percentile concept to perform the binary adaptation of the Sine-Cosine algorithm. To evaluate the results of this adaptation we will use the set covering problem (SCP). The experiments are designed with the objective of demonstrating the usefulness of the percentile concept in binarization. In addition, we verify the effectiveness of our algorithm through reference instances. The results indicate that the binary Percentile Sine-Cosine Optimization Algorithm (BPSCOA) obtains adequate results when evaluated with a combinatorial problem such as the SCP.

1 Introduction

In recent years, there is a research line of algorithms inspired by natural phenomena to solve optimization problems. As examples of these algorithms we have Cuckoo Search [1], Black Hole [2], Bat Algorithm [3] and Sine Cosine Algorithm [4] among others. Due to the type of phenomena on which these algorithms are inspired, many of these algorithms operate naturally in continuous spaces. On the other hand, the combinatorial problems have great relevance at a scientific and industrial level. We find them in different areas, such as Civil Engineering [5], Bio-Informatics [6], Operational Research [7,8], Big Data [9], resource allocation [10], scheduling problems [11,12], routing problems [13,14] among others. If we want to apply continuous algorithms to combinatorial problems, these algorithms must adapt. In the process of adaptation, the mechanisms of exploration and exploitation of the algorithm can be altered, having consequences in the efficiency of the algorithm.

Several binarization techniques have been developed to address this situation. In a bibliographic search, among the main methods of binarization used

R. Silhavy et al. (Eds.): CoMeSySo 2018, AISC 859, pp. 285–295, 2019.
https://doi.org/10.1007/978-3-030-00211-4_25

are transfer functions, angular modulation, and quantum approximation. In this article, we present a binarization method that uses the percentile concept to group solutions and then performs the binarization process. To verify the effectiveness of our method, we use the Sine-Cosine algorithm. This algorithm was proposed in [4] and applied to the test functions.

To check our binary percentile sine-cosine optimisation algorithm (BPSCOA), we use the well-known set covering problem. Experiments were developed using a random operator to validate the contribution of the percentile technique in the binarization process of the sine-cosine algorithm. In addition, to verify our results, the JPSO algorithm developed by [15] and MDBBH developed in [11] were chosen. The results show that BPSCOA algorithm obtain competitive results.

2 Set Covering Problem

The set covering problem corresponds to a classical problem in combinatorics, and complexity theory. The problem aims to find subsets of a set. Given a set and its elements, we want to find subsets that completely cover the set at a minimum cost. The SCP is one of the oldest and most studied optimization problems. It is well-known to be NP-hard [16].

SCP is an active problem because medium and large instances often become intractable and cannot be solved any more using exact algorithms. Additionally, SCP due to its large number of instances, is used to verify the behavior of proposed new algorithms. In recent years, SCP has been approached by various continuous Swarm intelligence metaheuristics and using different methods to binarize metaheuristics. In [17], they used the teaching-learning-based optimisation with a specific binarization scheme to solve medium and large size instances. An SCP in fault diagnosis application was developed by [18]. In this article, the Gravitational Search algorithm was used and transfer functions were applied to perform the binarization. A rail scheduling problem application using SCP was developed in [11]. In [15] a Jumping PSO was used to solve SCP and in [10] CS metaheuristics were binarized using a percentile algorithm.

SCP has many practical applications in engineer, e.g., vehicle routing, facility location, railway, and airline crew scheduling [7,19–21] problems.

The SCP can be formally defined as follows. Let $A = (a_{ij})$, be a $n \times m$ zero-one matrix, where a column j cover a row i if $a_{ij} = 1$, besides a column j is associated with a non-negative real cost c_j. Let $I = \{1, ..., n\}$ and $J = \{1, ...m\}$, be the row and column set of A, respectively. The SCP consists in searching a minimum cost subset $S \subset J$ for which every row $i \in I$ is covered by at least one column $j \in J$, i.e.,:

$$\text{Minimize } f(x) = \sum_{j=1}^{m} c_j x_j \tag{1}$$

$$\text{Subject to } \sum_{j=1}^{m} a_{ij} x_j \geq 1, \forall i \in I, \text{ and } x_j \in \{0, 1\}, \ \forall j \in J \tag{2}$$

where $x_j = 1$ if $j \in S$, $x_j = 0$ otherwise.

3 Sine-Cosine Algorithm

The standard sine-cosine algorithm (SCA) uses the sine and cosine functions to perform the exploration and exploitation of the search space. SCA is a population-based optimization algorithm. Let X_i be a solution where $i \in \{1, ..., N\}$ and $j \in \{1, ..., D\}$ then each solution starts in as described in Eq. 3

$$x_{i,j} = x_j^{min} + rand \times (x_j^{max} - x_j^{min}) \tag{3}$$

Where, N indicates the number of solutions, D indicates the size of the respective problem. x_j^{min} and x_j^{max} are the minimum and maximum value of the jth component. After initialization, the positions of each solution are updated using Eqs. 4 and 5.

$$X_i^{t+1} = X_i^t + r_1 \times sin(r_2) \times |r_3 P_i^t - X_i^t| \tag{4}$$

$$X_i^{t+1} = X_i^t + r_1 \times cos(r_2) \times |r_3 P_i^t - X_i^t| \tag{5}$$

where X_i^t is the position of current solution at t-th iteration. r_1, r_2 and r_3 are random numbers. P_i is the position of the destination point. These two equations are combined based on the random number r_4. This is shown in Eq. 6:

$$X_i^{t+1} = \begin{cases} X_i^t + r_1 \times sin(r_2) \times |r_3 P_i^t - X_i^t|, & \text{if } r_4 < 0.5 \\ X_i^t + r_1 \times cos(r_2) \times |r_3 P_i^t - X_i^t|, & \text{if } r_4, \geq 0.5 \end{cases} \tag{6}$$

In order to balance exploration and exploitation, the range of sine and cosine in Eqs. 4 and 5 is changed adaptively using the following Eq. 7

$$r_1 = a - t\frac{a}{T} \tag{7}$$

where t is the current iteration, T is the maximum number of iterations, and a is a constant.

4 Binary Percentile Sine-Cosine Algorithm

The application of data mining and machine learning techniques is used in many areas, such as transport [26], smart cities [24,25,25], agriculture [23] and computational intelligence [22]. In this article, we explore the percentile concept applied to the binarization of continuous metaheuristics. SCA is used as the algorithm to binarize. Once binarized the resulting algorithm BPSCA is applied to the combinatorial problem SCP. In the following subsections, the SCA binarization process is detailed.

4.1 Initialization

For initialization of a new solution, a column is randomly chosen. It is then queried whether the current solution covers all rows. In the case of the solution does not meet the coverage condition, the heuristic operator is called (Sect. 4.3). This operator aims to select a new column. This heuristic operation is iterated until all rows are covered. Once the coverage condition is met, the solution is optimized. The optimization consists of eliminating columns where all their rows are covered by more than one column. The detail of the procedure is shown in Algorithm 1.

Algorithm 1. Initialization Operator

1: **Function** Initialization()
2: **Input**
3: **Output** Initialized solution S_{out}
4: $S \leftarrow$ SRandomColumn()
5: **while** All row are not covered **do**
6: $\quad S$.append(Heuristic(S))
7: **end while**
8: $S \leftarrow$ dRepeatedItem(S)
9: $S_{out} \leftarrow S$
10: **return** S_{out}

4.2 Percentile Binary Operator

Since our SCA algorithm is continuous swarm intelligence metaheuristic, it works in an iterative way by updating the position and velocity of the particles in each iteration. As SCA is continuous, the update is done in $\mathbb{R}^n$. In Eq. 8, the position update is written in a general way. The $x(t+1)$ variable represents the x position of the particle at time $t+1$. This position is obtained from the position x at time t plus a Δ function calculated at time $t+1$. The function Δ is proper to each metaheuristic and produces values in $\mathbb{R}^n$. For example in Cuckoo Search $\Delta(x) = \alpha \oplus Levy(\lambda)(x)$, in Black Hole $\Delta(x) = \text{rand} \times (x_{bh}(t) - x(t))$ and in the Firefly, Bat and PSO algorithms Δ can be written in simplified form as $\Delta(x) = v(x)$.

$$x(t+1) = x(t) + \Delta_{t+1}(x(t)) \tag{8}$$

In the percentile binary operator, we considering the movements generated by the algorithm in each dimension for all particles. $\Delta^i(x)$ corresponds to the magnitude of the displacement $\Delta(x)$ in the i-th position for the particle x. Subsequently these displacement are grouped using $\Delta^i(x)$, the magnitude of the displacement. This grouping is done using the percentile list. In our case the percentile list used the values {20, 40, 60, 80, 100}.

The percentile operator has as entry the parameters percentile list (pList) and the list of values (vList). Given an iteration, the list of values corresponds to the magnitude $\Delta^i(x)$ of the displacements of the particles in each dimension. As a first step the operator uses the vList and obtains the values of the percentiles given in the pList. Later, each value in the vList is assigned the group of the smallest percentile to which the value belongs. Finally, the list of the percentile to which each value belongs is returned (pGroupValue). The algorithm is shown in Algorithm 2.

A transition probability through the function P_{tr} is assigned to each element of the vList. This assignment is done using the percentile group assigned to each value. For the case of this study, we particularly use the Step function given in Eq. 9.

$$P_{tr}(x^i) = \begin{cases} 0.1, & \text{if } x^i \in \text{group } \{0,1\} \\ 0.5, & \text{if } x^i \in \text{group } \{2,3,4\} \end{cases} \tag{9}$$

Afterwards the transition of each particle is performed. In the case of SCA search the Eq. 10 is used to perform the transition, where $\hat{x}^i$ is the complement of x^i. Finally, each solution is repaired using the heuristic operator.

$$x^i(t+1) := \begin{cases} \hat{x}^i(t), & \text{if } rand < P_{tg}(x^i) \\ x^i(t), & \text{otherwise} \end{cases} \tag{10}$$

Algorithm 2. Percentile binary operator

```
1: Function percentileBinary(vList, pList)
2: Input vList, pList
3: Output pGroupValue
4: percentileValue = getPValue(vList, pList)
5: for each value in vList do
6:     pGroupValue = getPGroupValue(pValue,vList)
7: end for
8: return pGroupValue
```

4.3 Heuristic Operator

The goal of the Heuristic operator is to select a new column for cases where a solution needs to be built or repaired. As input variables, the heuristic operator considers the incomplete solution S_{in} which must be completed. The operator obtains the columns that belong to S_in, then obtains the rows R that are not covered by the solution to S_in. With the set of rows not covered and using Eq. 11 we obtain in line 4 the best 5 rows to be covered. With this list of rows ($lRows$) on line 5 we obtain the list of the best columns according to the heuristic indicated in Eq. 12. Finally randomly in line 6 we obtain the column to incorporate.

$$WeightRow(i) = \frac{1}{L_i}. \tag{11}$$

Where L_i is the sum of all ones in row i

$$WeightColumn(j) = \frac{c_j}{|R \cap M_j|}. \quad (12)$$

Where M_j is the set of rows covered by Col j.

Algorithm 3. Heuristic operator

1: **Function** Heuristic(S_{in})
2: **Input** Input solution S_{in}
3: **Output** The new column *colOut*
4: *lRows* ← getBestRows(S_{in}, N=10)
5: *listcolumsnOut* ← getBestCols(*LRows*, M=5)
6: *colOut* ← getCol(*lcolsOut*)
7: **return** *colOut*

5 Results

In this section we detail the behavior of BPSCA when it is applied to SCP. The contribution of the binary percentile operator was studied when solving the different SCP instances. Additionally, BPSCA was compared with other algorithms that have recently resolved SCP. To solve the different SCP instances, a PC with Windows 10, core i7 processor and 16GB in RAM was used. The program was coded in Python 2.7. For the statistical analysis, the non-parametric Wilcoxon signed-rank test was used in addition to violin charts. The analysis of violin charts is performed by comparing the dispersion, median and the interquartile range of the distributions.

5.1 Insight of BPSCA Algorithm

In this module, we developed experiments that allowed us to study the contribution of the binary percentile operator with respect to the solutions obtained and the number of iterations used when solving SCP instances. To perform these evaluations, we used the instances of Balas and Carrera. A random operator was designed for the comparison of our binary percentile operator. This random operator uses a fixed value of 0.5 for executing the transitions. To compare the distributions of the results of the different experiments we use violin Chart. The horizontal X-axis corresponds to the problems, while Y-axis uses the measure % - Gap defined in Eq. 13

$$\% - Gap = 100 \frac{BestKnown - SolutionValue}{BestKnown} \quad (13)$$

Furthermore, a non-parametric test, Wilcoxon signed-rank test is carried out to determine if the results of BPSCA with respect to the random algorithm

Table 1. Setting of parameters for binary grasshopper search algorithm.

Parameters	Description	Value	Range
N	Number of solutions	25	[20, 25, 30]
G	Number of percentiles	5	[4, 5, 6]
Iteration number	Maximum iterations	500	[400, 500, 600]

have significant difference or not. The parameter settings and browser ranges are shown in Table 1.

BPSCA corresponds to our standard algorithm. *random.0.5* is the random variant. The results are shown in Table 2 and Fig. 1.

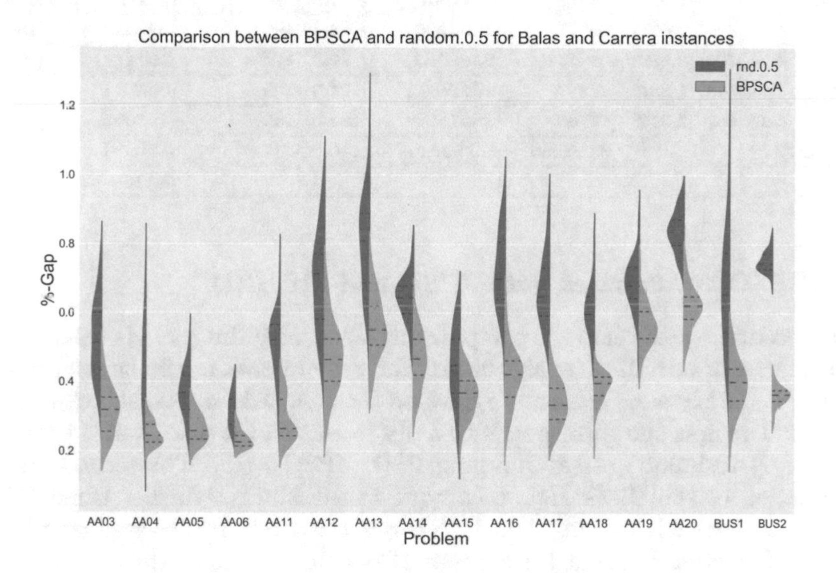

Fig. 1. Evaluation of percentile binary operator.

In Table 2, we compared the BPSCA and *rnd.0.5* . In the Average comparison, BPSCA outperforms *rnd.0.5* in all problems. The comparison of distributions is shown in Fig. 1. We analyzed de the dispersion of the *rnd.0.5* distributions, this is bigger than the dispersions of BPSCA in all problems. Therefore, this brings out the contributions of the percentile binary operator to the precision of the results. On the other hand, the BPSCA distributions are closer to zero than *rnd.0.5* distributions, indicating that BPSCA has consistently better results than *random.0.5.* When we evaluate the behavior of the algorithms through the Wilcoxon test, this test indicates that there is a significant difference between the two algorithms.

Table 2. Balas and Carrera instances.

Instance	Row	Col	Density	Best known	BPSCA (avg)	Std	(Secs) (Secs)	*rnd.0.5* (avg)	Std std
AA03	106	8661	4.05%	33155	**33332.3**	184.1	—196	33693.5	241.1
AA04	106	8002	4.05%	34573	**34665.1**	75.4	178	34931.0	231.2
AA05	105	7435	4.05%	31623	**31701.9**	48.9	238	31889.2	103.2
AA06	105	6951	4.11%	37464	**37489.5**	64.9	187	37651.3	97.1
AA11	271	4413	2.53%	35384	**35579.2**	128.5	231	35901.2	197.1
AA12	272	4208	2.52%	30809	**31245.1**	137.2	179	31631.3	210.2
AA13	265	4025	2.60%	33211	**33812.9**	141.1	199	34201.4	231.4
AA14	266	3868	2.50%	33219	**33659.4**	107.5	201	33967.3	121.1
AA15	267	3701	2.58%	34409	**34688.6**	117.3	210	34846.8	161.3
AA16	265	3558	2.63%	32752	**33116.4**	129.8	265	33601.3	183.9
AA17	264	3425	2.61%	31612	**31820.1**	73.2	204	32386.2	149.3
AA18	271	3314	2.55%	36782	**37128.8**	77.8	236	37417.6	148.1
AA19	263	3202	2.63%	32317	**32916.6**	87.4	199	33109.1	136.1
AA20	269	3095	2.58%	34912	**35693.5**	69.5	179	35963.1	97.8
BUS1	454	2241	1.88%	27947	**28200.1**	127.9	190	28756.1	231.6
BUS2	681	9524	0.51%	67760	**68218.7**	189.6	145	68678.9	101.2
Average				35495.56	35829.26	110.01	202.73	36164.1	165.1
p-value								3.41e−06	

5.2 BPSCA Compared with JPSO and MDBBH

In this section, we develop the comparisons made with the goal of evaluating the performance of our BPSCA algorithm. For the evaluation, the larger problems of the OR-library were chosen. To develop the comparison, two algorithms were selected. The first one corresponds to a discretization Particle Swarm Optimization (PSO) technique called Jumping PSO (JPSO), [15]. The second one is a binarization of the Black Hole technique called Multi Dynamic Binary Black Hole (MDBBH) Algorithm, [11]. JPSO uses a discrete PSO based on the frog jump. JPSO works without the concept of velocity, replacing this one by a component of the random jump which allows performing the movement in the discrete search space. On the other hand, MDBBH uses a binarization mechanism specific to the Black Hole algorithm. This is based on the concept of closeness to the Black Hole (BH). BH corresponds to the solution that has obtained the best value. When a solution is close to BH, the transitions are less likely to be performed. In the case of being away from the BH, the probability of transition is greater.

For instances E and F which are of medium size, the results of the MDBBH and BPSCA algorithms are similar when comparing their Best Value. Only, for instance, E.2 BPSCA was superior to MDBBH. When analyzing the average, JPSO obtained better results, however, the values of BPSCA were close, being MDBBH the one that obtained the worse performance. For the case of problems G and H, BPSCA was superior in problems G.2 and G.4 and H.3 when comparing their Best Value. In the case of averages, JPSO was superior in all cases. However,

BPSCA obtained results quite close, leaving MDBBH behind. In this point, we must emphasize that the percentile technique used in BPSCA allows binarizing any continuous swarm intelligence algorithm, unlike JPSO that is specific for PSO (Table 3).

Table 3. OR-Library benchmarks.

Instance	Row	Col	Density	Best known	JPSO Avg	MDBBH best	Avg	BPSCA best	Avg	Time(s)
E.1	500	5000	10%	29	29.0	29	29.0	29	29.4	11.3
E.2	500	5000	10%	30	30.0	31	31.6	30	30.4	14.4
E.3	500	5000	10%	27	27.0	27	27.4	27	27.2	19.7
E.4	500	5000	10%	28	28.0	28	29.1	28	28.2	13.6
E.5	500	5000	10%	28	28.0	28	28.0	28	28.1	13.9
F.1	500	5000	20%	14	14.0	14	14.1	14	14.0	18.1
F.2	500	5000	20%	15	15.0	15	15.3	15	15.0	17.7
F.3	500	5000	20%	14	14.0	14	14.8	14	15.3	18.9
F.4	500	5000	20%	14	14.0	14	14.9	14	14.2	18.8
F.5	500	5000	20%	13	13.0	14	14.1	14	14.5	21.3
G.1	1000	10000	2%	176	176.0	177	178.5	176	177.5	224.1
G.2	1000	10000	2%	154	155.0	157	160.6	156	155.8	201.7
G.3	1000	10000	2%	166	167.2	168	170.4	168	168.9	188.1
G.4	1000	10000	2%	168	168.2	169	170.9	168	170.1	206.5
G.5	1000	10000	2%	168	168.0	168	169.8	168	169.4	182.7
H.1	1000	10000	5%	63	64.0	64	64.9	65	65.7	202.7
H.2	1000	10000	5%	63	63.0	64	64	64	64.7	214.6
H.3	1000	10000	5%	59	59.2	59	60	59	60.5	185.7
H.4	1000	10000	5%	58	58.3	59	60.4	59	59.4	184.3
H.5	1000	10000	5%	55	55.0	55	56.4	55	55.8	167.9
Average				67.1	67.30	67.7	68.71	67.55	68.21	106.30

6 Conclusions

In this work, the percentile concept was used to binarize the SCA metaheuristic. The percentile concept can be applied in the binarization of any continuous metaheuristics. Experiments were designed to evaluate the contribution of the binary percentile operator, obtaining as a result that the operator contributes significantly to improve the accuracy and quality of the solutions. Later we compared BPSCA with the best metaheuristic algorithm that has solved SCP, our algorithm had a performance lower than 1.23 %, which is not a big difference considering that JPSO uses a particular adaptation mechanism for PSO.

As future works, we believe that the integration between machine learning techniques with metaheuristic is an area little explored and could help to improve binarization performance both in the quality of the solutions and in the convergence times. additionally, we are interested in exploring the binarization of another metaheuristic with the percentile technique. Finally, it is interesting to generate descriptors that allow us to understand how the percentile algorithm translates the exploration and exploitation properties of the continuous space into the binary space in order to better understand the binarization process.

References

1. Yang, X.S., Deb, S.: Cuckoo search via lévy flights. In: Nature Biologically Inspired Computing NaBIC 2009 World Congress on 2009, pp. 210–214, IEEE (2009)
2. Hatamlou, A.: Black hole: a new heuristic optimization approach for data clustering. Inf. Sci. **222**, 175–184 (2013)
3. Yang, X.S.: A new metaheuristic bat-inspired algorithm. In: Nature Inspired Cooperative Strategies For Optimization (NICSO 2010), pp. 65–74 (2010)
4. Mirjalili, S.: SCA: a sine cosine algorithm for solving optimization problems. Knowl.-Based Syst. **96**, 120–133 (2016)
5. Khatibinia, M., Yazdani, H.: Accelerated multi-gravitational search algorithm for size optimization of truss structures. Swarm Evol. Comput. **38** 109–119 (2017)
6. Barman, S., Kwon, Y.-K.: A novel mutual information-based boolean network inference method from time-series gene expression data. Plos One **12**(2), e0171097 (2017)
7. Crawford, B., Soto, R., Monfroy, E., Astorga, G., García, J., Cortes, E.: A meta-optimization approach for covering problems in facility location. In: Workshop on Engineering Applications, pp. 565–578, Springer (2017)
8. Crawford, B., Soto, R., Astorga, G., García, J.: Constructive metaheuristics for the set covering problem. In: International Conference on Bioinspired Methods and Their Applications, pp. 88–99, Springer (2018)
9. García, J., Altimiras, F., Fritz, A.P., Astorga, G., Peredo, O.: A binary cuckoo search big data algorithm applied to large-scale crew scheduling problems. Complexity **2018**, 15 (2018)
10. García, J., Crawford, B., Soto, R., Astorga, G., A percentile transition ranking algorithm applied to knapsack problem. In: Proceedings of the Computational Methods in Systems and Software, pp. 126–138, Springer (2017)
11. García, J., Crawford, B., Soto, R., García, P.: A multi dynamic binary black hole algorithm applied to set covering problem. In: International Conference on Harmony Search Algorithm, pp. 42–51, Springer (2017)
12. García, J., Crawford, B., Soto, R., Astorga, G., A percentile transition ranking algorithm applied to binarization of continuous swarm intelligence metaheuristics. In: International Conference on Soft Computing and Data Mining, pp. 3–13, Springer (2018)
13. Franceschetti, A., Demir, E., Honhon, D., Van Woensel, T., Laporte, G., Stobbe, M.: A metaheuristic for the time-dependent pollution-routing problem. Eur. J. Oper. Res. **259**(3), 972–991 (2017)
14. Crawford, B., Soto, R., Astorga, G., García, J., Castro, C., Paredes, F.: Putting continuous metaheuristics to work in binary search spaces. Complexity **2017**, 19 (2017)
15. Balaji, S., Revathi, N.: A new approach for solving set covering problem using jumping particle swarm optimization method. Nat. Comput. **15**(3), 503–517 (2016)
16. Gary, M.R., Johnson, D.S.: Computers And Intractability A Guide To The Theory of NP-Completeness. W. H. Freeman and Company, New York (1979)

17. Lu, Y., Vasko, F.J.: An or practitioner's solution approach for the set covering problem. Int. J. Appl. Metaheuristic Comput. (IJAMC) **6**(4), 1–13 (2015)
18. Li, Y., Cai, Z.: Gravity-based heuristic for set covering problems and its application in fault diagnosis. J. Syst. Eng. Electron. **23**(3), 391–398 (2012)
19. Kasirzadeh, A., Saddoune, M., Soumis, F.: Airline crew scheduling: models, algorithms, and data sets. Euro J. Transp. Logistics **6**(2), 111–137 (2017)
20. Horváth, M., Kis, T.: Computing strong lower and upper bounds for the integrated multiple-depot vehicle and crew scheduling problem with branch-and-price. Cent. Eur. J. Oper. Res. **8**, 1–29 (2017)
21. Stojković, M.: The operational flight and multi-crew scheduling problem. Yugoslav J. Oper. Res. **15**(1) (2016)
22. García, J., Crawford, B., Soto, R., Carlos, C., Paredes, F.: A k-means binarization framework applied to multidimensional knapsack problem. In: Applied Intelligence, pp. 1–24 (2017)
23. García, J., Pope, C., Altimiras, F.: A distributed k-means segmentation algorithm applied to lobesia botrana recognition. Complexity **2017**, 14 (2017)
24. Graells-Garrido, E., García, J.: Visual exploration of urban dynamics using mobile data. In: International Conference on Ubiquitous Computing and Ambient Intelligence, pp. 480–491, Springer (2015)
25. Graells-Garrido, E., Peredo, O., García, J.: Sensing urban patterns with antenna mappings: the case of santiago, chile. Sensors **16**(7), 1098 (2016)
26. Peredo, O.F., García, J.A., Stuven, R., Ortiz, J.M.: Urban dynamic estimation using mobile phone logs and locally varying anisotropy. In: Geostatistics Valencia 2016, pp. 949–964, Springer (2017)

Towards the Methodology for the Implementation of Sales: Forecast Model to Sustain the Sales and Operations Planning Process

Francisco Edson Santos Farias, Igor de Magalhães Carneiro Nascente, Milton Carlos de Almeida Lima, Plácido Rogerio Pinheiro(✉), and Tulio Fornari

Administration Department, University of Fortaleza, Fortaleza, Brazil
edson.magis@hotmail.com, igornascente@gmail.com, miltonclima@hotmail.com, placido@unifor.br, fornaritulio@yahoo.com.br

Abstract. In the current context change is part of most companies, so companies are in need of adjusting the speed of these changes and often anticipate them. It is timely the need for companies to seek methodologies, tools and techniques to achieve the objectives constituted by the company's strategy. The use of new methodologies and techniques has shown an abundance of data and information, which most of the time does not have an in-depth study and are often analyzed in an unaccompanied way of other information that end up leading to partial and often biased decisions, influencing other sectors and shaking the company's strategy. A methodology that proposes to mitigate these planning dysfunctions in the organization is "Sales and Operations Planning" or S&OP (Sales and Operations Planning). The objective of this work is to present the steps that correspond to the structure of the S&OP, with emphasis on a modeling proposal for the sales forecasting process, which is extremely critical to the entire S&OP process because it involves studies of mathematical models.

Keywords: Sales forecasting · Strategic planning and S&OP

1 Introduction

According to Silva (1996), the new world economic scenario and the growing awareness of the Brazilian people have forced Brazilian companies to review their position vis-à-vis consumers and society.

Organizations are pressured from all sides and compete for survival. Some organizations expect changes to occur in return, others anticipate change.

Nadler (1993) apud Monteiro (2003) point out that these pressures have been occurring since the 1980s, due to radical and continuous changes in the organizational environment to be highlighted: (i) the increasingly fierce competition with the advent of Internet companies do not only compete at regional levels, but at global levels, making them seek more and more industrial excellence, (ii) with accelerated technological

R. Silhavy et al. (Eds.): CoMeSySo 2018, AISC 859, pp. 296–306, 2019.
https://doi.org/10.1007/978-3-030-00211-4_26

innovation, companies are increasingly able to bargain by acquiring new equipment, (iii) new technologies, making the creation of methods and products constantly and dynamically always aiming for greater market share, (iv) excess supply on a worldwide basis, causing more suppliers than buyers worldwide.

In this new context, business success depends exclusively on a rapid response capability, whether small, medium or large.

According to Slack et al. (2002), "the world is a smaller place to do business", thus, it is suggested below the challenges of production in which companies have to fit, to avoid their decay in the globalized market: (a) Globalization (b) Globalization and operations decisions, (c) Ant globalization Movement, (d) International Location, (e) Headquarters Configuration, (f) Regional Configuration, (g) Global Coordinate Configuration, (h) Social Responsibility, (i) Environmental Responsibility and (j) Technology.

In Brazil since the 1990s, pressured mainly by the fiercest competition generated by the country's economic opening, the organizations installed here have been undergoing profound changes aimed at improving their quality of processes, products and services aiming at a growing competitiveness.

According to Kotter (2015), organizations around the world are finding it hard to keep up with the accelerating pace of change, and much more to get ahead of them.

Faced with this increasingly unpredictable and competitive world, it is increasingly difficult to grow and maintain the profitability of companies in a competitive way (Kotter 2015).

One of the ways companies seek to achieve this growing competitiveness, in line with the speed demanded by the market, and sometimes even to overcome difficulties is to ensure that everything that has been decided strategically, with short, medium and long term prospects through operational decisions.

The results of companies are influenced by the links between departments of the companies, mainly through the functions of production and sales. The first connection between the company and the market is the information on sales order or sales forecast that is calculated on the customer or the corresponding market. Waddell and Sohal (1994) calls this action "Sales Forecasting". An appropriate sales forecasting model for each company is imperative for effective targeting and is an important tool for decision making. So, S&OP is, in particular, a structured methodology, where the information collected by the different departments (sales, production and engineering) is discussed, analyzed, and in the end, the plans of each department are aligned.

In this article, it is proposed the structuring of a forecasting sales model inserted in the S&OP technique. The main objective is to deepen the quantitative techniques of demand forecasting (here called forecasting techniques) in support of management decision making inserted in the context of the S&OP technique. The proposed methodology aims to describe a proposal to build a forecast sales model, based on quantitative forecasting techniques. These structures are being applied as a decision base for a company in the food industry.

The proposed methodology brings as main contributions the steps that are necessary as well as the motivations to create an S&OP, in parallel it brings a structuring context of a forecasting system due to the quantity of sku's and its logistic complexity.

The definitions of S&OP as well as their applications are numerous in the context of manufacturing, Côrrea (2001) points out that S&OP "*seeks to identify from a horizon of the future, along with current knowledge and facts, factors that can influence the decisions that are being taken now and that aim at the achievement of certain objectives are in the short, medium and long term, being characterized by monthly reviews and monthly and continuous in the light of fluctuations in market demand*" Fig. 1.

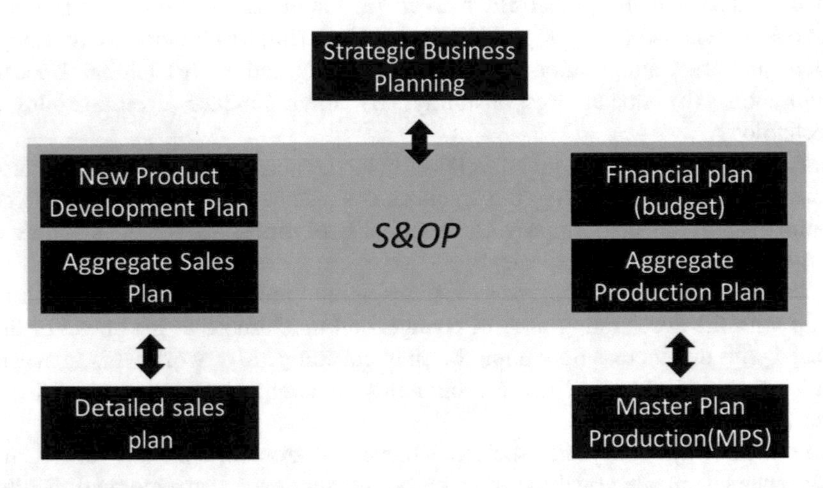

Fig. 1. S&OP in the global planning process Côrrea (2001)

One of the objectives of the S&OP implementation process is to minimize the actions resulting from decision-making vacuums, that is, sectors make decisions that directly impact other sectors and are not aware of these impacts, so S&OP is a means of orchestrating all departments.

Mac Gougan (2003), lists some points that should be avoided:

(a) Change of culture: first define capacities and resources;
(b) Integrity of the information: to know the relevant facts and not the details;
(c) Forecast at a high level: basis of forecast in quantity and values grouped and not in daily items;
(d) Limitation of systems: Process outputs are normally spreadsheets and decision-making graphs, which cannot serve as inputs for integrated MRP (Materials Requirement Planning) systems.

2 Methodology for Implementation of S&OP

Wallace (1999) suggests that for the implementation of the S&OP process the steps should be followed with discipline. It is interesting to establish a cycle of meetings that encompasses the company's planning process, Wallace (1999) exemplifies in 5 steps

(Fig. 2), steps that must be managed by multifunctional groups that must be composed of people from the areas of: Production, Sales and Marketing, Customer Service, Materials Management, Research and Development, Product Engineering, among others.

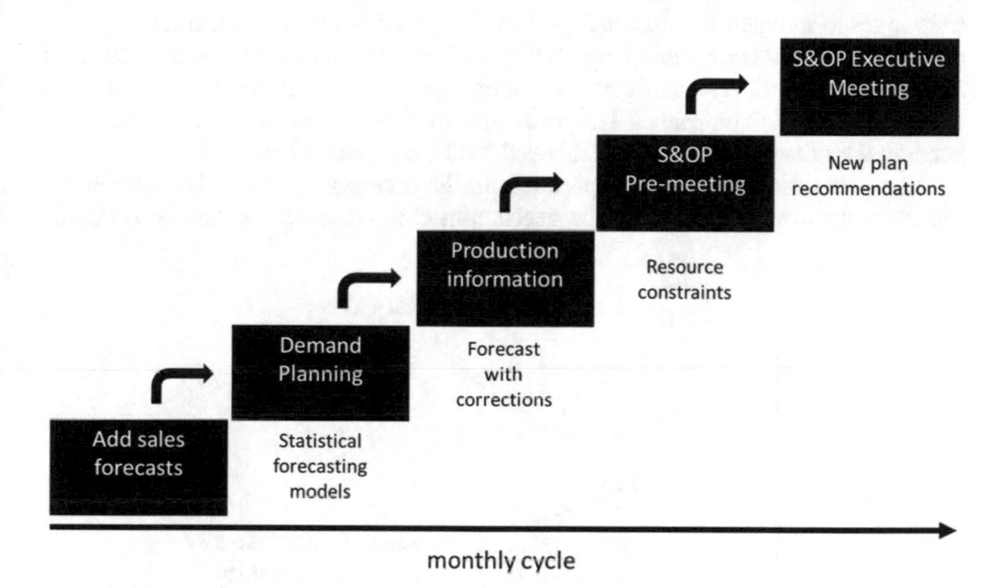

Fig. 2. Stages of construction of an S&OP model adapted by the author.

It is noticed that the first step proposed by Wallace (1999) is to aggregate the sales forecasts, so the creation of a forecasting system is fundamental to initiate a process of construction of S&OP.

It is understood that some companies consider implementing an S&OP system, however, it is necessary to have an established forecasting model.

3 Case Study

The object of study is a national company, with headquarters in the Northeast in the food, founded in the 60's, leader in the national market with 5 factories, 24 logistics centers, with approximately 550 sku's with sales in more than 300,000 establishments with approximately 4,100 employees with family management.

Despite major investments in technology and processes, the company faces major challenges such as: (i) significant increase in sales volume, (ii) acquisition of new businesses, (iii) introduction of new brands, (iv) new products and channels (v) increase in the number of factories and distribution centers and (vi) performance in new markets/regions.

In view of these scenarios over time, there have been some undesirable effects to be highlighted: (a) high deviations in sales forecasts, (b) lack and overshoot of products,

(c) customer order cuts (d) use of overtime and downtime to set up in the industry, (e) lack of alignment between areas, (f) consensus of root causes of problems, (g) discussion of areas with focus on justify inefficiencies, not treat them.

Such scenarios require a change approach and the proposal is to use the S&OP system as well as the implementation of a sales forecasting system based on forecasting techniques to mitigate the risks and problems exposed in the above items.

According to Mentzer and Cox (1997), qualitative methods are the most used for forecasting demand, although they have a low degree of precision; despite this, continue to be used in companies. The qualitative methods are related to the predictions generated by them to the goals established by the companies Dias (1999).

The cost of forecasting is proportional to the accuracy required. The greater the accuracy, the lower the losses of the decision-making processes, according to Fig. 3.

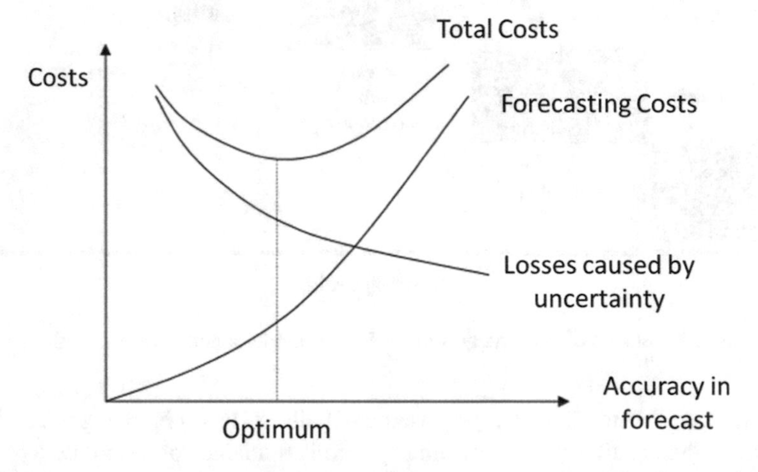

Fig. 3. Relation between accuracy and cost of forecasting. (Adapted from Montgomery et al. 1990).

According to Pellegrini (2001) a forecasting system requires a lot of knowledge and skill in 4 basic areas: (i) identification and definition of the problems to be dealt with in forecasting; (ii) application of forecasting methods; (iii) procedures for selecting the appropriate method for specific situations; and (iv) organizational support to adapt and use the required forecasting methods.

Makridakis et al. (1998) points out that a forecasting system depends on three conditions: (1) Information history; (2) the possibility of transforming this history into data; (3) assumption of the repetition of patterns observed in past data in the future time. Such conditions are considered essential for the use of a quantitative forecasting method.

3.1 Structuring the Forecasting System for a Company Studied

It is intended from the proposed structuring that this model allows the choice of the model for each type of product.

Due to the importance given by the company's management, it is considered appropriate to create a sales forecasting system allied to the S&OP system. The current forecasts are made based only on the opinion and sentiment of the managers and directors of sales, not using any statistical tool to do so. Therefore, it is suggested from Fig. 4, a step-by-step for the implementation of forecasting techniques in the company, with the aim of increasing the accuracy of forecasts, enabling the following objectives to be achieved: (i) higher forecast accuracy (ii) reduction of sales losses, (iii) integration between areas, (iv) assertiveness and speed of decision making.

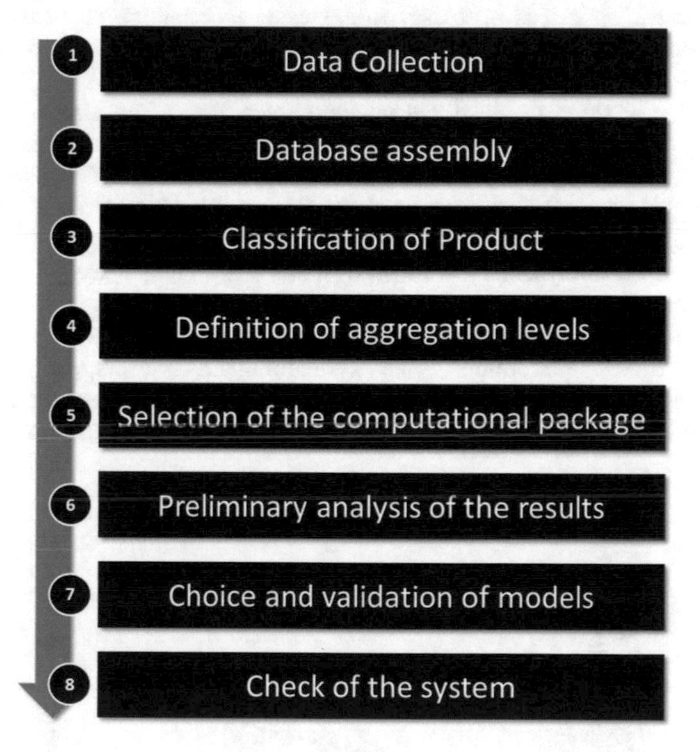

Fig. 4. Proposal of steps to build a forecasting system (Adapted by the author).

Figure 4 presents the 8-step proposal for the construction of a forecasting system where:

(1) **Data collection:** Makridakis et al. (1998) defined data collection in 2 slopes (1) numerical data, (2) subjective data from the perceptions and opinions of people involved in the process.
(2) **Database assembly:** It is suggested that the data be stored in easily accessible programs with easy reading comprehension, preferably in spreadsheets such as excel, or some similar program, the purpose of this step is to become the best data possible for analysis.

(3) **Classification of products:** Correa and Gianesi (2006) defined as product classification the division into 3 groups (A, B and C) to differentiate the products so that they have different controls and treatments, the products belonging to the category A products account for 80% of turnover and 20% of products sold by the company, category B products account for 15% of turnover of 30% of products sold by the company and products of category C represent 5% of turnover and 50% of products sold by the company. Such classification is extremely important to define particular strategies for each product belonging to each category.

(4) **Definition of aggregation levels:** it is often difficult to verify the behavior of a time series, so if it is necessary to aggregate its temporal elements to verify a behavior characteristic of the sample, failure to perform this step may present some in the modeling of data.

(5) **Selection of the computational package:** regardless of the size of the company in any occasion when talking about statistical calculations is very complex and if it is necessary to use a computational tool, the choice of the computational package is extremely important in view of the need to if you have reliable data and at an appropriate speed for the theme.

(6) **Preliminary analysis of the results:** In this step, the data are presented graphically to identify possible apocryphal values in the time series, which would cause a problem in the modeling. Such apocryphal values are caused by market variations, erroneous promotions, typos, system integration errors, etc.

(7) **Choice and validation of models:** Makridakis et al. (1998) suggests that the choice of model should be based on the following factors: (a) Aspects that can

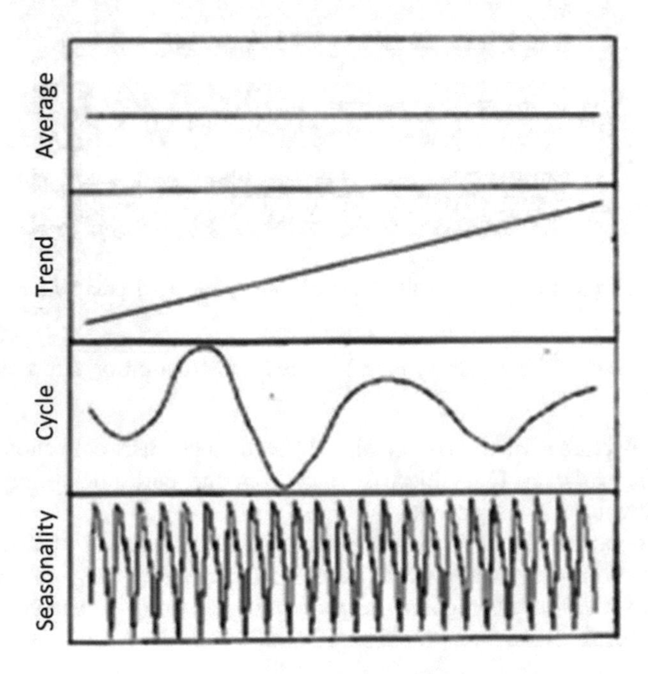

Fig. 5. Characteristics of a time series. (Adapted from Makridakis et al. 1998).

influence demand (promotions or promotional campaigns), (b) Characteristics of the time series, (c) Time aggregation of data, (d) Forecast interval.

(8) **Check of the system:** having the models and their parameters known satisfactorily, their use in forecasting future demand can be tested.

3.2 Main Models Used as Qualitative Methods for Forecasting

Demand forecasting can and should be done using mathematical models. The use of the model basically obeys the reaction of the time series to be analyzed. A temporal series can expose up to four different characteristics in its behavior: average, seasonality, cycle and trend Makridakis et al. (1998). These characteristics are exemplified in Fig. 5.

In the Table 1 presents some models used as quantitative methods for forecasting demand.

Table 1. Model and main feature - adapted by the author

Model	Main feature
Holt's model	Time series with linear trend
Winters models	Linear trend, plus seasonality
Seasonal models	Occurs when the series exhibits a periodic characteristic that repeats at each time intervals
Multiplicative seasonal model	Amplitude of the seasonal cycle varies over time
Seasonal additive model	The seasonal cycle amplitude is constant over time
Auto correlation	Describes the correlation between two values of the same time series at different time periods
Box-Jenkins models	The values of a time series are highly dependent
Stochastic and deterministic models	Is characterized by a family of random variables that describe the evolution of some phenomena of interest

4 Preliminary Analysis

The company counts its demand data weekly although its production planning is done monthly. The forecasting horizon considered ideal by the company is 12 weeks. It is understood that this horizon, the company would be able to adapt to fluctuations in the demand of the products it markets, with the maturity of the forecasting system the company intends to transform this scenario if expanded in more periods as Fig. 6.

To adapt the model was proposed through Table 2, indicating the main changes that must occur from the current process to the forecasting sales forecasting process.

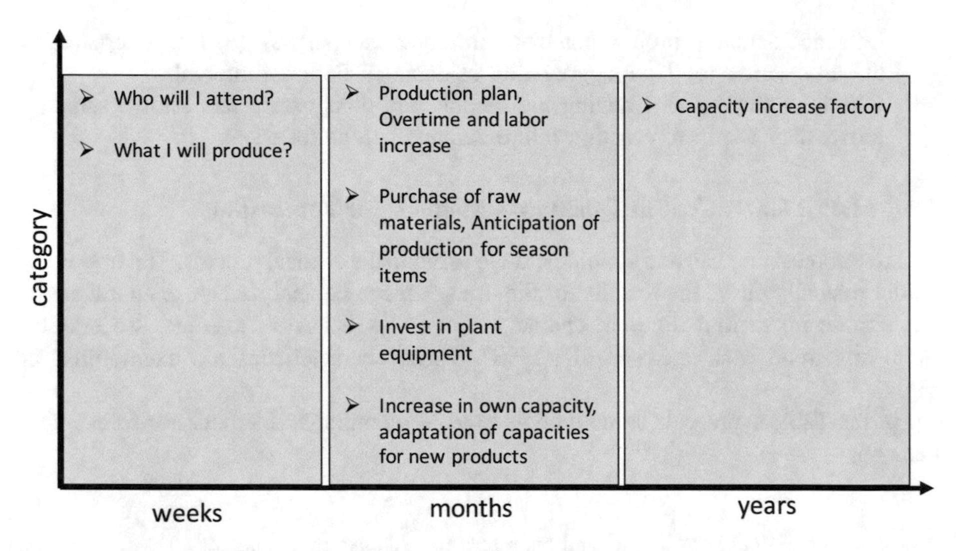

Fig. 6. Proposal of scenario for forecasting period

Table 2. Main changes of the current model × forecast forecasting model

Sub process	Main change	Current situation	Future situation	Benefits
General	Increased planning horizon	Planning for the next 3 months (focus on the current month)	Planning for the next 3–12 months (focus next month)	Can prepare for • New products • Promotional actions • Imported purchases • Seasonality
Forecast	Use of statistical methods according to the demand profile	Use of the three-month moving average	Use of classic decomposition and Holts method, as well as clearing outliers in historical	• Identification os seasonality and trends • Improves the baseline for planning
Plan for sales	Integration of marketing and trade into planning	Communication of marketing actions and new products in a disorganized way and at the time	Integrate marketing actions and new products into S&OP meetings and measure their impacts on the plan	Anticipate and amplify the preparation for promotional actions, product launches and discontinuations
	Integration in the number of the Budget × Plan of sale × Supply	Numbers are different in each process	Leverage existing sales meetings and integrate number with S&OP	Have the correct reading of the market and communicate this number to prepare the chain
	Use of sales-relevant information in planning execution	Not used in a structured way	Include in the file • Critical statistics and SKUs • Planning error • Market potential • Costs and margin for coffee	• Plan focus of action in a structured way • Focus on the most critical products

5 Conclusion

Despite their practical importance, forecasting methods are unprecedented by a large number of Mentzer and Cox companies (1997). This fact, although it expresses the Brazilian reality, does not represent what has been done in other countries, where these techniques are well known, including in the service sectors Winston (1994).

The proposal of structuring a forecasting process can efficiently assist in the improvement of the forecasting process, preparing the other processes of the company allowing a greater integration between sectors.

It is important to stress that forecasting methods can generate a competitive value for companies based on a defined process expectation, that is, they should not be seen in isolation, but in concomitant with other processes, especially S&OP.

It is important to highlight that the achievement of a model with a good understanding of the data does not always imply in obtaining a satisfactory prediction of the demand of the modeled product. The article presented, besides exemplifying the proposed methodology, highlights the importance of managerial analysis in the preliminary analysis stage.

Acknowledgment. The author's thanks to the National Council for Technological and Scientific Development (CNPq) through grants #304747/2014-9, Edson Queiroz Foundation/University of Fortaleza and FINEP (Funding Agency for Studies and Projects) for all the support provide.

References

da Silva, J.M.: The Environment of Quality in Practice-5S. Christiano Ottoni Foundation, Belo Horizonte (1996)

Slack, N., et al.: Production Administration, 2nd edn. Atlas S/A, São Paulo (2002)

Kotter, J.: Accelerate: Have Strategic Agility in a World in Constant Transformation (John P. Kotter, translated by Cristina Yamagami). HSM Editora, São Paulo (2015)

Waddell, D., Sohal, A.S.: Forecasting: the key to managerial decision making. Manag. Decis. **32** (1), 41–49 (1994)

Mac Gougan, G.: S&OP for top management in a small business. In: International Conference Proceedings, APICS, F-06, pp. 1–4 (2003)

Wallace, T.F.: Sales and Operations Planning. T. F. Wallace & Company, Cincinnati, Ohio (1999)

Mentzer, J.T., Cox Jr., J.E.: Familiarity, application, and performance of sales forecasting techniques. J. Forecast. **3**(1), 27–37 (1997)

Dias, G.P.P.: Proposed Sales Forecasting Process for Consumer Goods. Annals of the XIX ENEGEP - CD-ROM, Rio de Janeiro (1999)

Makridakis, S., Wheelwright, S.C., Hyndman, R.J.: Forecasting - Methods and Applications, 3rd edn. John Wiley, New York (1998)

Winston, W.L.: Operations Research: Applications and Algorithms, 3rd edn. Duxbury Press, Belmont, CA (1994)

Monteiro, W.R.: Implementation of Business Process Reengineering: Case Study of Organizations in Brazil 2003. 137 f. Thesis (Master in Administration) - Faculty of Economics, Administration and Accounting, Department of Administration, University of São Paulo (USP), São Paulo (2003)

Côrrea, H.L.: Planning, programming and production control: MRPII/ERP: concepts, use and implementation. In: Corrêa, H.L., Gianesi, I.G.N., Caon, M. (eds). 4. Atlas, São Paulo (2001)
Pellegrini, F.R.: Methodology for the Implementation of Demand Forecasting Systems. 146 f. Thesis (Master in Production Engineering) - Federal University of Rio Grande do Sul School of Engineering Post-Graduation Program in Production Engineering, Porto Alegre (2000)

The Quantum Computer Model Structure and Estimation of the Quantum Algorithms Complexity

Viktor Potapov(✉), Sergei Gushanskiy, Alexey Samoylov, and Maxim Polenov

Department of Computer Engineering, Southern Federal University, Taganrog, Russia
vitya-potapov@rambler.ru,
{smgushanskiy, asamoylov, mypolenov}@sfedu.ru

Abstract. Present paper describes the basics of quantum algorithms development and modeling of entangled quantum computations used in quantum algorithms. Since quantum algorithms assume the use of vector and matrix algebras, so the suggested structure of the quantum computer model (QCM) considers this specialty and reflects all functional features of such model. The basic tasks of the proposed model simulation are determined within the quantum algorithms execution taking into account the entanglement property. In accordance with the constructed system for determining the computational complexity of quantum algorithms, the entire set of necessary elements is shown: the time complexity, the number of operations and queries to the quantum oracle, the complexity class of the quantum algorithm.

Keywords: Qubit · Quantum register · Quantum computer model
Algorithms complexity · Quantum circuit

1 Introduction

Recently an interest in quantum computers has grown rapidly, especially after the appearance of operating quantum computers. The using of quantum computers makes it possible to significantly increase the speed of solving computational problems and, most important, to exponentially increase the speed of solving NP-complete problems [1], which can be solved on classical computers for unacceptable time. The study of entanglement as a property is one of the most important areas of research in the field of the theory of quantum information and quantum computing because the entanglement concept is a fundamental in applied algorithms. Entanglement is regarded as one of the main differences between quantum mechanical systems and classical ones, making the first interesting from the point of view of applications in the information processing and quantum communications.

Quantum computers (QC) can solve problems that are beyond the power of classical ones. These QC are not tools of improving the performance of any applications, however,

R. Silhavy et al. (Eds.): CoMeSySo 2018, AISC 859, pp. 307–315, 2019.
https://doi.org/10.1007/978-3-030-00211-4_27

one can distinguish the areas in which they could make a revolution. Among such areas there are quantum chemistry, material sciences, machine learning and cryptography.

Experimental devices containing about 50 qubits will be soon available on the market and will be able to solve well-defined computational tasks, which would require the most powerful classical supercomputers. However, in order not to repeat the mistakes of the past in the field of computing systems, it is already necessary at this particular preliminary stage to determine the optimization, theoretical and practical issues facing quantum computing devices and their components. Namely, quantum circuits, the optimization of the distribution of gates and qubit, which will reduce the number of links in the multi-node configuration, and the number of k-qubit cores with gates at the level of one node. An important advantage is also the reduction of the clock cycle of the quantum circuit.

2 Low-Level Random Quantum Schemes

A quantum circuit is a sequence of physical transformations from a finite set of basic elementary transformations – gates. At the input, the quantum circuit gets quantum bits and it has the probabilistic result. Consequently, any transformation is uniquely defined by values on the basis states and the k quantum bits transformation can be written in the form of a $2^k \times 2^k$ matrix.

Although one of the computational tasks proposed to demonstrate the computational power of a quantum computer − the execution of low-level random quantum circuits – is not scientifically useful itself, its implementation schemes are still very useful for calibrating, checking and testing temporary quantum devices. The low-level random quantum scheme, which is shown in Fig. 1, demonstrates quantum superiority [2]. The identical schemes are created using the following rules. A Hadamard gate is applied to each qubit at 0 time step. As a consequence, eight different templates (6×6 qubits) of controlled Z (CZ) gates are applied repeatedly until the desired depth of the circuit is reached. The CZ-gates, which are logical two-qubit gates and described by Pauli matrices, are represented by lines between two qubits (further we will identify such line with a node notion). This template ensures that all possible qubit interactions in this two-dimensional nearest neighbor architecture are performed every 8 cycles. Single qubit gates are selected randomly.

Therefore, in addition to testing of the quantum algorithms and conducting a study of their operation in the presence of noise, the model of the quantum circuit can provide the means for conducting calibrations and control measurements that allow obtaining the most productive computing structure. Quantum simulators [3] are comparable with the tools of structural modeling, allowing to implement the simulation of the classical calculator. The simulator of the quantum calculator developed earlier was realized and optimized through a multilevel approach.

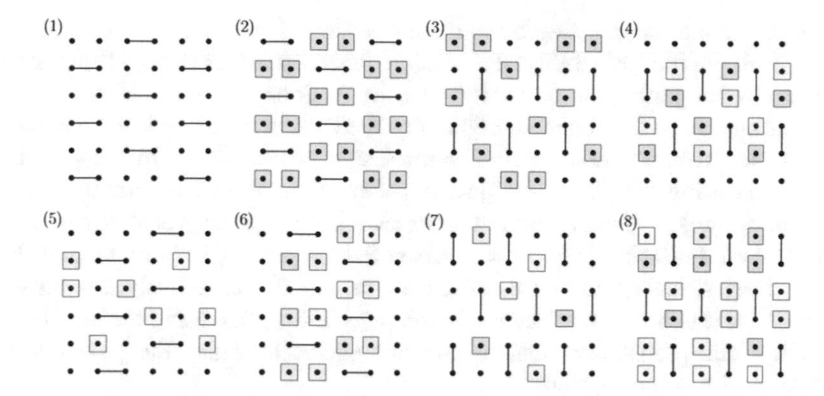

Fig. 1. Low-level random quantum scheme

3 The Structure of the Quantum Computer Model

The structure of the QCM, reflecting all its functional features, advantages and disadvantages is shown in Fig. 2.

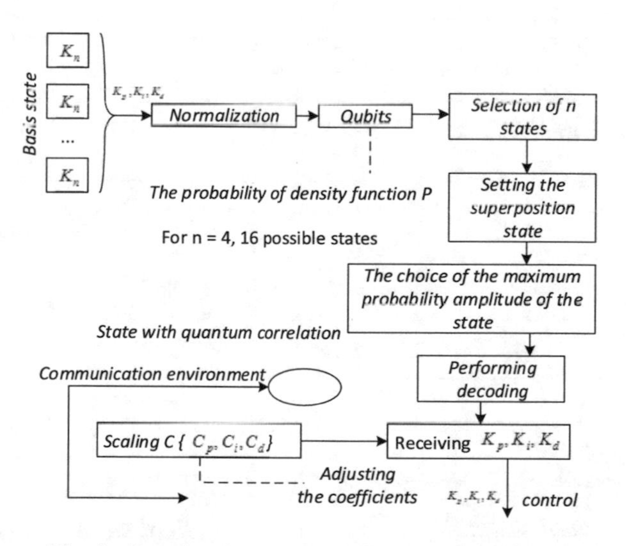

Fig. 2. Functional structure of the QCM in real-time

A control signal, generated from the basis state formed previously by the user, is entered into the QCM in real-time. At the next step of the entangled quantum computations simulation [4] the procedure for normalizing the signals, obtained at the previous stage is realized within the interval [0, 1], by separating the amplitudes of the trajectory of the control signals by the maximum amplitudes. The process of forming a set of qubits from the current values of the normalization control signals occurs after signal normalization. This process requires to determine the probability density function at the preliminary step by random selection of the trajectories of the control signals. Further, the probability distribution functions of an integral character are calculated by integrating the probability density function obtained earlier.

The advantage of the obtained probability distribution functions is the possibility of selection the "virtual" states $|1>$ of control signals capable of forming a state of superposition using the Hadamard gate transformations from the current state of the input control signals. These transformations are made using the probability distribution law of the form $P(|0>) + P(|1>) = 1$, where $P(|0>)$ and $P(|1>)$ are the probabilities of the real and virtual current states of the control signals, respectively. For the current real state $|1>$ it is possible to determine the probability value using the distribution of probability. The probability conservation law allows to define the probability of a virtual state of a control signal.

The computational process in the simulation of entangled quantum computations is shown in Fig. 3. It implies the use of various kinds of quantum computations described in the paper and the generation of qubits set for the quantum state at a particular time.

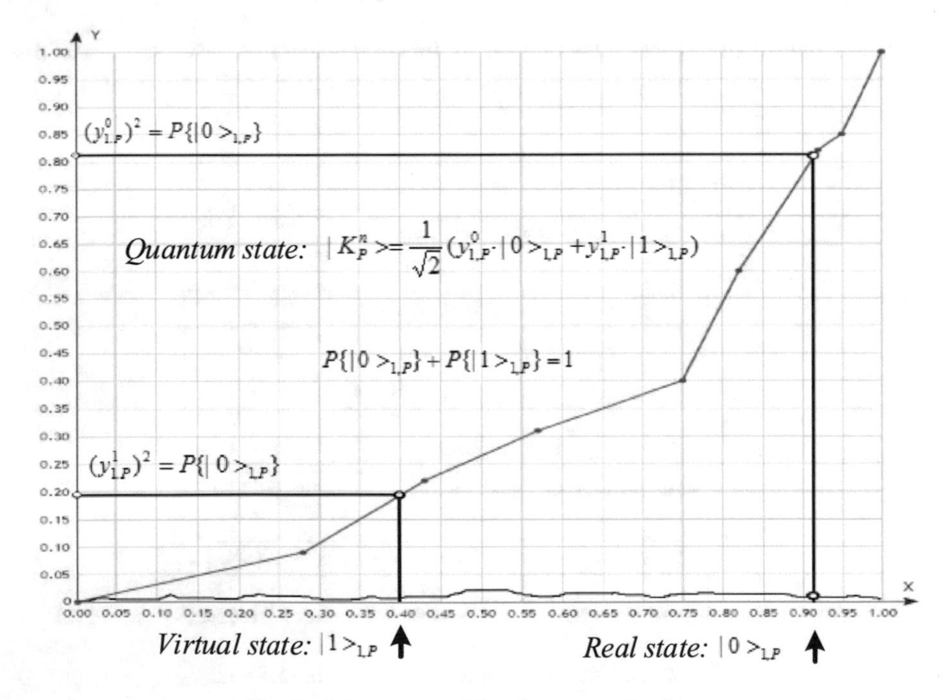

Fig. 3. The process of forming a set of qubits

Applying the tensor product between the Hadamard transformations [5], we obtain terms of the form $K^{n}_{P} \otimes K^{n_2}_{D}$ and analogous combinations of the gain factors. The example described above indicates the existence of sixteen probabilistic states that define variations among correlations in accordance with their type and law. Now we illustrate the process of calculating the amplitudes of quantum states under conditions of superposition of states with mixed types of quantum correlation and selection the maximum among them. Here, the process of choosing a quantum state is realized using the principle of the maximum value of the probability amplitude (the minimum of Shannon information entropy).

Quantum state of superposition:

$$|K_P^n > \otimes |K_P^{n_2} > \otimes |K_D^n > \otimes |K_D^{n_2} > = \frac{1}{\sqrt{2^n}}(\alpha_1 \cdot |0000\rangle + \alpha_2 \cdot |0001\rangle + \ldots + \\ + \alpha_{2^n-1} \cdot |1110\rangle + \alpha_{2^n} \cdot |1111\rangle), \tag{1}$$

where $|0001\rangle$ is the quantum correlation, α_{2^n-1} – the probability amplitude, $\alpha_{2^n} \cdot |1111\rangle$ is the classical state. Interference [6], selected state: $\alpha_k = |a_1, a_2, a_3, a_4 >$, oracle:

$$\begin{cases} \alpha_1 = y_{1,P}^0 \cdot y_{2,P}^0 \cdot y_{1,D}^0 \cdot y_{2,D}^0 \\ \alpha_2 = y_{1,P}^0 \cdot y_{2,P}^0 \cdot y_{1,D}^0 \cdot y_{2,D}^1 \\ \ldots \\ \alpha_{16} = y_{1,P}^1 \cdot y_{2,P}^1 \cdot y_{1,D}^1 \cdot y_{2,D}^1 \end{cases}$$

The iterative process of the QCM is performed using the standard decoding procedure identical to the inner product of vectors in the Hilbert space and choosing the scaling factors for the output values of the projected gain factors, described in the "Receiving K_p, K_i, K_d" block in Fig. 1.

4 Estimation of the Complexity of Quantum Algorithms by the Complexity Function

The algorithm complexity can be considered as the number of "elementary" operations performed by the algorithm to solve a particular problem in a given formal system. When using algorithms [7] to solve the practical problems, we are faced with the problem of rational choice of the algorithm for solving the problem. The solution of the choice problem is connected with the creation of a system of comparative estimates, which in turn significantly based on the formal model of an algorithm. The specific problem is given by N words of memory, thus at the input of the algorithm there are N_β qubits. The tool that implements the algorithm consists of M machine instructions for β_M qubits – $M_\beta = M * \beta_M$ qubits of information. Obviously, for different applications the weights of the resources will be different, which leads to the following complex evaluation of the algorithm:

$$\varphi_\gamma =_1 *F_a(N) + c_2 * M_\beta + c_3 * S_d + c_4 * S_\gamma, \tag{2}$$

where c_i are the weights of resources, S_d – the memory for storing intermediate results, S_γ is the memory for organizing the computational process.

All weights were determined according to the importance and significance of a particular resource for the successful algorithm execution and the whole device functioning. $F_a(N)$ is the number of "elementary" operations performed by an algorithm to solve a particular problem in a given formal system.

Counting the Number of Transactions
One way to estimate the complexity (T_n) is to calculate the number (N_o) of performed operations. Let us consider the Grover algorithm as an example. This algorithm will execute:

1. The register transfer into a superposition state; equalizing the probabilities of all N states.
2. Performing unitary transformations: R (phase rotation conversion) and D (diffusion transformation) O(logN) times. After each iteration the amplitude of the desired state will be changed by the amount $2 * M(\phi) + \phi_i$, where $M(\phi)$ is the average value of the vector components after the transformation R and ϕ_i – the amplitude of the quantum state until transformation R.
3. Measuring the state of the system. There is a number of iterations when the probability of the desired result will decrease.

5 Estimation of the Complexity of Quantum Algorithms Depending on Memory

One of the optimal measures for estimating quantum algorithms is estimation the number of queries for a quantum oracle [8] (a quantum analog of a "black box" type device) during the quantum algorithm implementation. Consequently, the algorithms for solving various problems are compared in terms of the number of requests. Thus, the fewer requests we have the more effective the algorithm is. In this case the complexity estimation is performed by the memory, because the query bit is a part of the memory of a quantum computing device. By giving this bit to the "black box", the algorithm receives the function value in response.

It is also worth to mention that the quantum algorithms complexity can also be determined by the processor performance, if we deal with quantum computers simulators.

5.1 A System for Determining the Computational Complexity/Performance of Quantum Algorithms

In accordance with the foregoing, it will be worthwhile to use the following system to determine the computational complexity/productivity of quantum algorithms, expressed by a certain ratio of the elements in the equation:

$$\sum_n S = \frac{O(n)}{N_o + N_q} * (P|NP|ZPP|BPP|BQP), \tag{3}$$

where $O(n)$ is the time complexity of the quantum algorithm, or, if simpler, the running time of operation of the particular algorithm, N_o – the number of operations performed in the algorithm, N_q – the number of queries to the quantum oracle during the quantum algorithm implementation, $P|NP|ZPP|BPP|BQP$ is the complexity classes of the quantum algorithms (| is the logical OR operation). The system below presents the mathematical formalizations of all the complexity classes:

$$\begin{cases} P : C_M(n) = \max\limits_{x:|x|=n} T_M(x)(1), C_M(n) < n^c \\ NP : \cup_{i=0}^{\infty} NTIME(i * n^i) = \cup_{i=0}^{\infty} \cup_{k=0}^{\infty} NTIME(i * n^k), \\ NTIME(f(n)) = \{L|\exists m : L(m) = L, T(m,x) \le f(|x|)\} \\ ZPP : 0|1 \\ BPP : P, error = (2\sqrt{p(1-p)})^n \\ BQP : P, 0 \le error \le \frac{1}{3} \end{cases} \quad (4)$$

The algorithm for polynomial time (P) is equivalent to a deterministic Turing machine (TM), which calculates the response for a word from the input alphabet given to the input tape. The complexity of the function f is the function C, which depends on the length of the input word and is equal to the maximum of the machine's operating time for all input words of a fixed length. NP is the class of tasks that can be solved for a time P, and the class NTIME (f) is the class of tasks for which a nondeterministic TM exists. TM work stops when the length of the input is exceeded.

Table 1. Estimating the time complexity of quantum algorithms

Name/Running time (O (n))	Examples of algorithms [9] and tasks
Cubic/O(n^3)	Shor's factorization algorithm, algorithm for solving the discrete logarithm problem, Pell's equation, principal ideal algorithm, unified grouping, Gaussian sums, Abelian subgroup, non-Abelian subgroup, algorithm for finding a subset, linear systems, connected tree, hidden nonlinear structures, quantum modeling, string overwriting
The polynomial/$2^{O(\log n)}$ = poly(n), poly(n) = $n^{O(1)}$	The algorithm for solving exponential consistency, the algorithm for verifying the product of matrices, the sum of the subset, the problem of constraint satisfaction, Grover's search algorithm, algorithm of Bernstein-Vazirani, the formula estimation, the algorithm for finding the contradictions and uniqueness of the elements, the graph contradiction, the algorithm for finding the subset, the algorithm for finding the period of the function, Simon's algorithm
Linear-logarithmic/O (n * log n)	Algorithms based on quantum Fourier transforms, Viterbi's convolution decoding algorithm
Exponential/twice exponential/$2^{poly(n)}/2^{2poly(n)}$	The Deutsch algorithm, the Deutsch-Jozsa algorithm
Variation	Decoding, quantum cryptanalysis, machine learning
Unknown	Adiabatic algorithms

5.2 Estimation of the Time Complexity of Quantum Algorithms

The following Table 1 summarizes some complexity classes usually considered in such cases. Time complexity (Fig. 4) of the last two types of algorithms (variable, unknown) depends on the size of the input parameters.

Also, we should mention a number of important types of time complexity: the constant – O(1); the Ackerman inverse function – $O(\alpha(n))$; the repeated logarithmic – O(log*n); the logarithmic/doubly logarithmic – O(log n)/O (log log n); poly-logarithmic – poly(log n); fractional degree – $O(n^c)$ for $0 < c < 1$; linear – O (n), quadratic – $O(n^2)$; quasi-polynomial – $2^{poly(\log n)}$; factorial – O(n!); subexponential – $2^{o(n)}$.

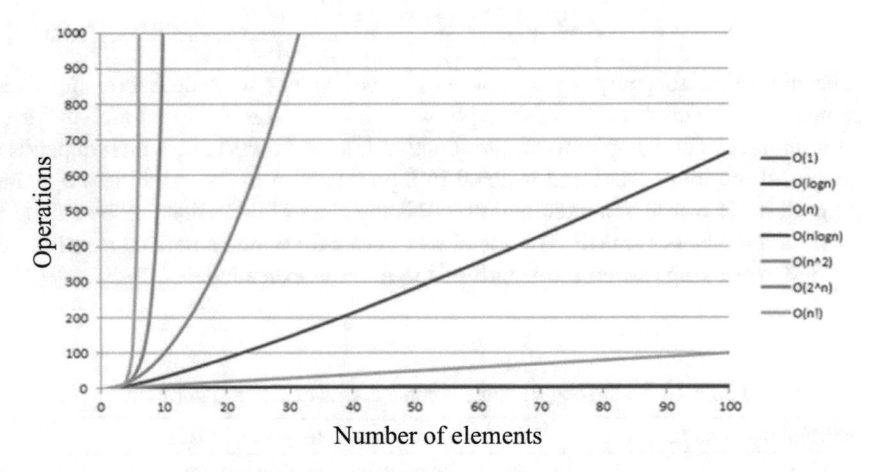

Fig. 4. Temporal complicity of algorithms

6 Conclusion

Simulation and implementation of quantum computations in the field of quantum algorithms allow:

- To predict and analyze the behavior of the quantum algorithm by calculation of the operations number, the process of a qubits set generation, and the estimation of the quantum algorithms complexity. Since the quantum system cannot be completely shielded from the environment, therefore, this prediction is relevant to any algorithm.
- Descriptively describe the universal methodology for implementing a system for determining the computational complexity/performance of quantum algorithms.
- Find new approaches to apply quantum algorithms to simulate any parameters for an executable task.

The classification of quantum algorithms in terms of an estimation of their time complexity has been performed and the main ideas and principles of their operations have been considered. As a consequence, the complexity estimation of a particular algorithm with respect to the complexity function and the derivation of a universal formula for its calculation have been determined. Thus, the computational process [10] is affected by: memory for storing intermediate results, memory for organizing the computational process, and means that implement an algorithm consisting of M machine instructions for β_M qubits of information.

The developed system for determining the computational complexity/performance of quantum algorithms describes the basics of the performance and computational complexity of quantum algorithms. It shows the entire set of necessary elements: the time complexity of the quantum algorithm, the number of operations executed by the algorithm, the number of queries to the quantum oracle, the complexity class of the quantum algorithm.

References

1. Wikipedia contributors: NP (complexity). In Wikipedia, The Free Encyclopedia. https://en.wikipedia.org/w/index.php?title = NP_(complexity)&oldid = 835998319 (2018)
2. Guzik, V., Gushanskiy, S., Polenov, M., Potapov, V.: Models of a quantum computer, their characteristics and analysis. In: Proceedings of the 9th International Conference on Application of Information and Communication Technologies (AICT 2015), pp. 583–587. IEEE Press (2015)
3. Pravilshchikov, P.: Quantum parallelism and a new model of computation. In: Proceedings of the XII All-Russian Meeting on the Control Problems, pp. 7319–7334 (2014) (in Russian)
4. Potapov, V., Guzik, V., Gushansky, S.: About the performance and computational complexity of quantum algorithms. In: Informatization and Communication, vol. 3, pp. 24–29. Moscow (2017) (in Russian)
5. Brassard, G., Høyer, P., Mosca, M., Tapp, A.: Quantum amplitude amplification and estimation. In: Contemporary Mathematics 305 Quantum Computation and Information, pp. 53–74. American Mathematical Society (2000)
6. Kornovan, D., Sheremet, A., Petrov, M.: Collective polaritonic modes in an array of two-level quantum emitters coupled to an optical nanofiber. Phys. Rev. B **94**, 16 (2016)
7. Smith, J., Mosca, M.: Algorithms for quantum computers. In: Handbook of Natural Computing, pp. 1451–1492. Springer (2010)
8. Potapov, V., Gushansky, S., Guzik, V., Polenov, M.: Architecture and software implementation of a quantum computer model. In: Software Engineering Perspectives and Application in Intelligent Systems Advances in Intelligent Systems and Computing, vol. 465, pp. 59–68. Springer (2016)
9. Potapov, V., Gushansky, S., Guzik, V., Polenov, M.: Development of methodology for entangled quantum calculations modeling in the area of quantum algorithms In: Software Engineering Perspectives and Application in Intelligent Systems Advances in Intelligent Systems and Computing. vol. 575, pp. 106–115. Springer (2017)
10. Ekert, A., Hayden, P., Inamori, H.: Basic Concepts in Quantum Computation. arXiv:quant-ph/0011013 (2008)

The Mean Square Stability of Semi-implicit Euler Method for the Model of Technology Innovation Network

Cheng Song[1(✉)], Longying Hu[1], and Haiyan Yuan[2]

[1] School of Management, Harbin Institute of Technology, Harbin 150001, China
hitsomsc@163.com

[2] Department of mathematics, Heilongjiang Institute of Technology, Harbin 150050, China

Abstract. This paper takes a model of the technology innovation network, deals with the mean-square stability of semi-implicit Euler method for the technology innovation network as a model. It is shown that the semi-implicit Euler method inherits the mean-square stability property of the analytic system and it is mean square stable with no restrictions on the step size and has important significance to maintain the network stability. Numerical experiments are presented to illustrate the results.

Keywords: Technology innovation network
Semi-implicit Euler method · Mean square stability

1 Introduction

Delay integro-differential equations (DIDEs) are often used to model some problems in biology, medicine, and many other fields. Taking random noise into account, we obtain stochastic delay integro-differential equations (SDIDEs). They are used in modeling the spread of infectious diseases and the financial mathematics, see [1–3]. Technology innovation network could be modeled as the neutral stochastic delay integro-differential equations (NSDIDEs) which take the influence of the change of the past on the present state into consideration. The analytical solution of stochastic differential equations is difficult to obtain, so, the numerical analysis becomes the study hot, see [4–8]. The study of the numerical methods for nonlinear stochastic delay integro differential equations just begin, so far only Li and Gan studied the exponential mean square stability of the stochastic θ-method for the nonlinear stochastic delay integro differential equations in [9] there is little numerical studies for the nonlinear neutral stochastic delay integro differential equations(NSDIDEs). The aim of our paper is to further study the semi-implicit Euler method for the model of the technology innovation network which has the following form.

R. Silhavy et al. (Eds.): CoMeSySo 2018, AISC 859, pp. 316–332, 2019.
https://doi.org/10.1007/978-3-030-00211-4_28

Let $N : R^d \mapsto R^d$, $f : R_+ \times R^d \times R^d \times R^d \mapsto R^d$ and $g : R_+ \times R^d \times R^d \times R^d \mapsto R^{d\times l}$ be the Borel measurable functions, where $R_+ = [0, +\infty)$. Consider d-dimensional NSDIDE of the form

$$\begin{cases} \mathrm{d}(y(t) - N(y(t-\tau))) & = f(t, y(t), y(t-\tau), \int_{t-\tau}^{t} k(t, s, y(s)\, ds))\mathrm{d}t \\ & +g(t, y(t), y(t-\tau), \int_{t-\tau}^{t} k(t, s, y(s)\, ds))\mathrm{d}\omega(t) \\ y(t) = \phi(t), \quad t \in [-\tau, 0]. \end{cases} \quad (1)$$

where the delay τ is a positive constants, and $\phi(t)$ is an F_0 - measurable, $\mathrm{C}([-\tau, 0]; R^d)$ - valued random variable satisfying

$$\sup_{-\tau \leq t \leq 0} \mathrm{E}[\phi^{\mathrm{T}}(t)\phi(t)] < +\infty \quad (2)$$

with the notation E denoting the mathematical expectation with respect to p.

Let the diffusion and drift terms satisfy the following conditions $(a1)$ and $(a2)$:

(a_1) (The local Lipschitz condition) There exists constants $K_L > 0$ and $L > 0$ such that the following inequality holds,

$$\begin{aligned} \|f(t, x_1, y_1, z_1) - f(t, x_2, y_2, z_2)\|^2 \vee \|g(t, x_1, y_1, z_1) - g(t, x_2, y_2, z_2)\|^2 \\ \leq K_L(\|x_1 - x_2\|^2 + \|y_1 - y_2\|^2 + \|z_1 - z_2\|^2) \end{aligned} \quad (3)$$

for all $\|x_1\| \vee \|x_2\| \vee \|y_1\| \vee \|y_2\| \vee \|z_1\| \vee \|z_2\|$ and $t \in R_+$.

(a_2) (The linear growth condition)

There exists constant K_G such that the following inequality holds,

$$\|f(t, x, y, z)\|^2 \vee \|g(t, x, y, z)\|^2 \leq K_G(1 + \|x\|^2 + \|y\|^2 + \|z\|^2) \quad (4)$$

for all $(t, x, y, z) \in \mathrm{R}_+ \times R \times R \times R$.

The conditions $(a1)$ and $(a2)$ are the sufficient condition for the existence and uniqueness of the solution for (1).

2 The Stability of NSDIDEs

Throughout this paper, we use the following notations. Let$\|.\|$ denote both the Euclidean norm in R^d and the norm in $R^{d\times l}$. The inner product of x, y is denoted by $\langle , \rangle$. Denote $\mathrm{C}([-\tau, 0]; R^d)$ the family of continuous functions from $[-\tau, 0]$ to equipped with the norm $\|\phi\| := \sup_{-\tau \leq t \leq 0} |\phi(t)|$. Let$\{\Omega, F, \{F_t\}_{t\geq 0}, \mathrm{P}\}$ be a complete probability space with a filtration $\{F_t\}_{t\geq 0}$ satisfying the usual condition (i.e., it is increasing and right continuous, and F_0 contains allP-null sets). Let $w(t) = (w_1(t), w_2(t), \ldots, w_l(t))^{\mathrm{T}}$ be standard l-dimensional Brownian motion defined on the probability space.

We assume that

$$f(t,0,0,0)=0, g(t,0,0,0)=0, k(t,s,0)=0, \tag{5}$$

for all $t \geq 0$, $t-\tau \leq s \leq t$.
And the initial data is bounded, that is that there is a finite M such that

$$\mathrm{E}\|\phi\|^2 \leq \mathrm{M}.$$

Condition (5) implies that (1) has a null solution $x(t) \equiv 0$.

Definition 1. *(see [6]) The trivial solution of* (5) *is said to be exponentially mean-square stable, if there exists a pair of constants $r>0$, and $C>0$, such that, whenever* $\sup_{-\tau \leq t \leq 0} \mathrm{E}|\phi(t)|^2 < +\infty$*, the following inequality holds*

$$\mathrm{E}|y(t)|^2 \leq C \sup_{-\tau \leq t \leq 0} \mathrm{E}|\phi(t)|^2 e^{-rt}, t \geq 0. \tag{6}$$

By through weakening the conclusion, we give the following definition.

Definition 2. *The trivial solution of* (5) *is said to be mean-square bounded, if there exists a pair of constants $r>0$, and $C>0$, such that, whenever* $\sup_{-\tau \leq t \leq 0} \mathrm{E}|\phi(t)|^2 < +\infty$*, the following inequality holds*

$$\mathrm{E}|y(t)|^2 \leq M, t \geq 0.$$

For the convenience, we make assumptions as followed:
(E1) If there is a positive constant $\lambda \in (0,1)$ such that

$$|N(x)| \leq \lambda |x|$$

(E2) If there are positive constants a_1,a_2, such that for any $(t,x,y) \in R_+ \times R^d \times R^d$ satisfying

$$(x-N(y))^{\mathrm{T}} f(t,x,y,0) \leq -a_1|x|^2 + a_2|y|^2$$

(E3) There exist nonnegative real numbers a_3, β_1, β_2, β_3, γ, such that for any $t>0$, x, y, $z \in R^d$, we have

$$|f(t,x,y,z) - f(t,x,y,0)| \leq a_3|z|$$

$$|g(t,x,y,z)|^2 \leq \beta_1|x|^2 + \beta_2|y|^2 + \beta_3|z|^2$$

$$|k(t,s,x) - k(t,s,y)| \leq \gamma|x-y|$$

(E4) The constants a_1 satisfies

$$1 + a_2 + \lambda^2 + \frac{(\beta_1 + \beta_2 + 4(a_3+\beta_3)\gamma^2\tau^2)}{2} < a_1$$

Lemma 1. *Assume that (1) satisfies the above conditions of the (E1), (E2), (E3) and (E4), then the trivial solution of (1) is mean square bounded.*

Proof. By the Itô formula we can derive that, for any $t > 0$,

$$
\begin{aligned}
|x(t) - Nx(t-\tau)|^2 =& |x(0) - N(-\tau)|^2 \\
&+ 2\int_0^t \langle x(u) - Nx(u-\tau), f(u, x(u), x(u-\tau), \\
&\int_{u-\tau}^u k(u,s,x(s))ds)\rangle du \\
&+ 2\int_0^t \langle x(u) - Nx(u-\tau), g(u, x(u), x(u-\tau), \\
&\int_{u-\tau}^u k(u,s,x(s))ds)\rangle d(u) \\
&+ \int_0^t |g(u, x(u), x(u-\tau), \int_{u-\tau}^u k(u,s,x(s))ds)|^2 du
\end{aligned}
\tag{7}
$$

Using (E1), (E2), (E3) and (E4), we obtain

$$
\begin{aligned}
&\langle x(u) - Nx(u-\tau), f(u, x(u), x(u-\tau), \int_{u-\tau}^u k(u,s,x(s))ds)\rangle \\
&= \langle x(u) - Nx(u-\tau), f(u, x(u), x(u-\tau), 0) \\
&+ f(u, x(u), x(u-\tau), \int_{u-\tau}^u k(u,s,x(s))ds) - f(u, x(u), x(u-\tau), 0)\rangle \\
&\leq -a_1|x(u)|^2 + a_2|x(u-\tau)|^2 + |x(u) - Nx(u-\tau)||a_3\int_{u-\tau}^u k(u,s,x(s))ds| \\
&\leq -a_1|x(u)|^2 + a_2|x(u-\tau)|^2 + \frac{|x(u) - Nx(u-\tau)|^2}{2} + \frac{a_3\gamma^2\tau}{2}\int_{u-\tau}^u |x(s)|^2 ds \\
&\leq -(a_1 - 1)|x(u)|^2 + (a_2 + \lambda^2)|x(u-\tau)|^2 + a_3\gamma^2\tau\int_{u-\tau}^u |x(s)|^2 ds
\end{aligned}
\tag{8}
$$

By the conditions (6), we have

$$
\begin{aligned}
&\left|g(u, x(u), x(u-\tau), \int_{u-\tau}^u k(u,s,x(s))ds)\right|^2 \\
&\leq \beta_1|x(u)|^2 + \beta_2|x(u-\tau)|^2 + \beta_3|\int_{u-\tau}^u k(u,s,x(s))ds|^2 \\
&\leq \beta_1|x(u)|^2 + \beta_2|x(u-\tau)|^2 + \beta_3|\gamma^2\tau\int_{u-\tau}^u |x(s)|^2 ds
\end{aligned}
\tag{9}
$$

Substituting (8) and (9) into (7) and rearranging terms yield that

$$
\begin{aligned}
|x(t) - Nx(t-\tau)|^2 = |x(0) - Nx(-\tau)|^2 + \int_0^t (2 - 2a_1 + \beta_1)|x(u)|^2 du \\
+ \int_0^t (2a_2 + 2\lambda^2 + \beta_2)|x(u-\tau)|^2 du \\
+ \int_0^t (a_3 + \beta_3)\gamma^2 \tau \int_{u-\tau}^u (|x(s)|^2) ds du \\
+ 2\int_0^t \langle x(u) - Nx(u-\tau), g(u, x(u), x(u-\tau), \\
\int_{u-\tau}^u k(u,s,x(s))ds)\rangle dw(u) \qquad (10)
\end{aligned}
$$

Take the mathematical expectation on both sides, and noticing

$$
\begin{aligned}
\mathrm{E}\big\langle X_n, g(t_n, X_n, X_{n-m}, \widetilde{X}_n)\Delta w_n \big\rangle = 0, \\
\mathrm{E}\big\langle f(t_n, X_n, X_{n-m}, \widetilde{X}_n), g(t_n, X_n, X_{n-m}, \widetilde{X}_n)\Delta w_n \big\rangle = 0, \\
\mathrm{E}\big|g(t_n, X_n, X_{n-m}, \widetilde{X}_n)\Delta w_n\big|^2 = h\mathrm{E}\big|g(t_n, X_n, X_{n-m}, \widetilde{X}_n)\big|^2
\end{aligned}
$$

it then follows from (10) that

$$
\begin{aligned}
\mathrm{E}\big|x(t) - Nx(t-\tau)\big|^2 = \mathrm{E}\big|x(0) - Nx(-\tau)\big|^2 + (2 - 2a_1 + \beta_1)\int_0^t \mathrm{E}|x(u)|^2 du \\
+ (2a_2 + 2\lambda^2 + \beta_2)\int_0^t \mathrm{E}|x(u-\tau)|^2 du \\
+ (a_3 + \beta_3)\gamma^2\tau \int_0^t \int_{u-\tau}^u \mathrm{E}|x(s)|^2 ds du
\end{aligned}
$$

Let $x(t) - Nx(t-\tau) = y(t)$, hence

$$
\begin{aligned}
\mathrm{E}\big|y(t)\big|^2 \le \mathrm{E}\big|y(0)\big|^2 + (2 - 2a_1 + \beta_1)\int_0^t \mathrm{E}|x(u)|^2 du \\
+ (2a_2 + 2\lambda^2 + \beta_2) + (a_3 + \beta_3)\gamma^2\tau^2 \int_0^t \sup_{u-\tau \le r \le u} \mathrm{E}|x(r)|^2 du \\
\le \mathrm{E}\big|y(0)\big|^2 + (2a_2 + 2\lambda^2 + \beta_2 + (a_3 + \beta_3)\gamma^2\tau^2)\tau\mathrm{E}|\phi|^2 \\
+ \kappa \int_0^t \sup_{0 \le r \le u} \mathrm{E}|x(r)|^2 du
\end{aligned}
$$

where $\kappa = 2 - 2a_1 + \beta_1 + 2a_2 + 2\lambda^2 + \beta_2 + (a_3 + \beta_3)\gamma^2\tau^2) < 0$, using (7) we then derive that

$$
\mathrm{E}|y(t)|^2 \le \mathrm{E}|y(0)|^2 + (2a_2 + 2\lambda^2 + \beta_2) + (a_3 + \beta_3)\gamma^2\tau^2\tau\mathrm{E}|\phi|^2
$$

with the assumption $x(t) - Nx(t-\tau) = y(t)$, we have the following inequality

$$\mathrm{E}|y(t)|^2 \leq ((2a_2 + 2\lambda^2 + \beta_2 + (a_3 + \beta_3)\gamma^2\tau^2)\tau + 2\lambda^2 + 2)\mathrm{E}|\phi|^2 \tag{11}$$

On the other hand, we know that

$$|x(t)| = |x(t) - Nx(t-\tau) + Nx(t-\tau)| \leq |y(t)| + |Nx(t-\tau)|$$

then we get that

$$\mathrm{E}|x(t)|^2 \leq 2\mathrm{E}|y(t)|^2 + 2\lambda^2\mathrm{E}|x(t-\tau)|^2,$$

define

$$\epsilon_0 = (2a_2 + 2\lambda^2 + \beta_2 + (a_3 + \beta_3)\gamma^2\tau^2)\tau + 2\lambda^2 + 2,$$

the following inequality could be deduced from (8)

$$\mathrm{E}|x(t)|^2 \leq \Big(\frac{2}{1-2\lambda^2}\epsilon_0 + (2\lambda^2)^{\lfloor \frac{n}{m} \rfloor + 1}\Big)M,$$

which implies that the method is mean square bounded.

Now, we state a theorem on mean-square exponential stability property for the analytical solution of (1).

Theorem 1. *Let Assumption (5) hold, if the conditions and (E1), (E2), (E3) and (E4) are satisfied, then the trivial solution of (1) is mean-square exponentially stable.*

Proof. Let $t \geq 0$, $\delta > 0$, by the Itô formula we can derive that

$$\begin{aligned} |x(t+\delta) - Nx(t+\delta-\tau)|^2 = {} & |x(t) - Nx(t-\tau)|^2 + 2\int_t^{t+\delta} \langle x(u) - Nx(u-\tau), \\ & f(u, x(u), x(u-\tau), \int_{u-\tau}^{u} k(u,s,x(s))ds)\rangle du \\ & + 2\int_t^{t+\delta} \langle x(u) - Nx(u-\tau), g(u, x(u), x(u-\tau), \\ & \int_{u-\tau}^{u} k(u,s,x(s))ds)\rangle dw(u) \\ & + \int_t^{t+\delta} |g(u, x(u), x(u-\tau), \int_{u-\tau}^{u} k(u,s,x(s))ds)|^2 du \end{aligned} \tag{12}$$

Taking expectations on the both sides of (7), using (8), (9) and the properties of the Itô integral, one has

$$\begin{aligned}
\mathrm{E}|x(t+\delta)-Nx(t+\delta-\tau)|^2 =& \mathrm{E}|x(t)-Nx(t-\tau)|^2 \\
&+2\int_t^{t+\delta}\mathrm{E}\langle x(u)-Nx(u-\tau), \\
&f(u,x(u),x(u-\tau),\int_{u-\tau}^{u}k(u,s,x(s))ds)\rangle du \\
&+\int_t^{t+\delta}\mathrm{E}|g(u,x(u),x(u-\tau), \\
&\int_{u-\tau}^{u}k(u,s,x(s))ds)|^2 du \\
\leq& \mathrm{E}|x(t)-Nx(t-\tau)|^2 \\
&+(2-2a_1+\beta_1)\int_t^{t+\delta}\mathrm{E}|x(u)|^2du \\
&+(2a_2+2\lambda^2+\beta_2+(a_3+\beta_3)\gamma^2\tau^2) \\
&\int_t^{t+\delta}\sup_{u-\tau\leq r\leq u}\mathrm{E}|x(r)|^2du
\end{aligned}$$

Let $\mu(t)=\mathrm{E}|x(t)-Nx(t-\tau)|^2$, we have

$$\begin{aligned}
|\mu(t+\delta)-\mu(t)| =& 2\int_t^{t+\delta}\mathrm{E}\langle x(u)-Nx(u-\tau), \\
&f(u,x(u),x(u-\tau),\int_{u-\tau}^{u}k(u,s,x(s))ds)\rangle du \\
&+\int_t^{t+\delta}\mathrm{E}|g(u,x(u),x(u-\tau),\int_{u-\tau}^{u}k(u,s,x(s))ds)|^2du
\end{aligned} \tag{13}$$

Combined with (E4), we obtained that

$$|\mu(t+\delta)-\mu(t)| \leq (2a_2+2\lambda^2+\beta_2+(a_3+\beta_3)\gamma^2\tau^2)\delta \tag{14}$$

therefore $\mu(t)$ is continuous. We then derive from (13) that

$$\mathrm{D}^+\mu(t) \leq (2-2a_1+\beta_1)\mathrm{E}|x(t)|^2+(2a_2+2\lambda^2+\beta_2+(a_3+\beta_3)\gamma^2\tau^2)\sup_{t-\tau\leq s\leq t}\mathrm{E}|x(s)|^2 \tag{15}$$

where $\mathrm{D}^+\mu(t)=\lim_{\delta\to 0}\sup\frac{\mu(t+\delta)-\mu(t)}{\delta}$.
Because

$$\begin{aligned}
\mathrm{E}|x(t)|^2 &= \mathrm{E}|x(t)-Nx(t-\tau)+Nx(t-\tau)|^2 \\
&\leq 2\mathrm{E}|x(t)-Nx(t-\tau)|^2+2\mathrm{E}|Nx(t-\tau)|^2 \\
&\leq 2\mu(t)+2\lambda^2\mathrm{E}|x(t-\tau)|^2
\end{aligned}$$

Take it into (15) and considering (E4), one can get

$$\begin{aligned}
\mathrm{D}^+\mu(t) \leq& 2(2-2a_1+\beta_1)\mu(t)+2(2a_2+2\lambda^2+\beta_2+(a_3+\beta_3)\gamma^2\tau^2 \\
&+2\lambda^2(2-2a_1+\beta_1))\sup_{t-\tau\leq s\leq t}\mu(s)
\end{aligned} \tag{16}$$

On the other hand, the condition (E4) implies

$$2a_1 - 2 - \beta_1 > 2a_2 + 2\lambda^2 + \beta_2 + (a_3 + \beta_3)\gamma^2\tau^2 + 2\lambda^2(2 - 2a_1 + \beta_1) \geq 0$$

By Theorem 7 in [6], we have

$$\mu(t) \leq e^{-v^* t} \sup_{-\tau \leq s \leq 0} \mu(s), t \geq 0,$$

where

$$v^* \in (0, 2(2a_1(1 + 2\lambda^2) - 2 - \beta_1(1 + 2\lambda^2) - 2a_2 - 6\lambda^2 - \beta_2 - (a_3 + \beta_3)\gamma^2\tau^2)]$$

is the zero root of

$$L(v) = v + 2(2 - 2a_1 + \beta_1) + 2(2a_2 + 2\lambda^2 + \beta_2 + (a_3 + \beta_3)\gamma^2\tau^2 + 2\lambda^2(2 - 2a_1 + \beta_1))e^{v\tau}$$

The proof is completed.

3 Mean-Square Stability of the Semi-implicit Euler Method

By applying the semi-implicit Euler method to nonlinear NSDIDE (1), then we have

$$\begin{aligned} y_{n+1} - Ny_{n+1-m} = y_n - Ny_{n-m} + \Delta t f(t_{n+1}, y_{n+1}, \bar{y}_{n+1}, \tilde{y}_{n+1}) \\ + g(t_n, y_n, \bar{y}_n, \tilde{y}_n)\Delta w_n \end{aligned} \tag{17}$$

where the step size Δt satisfies $\Delta t = \frac{\tau}{m}$ for a integer m, y_i is an approximation to $y(t_i)$, $\bar{y}_n$ is an approximation to $y(t_i - \tau)$ and $\bar{y}_i = y_{i-m}$, $\tilde{y}_i$ is an approximation to $\int_{t_i-\tau}^{t} {}_i k(t, s, x(s))ds$, for $t_i = i\Delta t, i = 1, 2, \ldots$, which are obtained by a convergent compound quadrature (CQ) formula:

$$\tilde{y}_i = \Delta t \sum_{j=0}^{m} q_j k(t_i, t_{i-j}, y_{i-j}), i = 1, 2, \ldots, \tag{18}$$

and $y_k = \phi(k\Delta t)$ for $k = -m, -m + 1, \ldots, 0$. $\Delta w_k := w((k + 1)\Delta t) - w(k\Delta t)$ is the Brownian increment.

3.1 Some Assumptions

For the stability analysis, we need the compound quadrature formula (18) that satisfies the following condition:

$$\Delta^2(m + 1) \sum_{j=0}^{m} |q_j|^2 = \tilde{q} \tag{19}$$

where $\tilde{q}$ is a positive constant, and $\tilde{q} \leq 4\tau^2$.

Now we recall some stability concepts for numerical methods.

Definition 3. *For a given step size Δt, a numerical method is said to be exponentially mean square stable if there is a pair of positive constants γ and C such that for any initial data $\phi(t)$ the numerical solution y_n produced by the method satisfies*

$$\mathrm{E}[y_n^{\mathrm{T}} y_n] \leq e^{-\gamma t_n} \sup_{-\tau \leq t \leq 0} \mathrm{E}[\phi(t)^{\mathrm{T}} \phi(t)], \forall n \geq 0.$$

Definition 4. *For a given step size Δt, a numerical method is said to be asymptotically mean square stable if for any initial data $\phi(t)$ the numerical solution y_n produced by the method satisfies*

$$\lim_{n \to \infty} \mathrm{E}[y_n^{\mathrm{T}} y_n] = 0.$$

Theorem 2. *Assume that system (1) satisfies (E2) with $-\mu_1 + \mu_2 < 0$, then the semi-implicit Euler method (17) is asymptotically mean square stable for all $\Delta t > 0$.*

Proof. From (17) it follows that

$$y_{n+1} - \mathrm{N} y_{n+1-m} - \Delta t f(t_{n+1}, y_{n+1}, \bar{y}_{n+1}, \tilde{y}_{n+1}) = y_n - \mathrm{N} y_{n-m} + g(t_n, y_n, \bar{y}_n, \tilde{y}_n) \Delta w_n$$

take square on both sides, we obtain

$$\begin{aligned}
&(y_{n+1} - \mathrm{N} y_{n+1-m})^{\mathrm{T}} (y_{n+1} - \mathrm{N} y_{n+1-m}) \\
&\quad + \Delta^2 f(t_{n+1}, y_{n+1}, \bar{y}_{n+1}, \tilde{y}_{n+1})^{\mathrm{T}} f(t_{n+1}, y_{n+1}, \bar{y}_{n+1}, \tilde{y}_{n+1}) \\
&\quad - 2(y_{n+1} - \mathrm{N} y_{n+1-m})^{\mathrm{T}} \Delta f(t_{n+1}, y_{n+1}, \bar{y}_{n+1}, \tilde{y}_{n+1}) \\
&\quad = (y_n - N y_{n-m})^2 + \Delta w_n^{\mathrm{T}} g(t_n, y_n, \bar{y}_n, \tilde{y}_n)^{\mathrm{T}} g(t_n, y_n, \bar{y}_n, \tilde{y}_n) \delta w_n \\
&\quad + 2(y_n - N y_{n-m})^{\mathrm{T}} g(t_n, y_n, \bar{y}_n, \tilde{y}_n) \delta w_n
\end{aligned} \tag{20}$$

since $w(t) = (w_1(t), w_2(t), \ldots, w_l(t))^{\mathrm{T}}$ is a standard l -dimensional Brownian motion we have

$$\begin{aligned}
&\mathrm{E}[\Delta w_n^{\mathrm{T}} g^{\mathrm{T}}(t_n, y_n, \bar{y}_n, \tilde{y}_n) g(t_n, y_n, \bar{y}_n, \tilde{y}_n) \Delta w_n] \\
&= \Delta t \mathrm{E}[g^{\mathrm{T}}(t_n, y_n, \bar{y}_n, \tilde{y}_n) g(t_n, y_n, \bar{y}_n, \tilde{y}_n)]
\end{aligned}$$

and

$$\mathrm{E}[(y_n - N y_{n-m})^{\mathrm{T}} g(t_n, y_n, \bar{y}_n, \tilde{y}_n) \Delta w_n] = 0$$

let $x_n = y_n - N y_{n-m}, n = 0, 1, \ldots$, substituting it into the above equality and then taking expectation on both sides, one receives

$$\begin{aligned}
\mathrm{E}[x_{n+1}^{\mathrm{T}} x_{n+1}] &= \mathrm{E}[x_n^{\mathrm{T}} x_n] + 2\Delta t \mathrm{E}[x_{n+1}^{\mathrm{T}} f(t_{n+1}, y_{n+1}, \bar{y}_{n+1}, \tilde{y}_{n+1})] \\
&\quad - \Delta t^2 \mathrm{E}[f^{\mathrm{T}}(t_{n+1}, y_{n+1}, \bar{y}_{n+1}, \tilde{y}_{n+1}) f(t_{n+1}, y_{n+1}, \bar{y}_{n+1}, \tilde{y}_{n+1})] \\
&\quad + \Delta t \mathrm{E}[g^{\mathrm{T}}(t_n, y_n, \bar{y}_n, \tilde{y}_n) g(t_n, y_n, \bar{y}_n, \tilde{y}_n)]
\end{aligned} \tag{21}$$

which gives

$$\begin{aligned} \mathrm{E}[x_{n+1}^{\mathrm{T}} x_{n+1}] \le \mathrm{E}[x_n^{\mathrm{T}} x_n] + 2\Delta t \mathrm{E}[x_{n+1}^{\mathrm{T}} f(t_{n+1}, y_{n+1}, \bar{y}_{n+1}, \tilde{y}_{n+1})] \\ + \Delta t \mathrm{E}[g^{\mathrm{T}}(t_n, y_n, \bar{y}_n, \tilde{y}_n) g(t_n, y_n, \bar{y}_n, \tilde{y}_n)] \end{aligned} \tag{22}$$

A combination of (E2), (E3) and (E4) gives

$$\begin{aligned} \mathrm{E}[x_{n+1}^{\mathrm{T}} x_{n+1}] &\le \mathrm{E}[x_n^{\mathrm{T}} x_n] + 2\Delta t(-a_1 \mathrm{E}|y_{n+1}|^2 + a_2 \mathrm{E}|\bar{y}_{n+1}|^2 + 2\Delta t(\mathrm{E}|y_{n+1}|^2 \\ &\quad + \lambda^2 \mathrm{E}|\bar{y}_{n+1}|^2) + a_3 \Delta t \mathrm{E}|\tilde{y}_{n+1}|^2 + \Delta t(\beta_1 \mathrm{E}|y_n|^2 + \beta_2 \mathrm{E}|\bar{y}_n|^2 + \beta_3 \mathrm{E}|\tilde{y}_n|^2 \\ &= \mathrm{E}[x_n^{\mathrm{T}} x_n] + 2\Delta t((1-a_1)\mathrm{E}|y_{n+1}|^2 + (a_2 + \lambda^2)\mathrm{E}|\bar{y}_{n+1}|^2) \\ &\quad + \Delta t a_3 \mathrm{E}|\tilde{y}_{n+1}|^2 + \Delta t(\beta_1 \mathrm{E}|y_n|^2 + \beta_2 \mathrm{E}|\bar{y}_n|^2 + \beta_3 \mathrm{E}|\tilde{y}_n|^2) \end{aligned} \tag{23}$$

using $\bar{y}_i = y_i - m$ and (18), we can get the following inequality

$$\begin{aligned} \mathrm{E}[x_{n+1}^{\mathrm{T}} x_{n+1}] &\le \mathrm{E}[x_n^{\mathrm{T}} x_n] + 2\Delta t(-a_1 \mathrm{E}|y_{n+1}|^2 + a_2 \mathrm{E}|y_{n+1-m}|^2 \\ &\quad + 2\Delta t(\mathrm{E}|y_{n+1}|^2 + \lambda^2 \mathrm{E}|y_{n+1-m}|^2) \\ &\quad + a_3 \Delta t^3 \mathrm{E}|\sum_{j=0}^{m} q_j k(t_{n+1}, t_{n+1-j}, y_{n+1-j})|^2 \\ &\quad + \beta_3 \Delta t^3 \mathrm{E}|\sum_{j=0}^{m} q_j k(t_n, t_{n-j}, y_{n-j})|^2 \\ &\quad + \Delta t \beta_1 \mathrm{E}|y_n|^2 + \Delta t \beta_2 \mathrm{E}|y_{n-m}|^2 \end{aligned}$$

So that

$$\begin{aligned} \mathrm{E}[x_{n+1}^{\mathrm{T}} x_{n+1}] - \mathrm{E}[x_n^{\mathrm{T}} x_n] &\le 2\Delta t(1-a_1)\mathrm{E}|y_{n+1}|^2 + 2\Delta t(a_2 + \lambda^2)\mathrm{E}|y_{n+1-m}|^2 \\ &\quad + a_3 \Delta t^3 \mathrm{E}|\sum_{j=0}^{m} q_j k(t_{n+1}, t_{n+1-j}, y_{n+1-j})|^2 \\ &\quad + \beta_3 \Delta t^3 \mathrm{E}|\sum_{j=0}^{m} q_j k(t_n, t_{n-j}, y_{n-j})|^2 \\ &\quad + \Delta t \beta_1 \mathrm{E}|y_n|^2 + \Delta t \beta_2 \mathrm{E}|y_{n-m}|^2 \end{aligned} \tag{24}$$

By using (24) in a recurrence way, we can obtain

$$\begin{aligned} \mathrm{E}[x_n^{\mathrm{T}} x_n] - \mathrm{E}[x_{n-1}^{\mathrm{T}} x_{n-1}] &\le 2\Delta t(1-a_1)\mathrm{E}|y_n|^2 + 2\Delta t(a_2 + \lambda^2)\mathrm{E}|y_{n-m}|^2 \\ &\quad + a_3 \Delta t^3 \mathrm{E}|\sum_{j=0}^{m} q_j k(t_n, t_{n-j}, y_{n-j})|^2 \\ &\quad + \beta_3 \Delta t^3 \mathrm{E}|\sum_{j=0}^{m} q_j k(t_{n-1}, t_{n-1-j}, y_{n-1-j})|^2 \\ &\quad + \Delta t \beta_1 \mathrm{E}|y_{n-1}|^2 + \Delta t \beta_2 \mathrm{E}|y_{n-1-m}|^2 \end{aligned}$$

$$\begin{aligned}
\mathrm{E}[x_{n-1}^{\mathrm{T}}x_{n-1}] - \mathrm{E}[x_{n-2}^{\mathrm{T}}x_{n-2}] \leq\; & 2\Delta t(1-a_1)\mathrm{E}|y_{n-1}|^2 + 2\Delta t(a_2+\lambda^2)\mathrm{E}|y_{n-1-m}|^2 \\
& + a_3\Delta t^3 \mathrm{E}|\sum_{j=0}^{m} q_j k(t_{n-1}, t_{n-1-j}, y_{n-1-j})|^2 \\
& + \beta_3\Delta t^3 \mathrm{E}|\sum_{j=0}^{m} q_j k(t_{n-2}, t_{n-2-j}, y_{n-2-j})|^2 \\
& + \Delta t\beta_1\mathrm{E}|y_{n-2}|^2 + \Delta t\beta_2\mathrm{E}|y_{n-2-m}|^2 \\
& \vdots \\
\mathrm{E}[x_1^{\mathrm{T}}x_1] - \mathrm{E}[x_0^{\mathrm{T}}x_0] \leq\; & 2\Delta t(1-a_1)\mathrm{E}|y_1|^2 + 2\Delta t(a_2+\lambda^2)\mathrm{E}|y_{1-m}|^2 \\
& + a_3\Delta t^3 \mathrm{E}|\sum_{j=0}^{m} q_j k(t_1, t_{1-j}, y_{1-j})|^2 \\
& + \beta_3\Delta t^3 \mathrm{E}|\sum_{j=0}^{m} q_j k(t_0, t_{-j}, y_{-j})|^2 \\
& + \Delta t\beta_1\mathrm{E}|y_0|^2 + \Delta t\beta_2\mathrm{E}|y_{-m}|^2
\end{aligned} \tag{25}$$

which on substitution into (24) gives

$$\begin{aligned}
\mathrm{E}|x_{n+1}|^2 - \mathrm{E}|x_0|^2 \leq\; & 2\Delta t(1-a_1)\sum_{i=0}^{n}\mathrm{E}|y_{i+1}|^2 + 2\Delta t(a_2+\lambda^2)\sum_{i=0}^{n}\mathrm{E}|y_{i+1-m}|^2 \\
& + a_3\Delta t^3 \sum_{i=0}^{n}\mathrm{E}|\sum_{j=0}^{m} q_j k(t_{i+1}, t_{i+1-j}, y_{i+1-j})|^2 \\
& + \beta_3\Delta t^3 \sum_{i=0}^{n}\mathrm{E}|\sum_{j=0}^{m} q_j k(t_i, t_{i-j}, y_{i-j})|^2 \\
& + \Delta t\beta_1 \sum_{i=0}^{n}\mathrm{E}|y_i|^2 + \Delta t\beta_2 \sum_{i=0}^{n}\mathrm{E}|y_{i-m}|^2 \\
\leq\; & 2\Delta t(1-a_1)\sum_{i=1}^{n+1}\mathrm{E}|y_i|^2 + 2\Delta t(a_2+\lambda^2)\sum_{i=1-m}^{n+1-m}\mathrm{E}|y_i|^2 \\
& + a_3\Delta t^3 \sum_{i=1}^{n+1}\mathrm{E}|\sum_{j=0}^{m} q_j k(t_i, t_{i-j}, y_{i-j})|^2 \\
& + \beta_3\Delta t^3 \sum_{i=0}^{n}\mathrm{E}|\sum_{j=0}^{m} q_j k(t_i, t_{i-j}, y_{i-j})|^2 \\
& + \Delta t\beta_1 \sum_{i=0}^{n}\mathrm{E}|y_i|^2 + \Delta t\beta_2 \sum_{i=0}^{n}\mathrm{E}|y_{i-m}|^2
\end{aligned}$$

$$\leq 2\Delta t(1-a_1)\sum_{i=1}^{n+1}\mathrm{E}|y_i|^2 + 2\Delta t(a_2+\lambda^2)\sum_{i=1-m}^{n+1-m}\mathrm{E}|y_i|^2$$

$$+ a_3\Delta t^3\sum_{i=1}^{n+1}\mathrm{E}|(m+1)\sum_{j=0}^{m}q_j^2k^2(t_i,t_{i-j},y_{i-j})|$$

$$+ \beta_3\Delta t^3\sum_{i=0}^{n}\mathrm{E}|(m+1)\sum_{j=0}^{m}q_j^2k^2(t_i,t_{i-j},y_{i-j})|$$

$$+ \Delta t\beta_1\sum_{i=0}^{n}\mathrm{E}|y_i|^2 + \Delta t\beta_2\sum_{i=0}^{n}\mathrm{E}|y_{i-m}|^2$$

Combined with (E4) and (19), one arrives

$$\begin{aligned}\mathrm{E}|x_{n+1}|^2 - \mathrm{E}|x_0|^2 \leq \big(&2\Delta t(1-a_1+a_2+\lambda^2) + (a_3+\beta_3)\Delta t\gamma^2\tilde{q} \\ &+ \Delta t(\beta_1+\beta_2)\big)\sum_{i=0}^{n}\mathrm{E}|y_i|^2 \\ &+ \big(\Delta t(\beta_1+\beta_2) + (a_3+\beta_3)\Delta t\gamma^2\tilde{q}\big)\sum_{i=-m}^{-1}\mathrm{E}|y_i|^2\end{aligned}$$

$$\begin{aligned}\mathrm{E}|x_{n+1}|^2 \quad \mathrm{E}|x_0|^2 < \big(&2\Delta t(1-a_1+a_2+\lambda^2) + (a_3+\beta_3)\Delta t\gamma^2\tilde{q} \\ &+ \Delta t(\beta_1+\beta_2)\big)\sum_{i=i}^{n}\mathrm{E}|y_i|^2 + \big(\beta_1+\beta_2+(a_3+\beta_3)\gamma^2\tilde{q}\big)\tau M\end{aligned}$$

We know that

$$2\Delta t(1-a_1+a_2+\lambda^2) + (a_3+\beta_3)\Delta t\gamma^2\tilde{q} + \Delta t(\beta_1+\beta_2) < 0$$

so one arrives

$$\mathrm{E}|x_{n+1}|^2 \leq \mathrm{E}|x_0|^2 + \big(\beta_1+\beta_2+(a_3+\beta_3)\gamma^2\tilde{q}\big)\tau M$$

then

$$\begin{aligned}\sum_{i=1}^{n}\mathrm{E}|y_i|^2 &\leq \frac{\mathrm{E}|x_0|^2 + \big(\beta_1+\beta_2+(a_3+\beta_3)\gamma^2\tilde{q}\big)\tau M - \mathrm{E}|x_{n+1}|^2}{-(2\Delta t(1-a_1+a_2+\lambda^2) + (a_3+\beta_3)\Delta t\gamma^2\tilde{q} + \Delta t(\beta_1+\beta_2))} \\ &\leq \frac{\mathrm{E}|x_0|^2 + \big(\beta_1+\beta_2+(a_3+\beta_3)\gamma^2\tilde{q}\big)\tau M}{-(2\Delta t(1-a_1+a_2+\lambda^2) + (a_3+\beta_3)\Delta t\gamma^2\tilde{q} + \Delta t(\beta_1+\beta_2))}\end{aligned}$$

Therefore

$$\lim_{k\to\infty}\mathrm{E}|y_k|^2 = 0.$$

It is shown that the semi-implicit Euler method is MS-stable.

4 The Numerical Experiment

The purpose of this section is to illustrate our theoretical results of the stability and obtained in Sects. 2 and 3. Let us consider the following technology innovation network model,

$$\begin{cases} \mathrm{d}(y(t)-\frac{1}{4}y(t-1)) & =(-14y(t)+y(t-1)sin(t-1)+\int_{t-1}^{t}cos(y(s))\,ds)\mathrm{d}t \\ & +(sin(y(t))+y(t-1)+\int_{t-1}^{t}cos(y(s)))\,ds)\mathrm{d}\omega(t) \\ y(t)=1, \quad t\in[-1,0]. \end{cases} \tag{26}$$

Consider the semi-implicit Euler method (17) for the nonlinear neutral stochastic delay integro-differential equation (26). For the test Eq. (26), we have that $\tau = 1$, $\lambda = \frac{1}{4}$, $a_1 = -14$, $\gamma = 1$, $a_2 = a_3 = 1$, $\beta_1 = 3$, $\beta_2 = \beta_3 = 3$. Therefore,

$$\begin{aligned} & 1+a_2+\lambda^2+\frac{\beta_1+\beta_2+4(a_3+\beta_3)\gamma^2\tau^2}{2} \\ & =1+1+\frac{1}{16}+\frac{3+3+4\times(1+3)\times 1}{2} \\ & =13\frac{1}{16} \\ & < a_1 = 14 \end{aligned}$$

By Theorem 1, the trivial solution of (26) is mean-square exponentially stable. It is easy to verify that nonlinear NSDIDE (26) satisfies the conditions of Theorem 2. In the following tests, we show the influence of stepsize Δt on MS-stability of the semi-implicit Euler method. We generate 10^3 numerical sample paths using the semi-implicit Euler method, denoting $Y_n^{(i)}$ as the numerical approximation to $y^{(i)}(t_n)$ at step point t_n in i th simulation of all 1000 simulations, and take $\frac{1}{1000}\sum_{i=1}^{1000}|Y_n^{(i)}|^2$ to approximate $\mathrm{E}|y_n|^2$.Taking step size $\Delta t = 0.1$, $\Delta t = 0.25$, $\Delta t = 0.5$ and $\Delta t = 1$, we obtain four groups of numerical solutions of Eq. (26) on interval [0,30], which are displayed in Figs. 1, 2, 3 and 4 respectively, and the curve is formed by $\mathrm{E}|y_n|^2$ in each figure.

Now, we will consider the following nonlinear vector NSDIDEs,

$$\begin{aligned} \begin{bmatrix} d(y_1(t)-\frac{1}{4}cos(y_1(t-\tau))) \\ d(y_2(t)-\frac{1}{8}sin(y_2(t-\tau))) \end{bmatrix} = & (A\begin{bmatrix} y_1(t) \\ y_2(t) \end{bmatrix} + B\begin{bmatrix} sin(y_1(t-\tau)) \\ cos(y_2(t-\tau)) \end{bmatrix} \\ & + C\begin{bmatrix} \int_{t-\tau}^{t} cos(y_1(s))ds \\ \int_{t-\tau}^{t} sin(y_2(s))ds \end{bmatrix})dt \\ & + D\begin{bmatrix} y_1(t)dw_1(t) \\ y_2(t-\tau)dw_2(t) \end{bmatrix}, t\geq 0, \end{aligned} \tag{27}$$

where

$$A=\begin{bmatrix} -28 & 0 \\ 0 & -30 \end{bmatrix}, B=\begin{bmatrix} 2 & -\frac{1}{2} \\ \frac{1}{4} & 1 \end{bmatrix}, C=\begin{bmatrix} 1 & 0 \\ 0 & 1 \end{bmatrix}, D=\begin{bmatrix} 1 & \frac{3}{2} \\ \frac{5}{2} & -\frac{1}{2} \end{bmatrix},$$

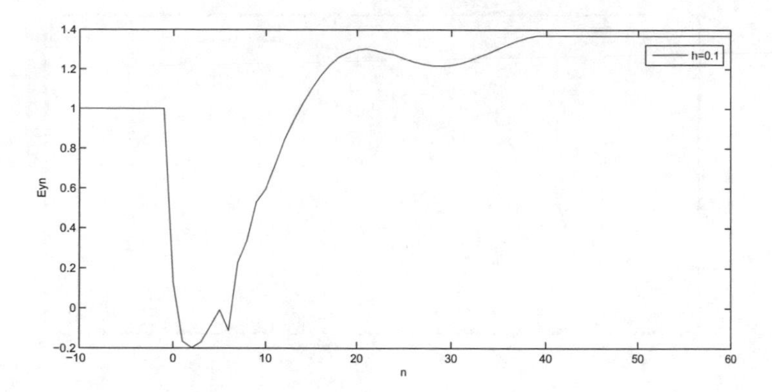

Fig. 1. Mean square stability of semi-implicit Euler method with $h = 0.1$ when applied to problem (26).

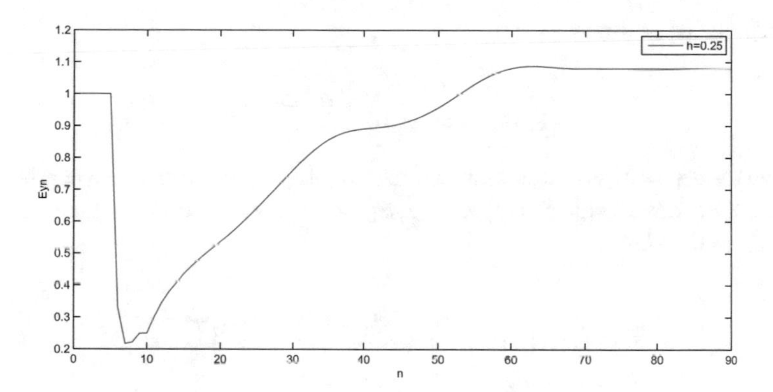

Fig. 2. Mean square stability of semi-implicit Euler method with $h = 0.25$ when applied to problem (26).

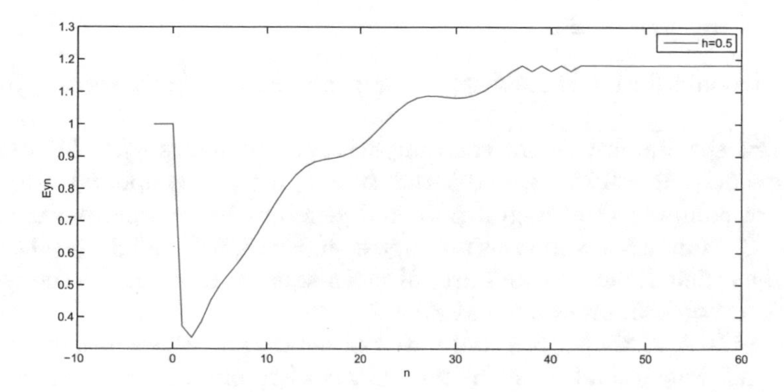

Fig. 3. Mean square stability of semi-implicit Euler method with $h = 0.5$ when applied to problem (26).

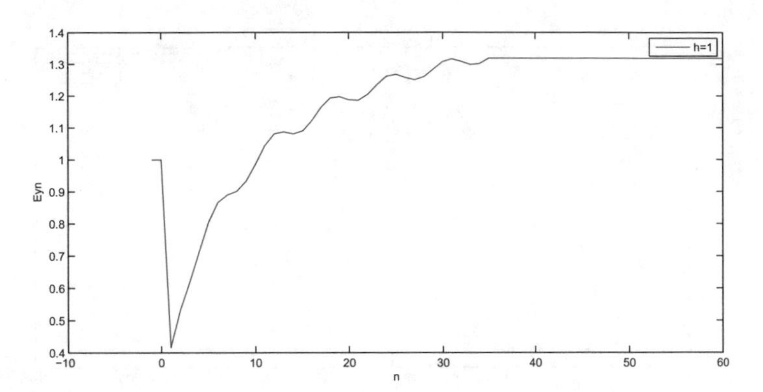

Fig. 4. Mean square stability of semi-implicit Euler method with $h = 1$ when applied to problem (26).

initial condition is given by

$$\begin{bmatrix} y_1(t) = t + \tau \\ y_2(t) = e^t + 2 \end{bmatrix}, t \in [-\tau, 0],$$

where we take $\tau = 1$. For this test, we use the $\|.\|_1$ to compute the correlate coefficients, so we have that $\tau = 1, \lambda = \frac{1}{4}, a_1 = -26, \gamma = 1, a_2 = 4, a_3 = 1, \beta_1 = \frac{7}{2}$, $\beta_2 = 2, \beta_3 = 0$, where,

$$\begin{aligned} &1 + a_2 + \lambda^2 + \frac{\beta_1 + \beta_2 + 4(a_3 + \beta_3)\gamma^2\tau^2}{2} \\ &= 1 + 4 + \frac{1}{16} + \frac{\frac{7}{2} + 2 + \frac{1}{4}}{2} \\ &= 5 + \frac{47}{16} \\ &< 26, \end{aligned}$$

then the trivial solution is mean-square exponentially stable according to Theorem 1.

We test the stability of the semi-implicit Euler schemes with different stepsizes $\Delta t = 0.1$, $\Delta t = 0.25$, $\Delta t = 0.5$ and $\Delta t = 1$, use the compound Trapezoidal formula to compute the integral part and generate 10^3 sample paths for each stepsize. The numerical solutions are plotted in Fig. 5, 6, 7 and 8, they show that the semi-implicit Euler schemes are all mean square stable under the stepsizes $\Delta t = 0.1$, $\Delta t = 0.25$, $\Delta t = 0.5$ and $\Delta t = 1$.

From Figs. 1, 2, 3, 4, 5, 6, 7 and 8 we can see that for the semi-implicit Euler method and Trapezoidal formula, the numerical results are in good agreement with the theoretical prediction (19). Taking into account the above two aspects, we can say that the semi-implicit Euler method is a competitive method for the technology innovation network model,it could maintain the network stability.

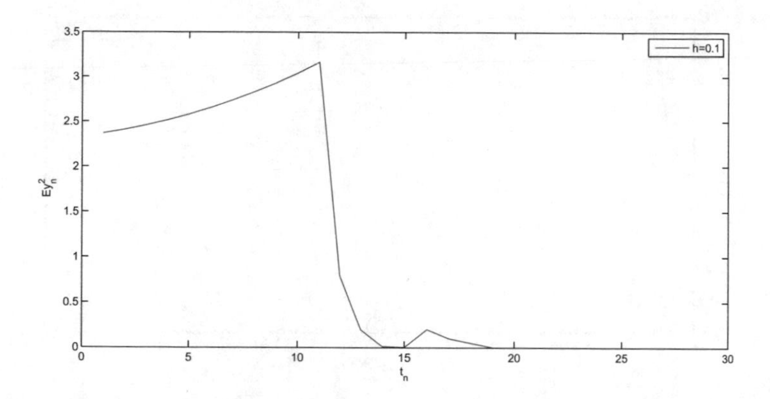

Fig. 5. Mean square stability of semi-implicit Euler method with $h = 0.1$ when applied to problem (27).

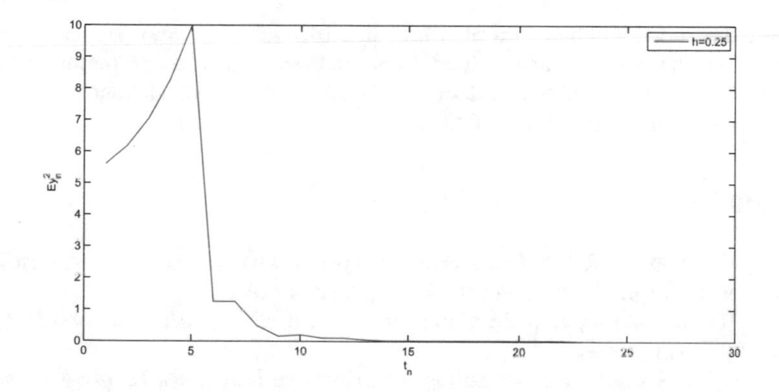

Fig. 6. Mean square stability of semi-implicit Euler method with $h = 0.25$ when applied to problem (27).

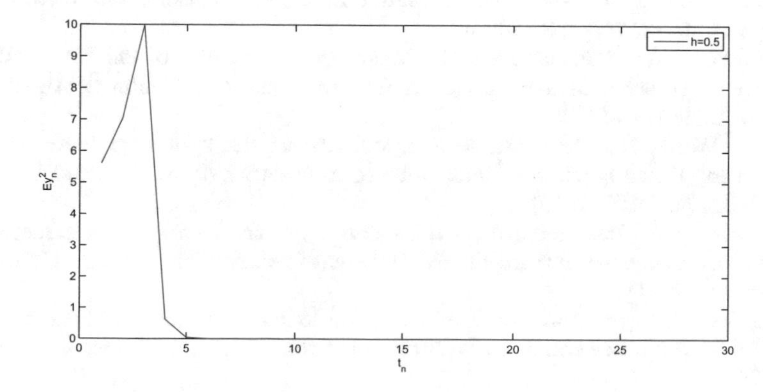

Fig. 7. Mean square stability of semi-implicit Euler method with $h = 0.5$ when applied to problem (27).

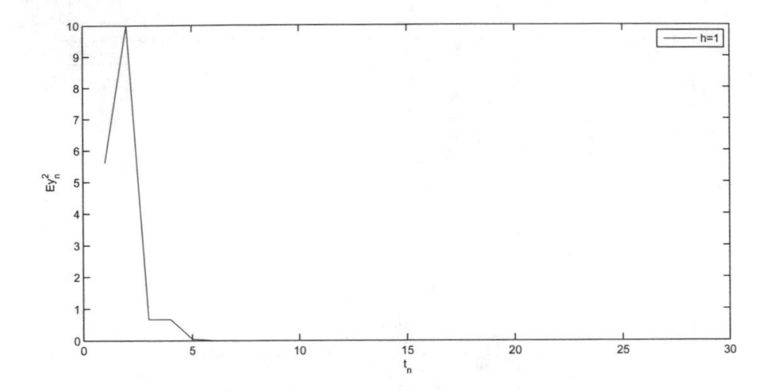

Fig. 8. Mean square stability of semi-implicit Euler method with $h = 1$ when applied to problem (27).

Acknowledgements. The authors would like to thank the anonymous referees for their valuable comments which helped us to improve the present paper. This work was supported by the Creative Talent Project Foundation of Heilongjiang Province Education Department (UNPYSCT-2015102).

References

1. Chang, M., Youree, R.K.: The European option with hereditary price structures: basic theory. Appl. Math. Comput. **102**, 279–296 (1999)
2. Hobson, D.G., Rogers, L.C.G.: Complete models with stochastic volatility. Math. Finan. **8**, 27–48 (1998)
3. Shaikhet, L., Roberts, J.: Reliability of difference analogues to preserve stability properties of stochastic volterra integro differential equations. Adv. Differ. Eqn. **1**, 1–22 (2006)
4. Ding, X., Wu, K., Liu, M.: Convergence and stability of the semi-implicit Euler method for linear stochastic delay integro differential equations. Int. J. Appl. Math. Comput. Sci. **83**, 753–763 (2006)
5. Rathinasamy, A., Balachandran, K.: Mean square stability of milstein method for linear hybrid stochastic delay integro differential equations. Nonlinear Anal.: Hybrid Syst. **2**, 1256–1263 (2008)
6. Tan, J., Wang, H.: Convergence and stability of the split step backward Euler method for linear stochastic delay integro differential equations. Math. Comput. Model. **51**, 504–515 (2010)
7. Rathinasamy, A., Balachandran, K.: T- stability of the split step theta methods for linear stochastic delay integro differential equations. Nonlinear Anal.: Hybrid Syst. **5**, 639–646 (2011)
8. Sytch, M., Tatarynowicz, A.: Exploring the locus of invention: the dynamics of network communities and firms invention productivity. Acad. Manag. J. **57**(1), 249–279 (2014)
9. Li, Q., Gan, S.: Mean square exponential stability of stochastic theta methods for nonlinear stochastic delay integro differential equations. J. Appl. Math. Comput. **39**, 69–87 (2012)

Stepwise Regression Clustering Method in Function Points Estimation

Petr Silhavy[(✉)], Radek Silhavy, and Zdenka Prokopova

Faculty of Applied Informatics, Tomas Bata University in Zlin, nam. T.G.M 5555, 76001 Zlin, Czech Republic
{psilhavy, rsilhavy, prokopova}@utb.cz

Abstract. This study proposed a stepwise regression clustering method for software development effort estimation. The proposed algorithm is based on functional points analysis and is used for forming clusters, which contains analogical projects. Furthermore, it is expected that clusters will be shaped well for the regression prediction models. The proposed models are based on Cook distance, which is used for elimination project from clusters. Model performance is proved for selected clusters. Overall model performance influenced by selected clusters, therefore, there is no statistically significant difference between regression models based on clustered and un-clustered datasets.

Keywords: Regression clustering · Function point analysis
Stepwise regression · Development effort prediction

1 Introduction

Estimation workload and software development effort is the crucial task in the software project management. Software industry suffers from poor project management and budgeting. Software development effort and budgeting are correlated issues. Estimation approaches and methods are under research nowadays. Models from computational statistics [1] and machine learning [2, 3] have been under the investigation.

Function Point Analysis (FPA) is the response of a need of formal approach in estimation; FPA is a standardised method in ISO standardisation. FPA depends on the ability and experiences of an analyst, who is responsible for evaluating parameters. The system is described as a set of functions, but the analyst must understand the system in great detail.

A tendency of personal influence on the estimation can be described in the case of each analyst. The personal opinion of the analyst, which influences the estimation, makes this method unsuitable for comparing productivity in system or software development.

In effort estimation research this personal influence and estimation error can be minimised.

In this paper, the International Function Point Users Group (IFPUG) [4, 5] method is used as the basis for the research. Albrecht [6] introduces FPA method as method base on functional measurement. The original approach is still in use. Data and transaction are taken in use. Both data and transactions are evaluated at the conceptual

R. Silhavy et al. (Eds.): CoMeSySo 2018, AISC 859, pp. 333–340, 2019.
https://doi.org/10.1007/978-3-030-00211-4_29

level, i.e., they represent data and operations that are relevant to the user, based on requirements specification. The boundary indicates the border between the application being measured and the external applications and user domain.

In publications [2, 7, 8] authors declare that selection of data subset - selection of similar past projects can significantly increase estimation ability of models.

The previously published study investigates the significance of using subset selection methods for the prediction accuracy of Multiple Linear Regression (MLR) models, obtained by the stepwise approach.

The results show that clustering techniques decrease estimation errors significantly when compared to unadjusted methods.

It can be concluded that these studies prove subset selection techniques as a significant method for improving the prediction ability of linear regression models - which are used for software development effort estimation.

In this study, a regression clustering method based on stepwise multiple linear regression (RCMLR) is proposed for software development effort estimation, when IFPUG method is used for baseline estimation. This study will answer a research question if regression clustering can significantly improve an unadjusted IFPUG method prediction using MLR. Only the core elements of IFPUG method are used as predictors, and known values of the normalised effort are used as dependent variable.

The rest of the paper is organised as follows: Sect. 2 is an experiment design and problem definition, Sect. 3 describes research question and hypothesis formulation, Sect. 4 brings results and discussion, and Sect. 5 is a conclusion.

2 Experiment Design and Problem Statement

The proposed method of regression clustering in stepwise linear regression (RCMLR) is used as follows:

For this study, the Best Performing Regression Model (BPRM) [1, 2, 9, 10] model is used. The model contains an intercept, linear terms, and squared terms.

For experimental purposes, an International Software Benchmarking Standards Group (ISBSG) dataset was adopted [11]. A total of 1650 data entries were selected (DS). Only the entries which were completed in all needed fields were used. For this study following variables, from IFPUG method were adopted:

The IFPUG which is used in this study introduces 6 parameters, which are derivate from functional requirements. This study used following parameters:

- External Inputs (EI, InputCount), which describes basic data processing or controlling processing incoming to the application over a boundary.
- External Outputs (EO, OutputCount), which is used for sending data or control processes outside an application boundary after processing is performed.
- External Inquiry (EC, EnquiryCount), which sends a data or control processes outside an application boundary, but no processing is performed.
- Internal Logical Files (ILF, FileCount), which describes data processing in relation form, it should be a data table.

- External Interface File (EIF, InterfaceCount), which represents a logically connect data, control information, which are maintained by the external system - outside application boundary.
- Normalised Work Effort – (NWE, NormalisedWorkEffort), which represents a full life-cycle effort for all teams reported. If the full life-cycle effort was not reported, then this is an estimation by ISBSG.

The dataset DS was used to create training and testing sets. Training and testing parts were created using the hold-out approach in ratio 2:1. In Table 1 statistical characteristics of training and testing, the dataset is presented.

Table 1. Dataset statistical characteristics (person-hours)

	n	Median NWE	SD NWE	Min NWE	Max NWE	Range NWE
Training	1,100	2,390	7,003	21	66,600	66579
Testing	550	2,404	11,973	31	150040	150009

In Fig. 1 there can be seen a histogram of the relative sizes training dataset. All data entries are used in functional points, which were count using IFPUG approach. Relative size scales origin in ISBSG dataset and is used as in Table 2 [11]. In Fig. 2, there can be seen a histogram of relative sizes of testing part of the dataset.

Table 2. Relative project size in functional points

Size	Min	Max
XXS	0	10
XS	10	30
S	30	100
M1	100	300
M2	300	1,000
L	1,000	3,000
XL	3,000	9,000
XXL	9,000	18,000
XXXL	18,000	–

2.1 Proposed Algorithm

The proposed RCMLR algorithm is based on stepwise regression approach, which is used for two tasks. Firstly it is used for clustering, where clusters are determined based on Cook distance [12]. Using the Cook distance (1) as a cluster cut-off is possible and influential data entries are identified for regression analysis. A similar approach was introduced by [13].

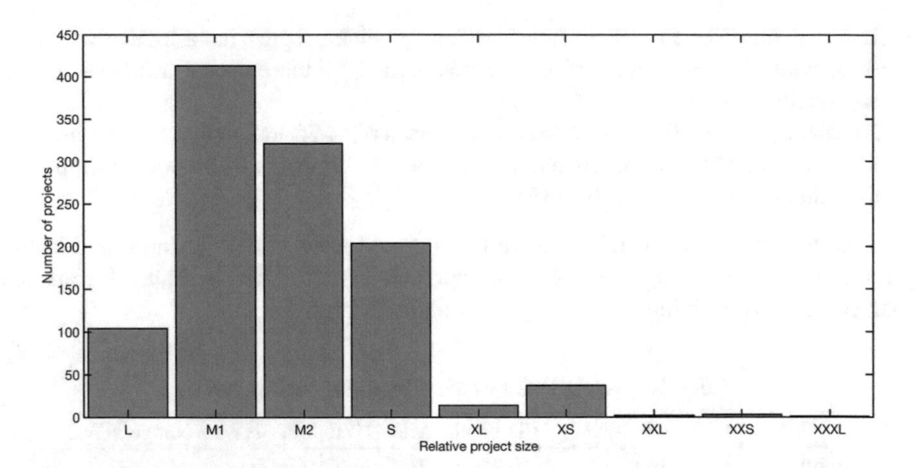

Fig. 1. Project relative sizes in training dataset

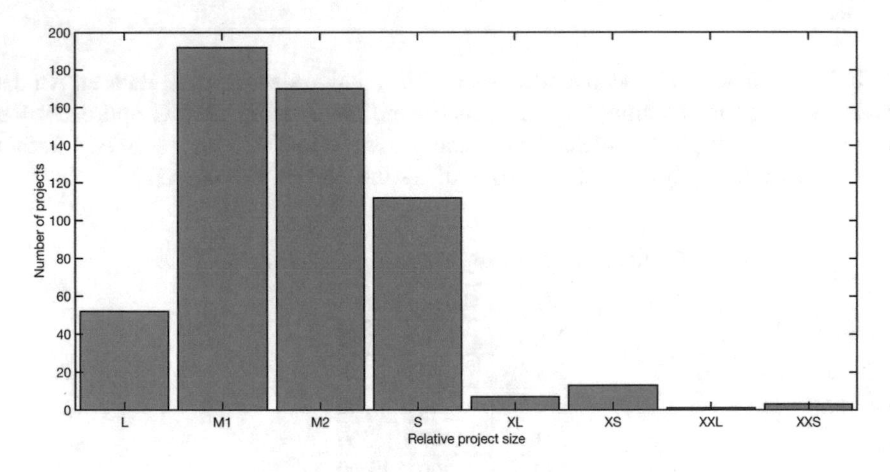

Fig. 2. Project relative sizes in testing dataset

$$D_i = \frac{\sum_{j=1}^{n}\left(\widehat{Y}_j - \widehat{Y_{j(i)}}\right)^2}{(p+1)\hat{\sigma}^2} \tag{1}$$

Where D is Cook Distance and is defined as a sum of all changes in the regression model when specific observation is removed from the regression model. $\widehat{Y_{j(i)}}$ stands for fitthe ted response, p is a number of predictors and $\hat{\sigma}^2$ is an estimated error variance for all observations.

The proposed RCMLR approach works as follows:

1. Creating training and testing sets by using the hold-out (2:1) method.
2. Setting a feature list - EI, EO, EC, ILF, EIF. Dependant variable was set to NWE.
3. Applying a stepwise MLR regression to all projects in training set.

4. Removing all outliners from a cluster.
5. Computing a regression model again, while all data entries in the cluster meet the condition of 3 times of Cook Distance median.
6. Repeat all process as many times as no outliners are identified.
7. Classification observation in testing dataset into clusters, using discriminant analysis.
8. Estimation of work effort in hours is performed by using a cluster-specific model.
9. Prediction error (PE) for projects in the testing set is computed.
10. MAPE and other evaluation criteria are computed.

3 Research Questions and Hypothesis Formulation

RQ1: Can regression clustering improve a prediction ability of stepwise MLR (RCMLR model) when compared to non-clustered MLR?

Let us assume that, the mean prediction error of proposed model (RCMLR) will be significantly lower than stepwise regression model on non-clustered testing data (MLR). To decide whether the employing regression clustering improve a prediction capability. A statistical hypothesis was tested

H_0: $\mu PE_{MLR} = \mu PE_{RCMLR}$; there is no difference in prediction capability between MLR and RCMLR. No difference in the mean of prediction errors.

Alternative Hypothesis

H_1: $\mu PE_{RCMLR} < \mu PE_{MLR}$; there is difference in prediction capability between MLR and the RCMLR. The mean prediction error is significantly lower for RCMLR than for the MLR method.

The statistical significance of the results was validated by using the pair t-test for two samples. It is used as a test of the null hypothesis - that the means of two normally distributed populations (two sample) are equal. The t-test will be used for the evaluation of prediction errors. Usage of t-test was proofed by analysis in [1].

3.1 Evaluation Criteria

All the tested methods were evaluated according to (2) Mean Absolute Percentage Error (MAPE), (3) the Sum of Squared Errors (SSE), (4) Mean Squared Error (MSE) and (5) Prediction Error. MAPE was selected, because of in [14] authors proofs that MAPE has practical and theoretical relevance for evaluation of regression models and its intuitive interpretation regarding the relative error.

$$\text{MAPE} = \frac{1}{n}\sum_{i=1}^{n}\frac{|y_i - \widehat{y_i}|}{y_i} \times 100 \tag{2}$$

$$SSE = \sum_{i=1}^{n} \varepsilon_i^2 \quad (3)$$

$$MSE = \frac{1}{n}\sum_{i=1}^{n} \varepsilon_i^2 \quad (4)$$

$$PE = \varepsilon = y_i - \widehat{y_i} \quad (5)$$

Where n is the number of observations, y_i is the known real value, $\widehat{y_i}$ is the predicted value.

4 Results and Discussion

In this study an RCMLR and MLR models where compared. In the Table 3 are summarised results for the testing dataset (n = 550).

Table 3. RCMLR and MLR models comparison

	n	MAPE	μPE	SSE	MSE	SD
RCMLR	550	542.48	−94.00	1.0936e+12	1.9883e+09	4.4433e+04
MLR	550	227.01	−939.96	8.0596e+10	1.4654e+08	1.2078e+04

In Fig. 3, the histograms of prediction errors can be seen. Both models produce a normally distributed residuals.

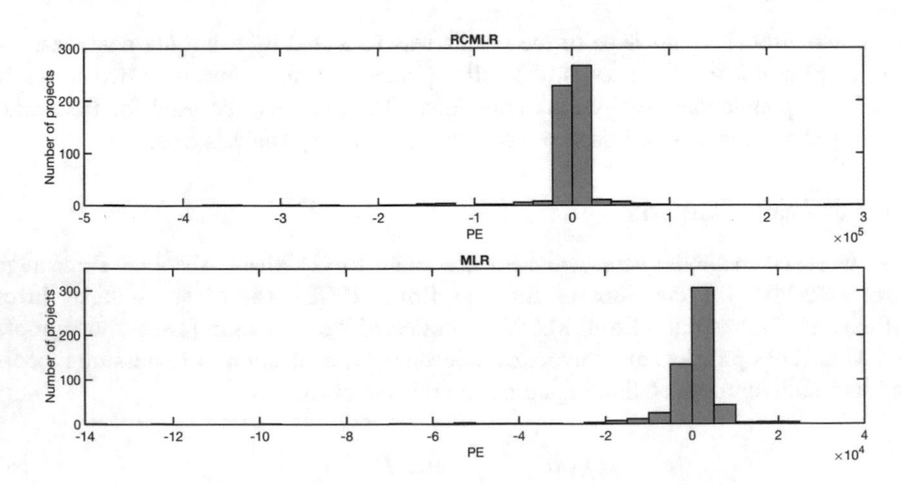

Fig. 3. Prediction errors (PE) for RCMLR and MLR

For RCMLR model the behaviour in clusters shows that performance is not equal. See Fig. 4 for median values of each identified cluster.

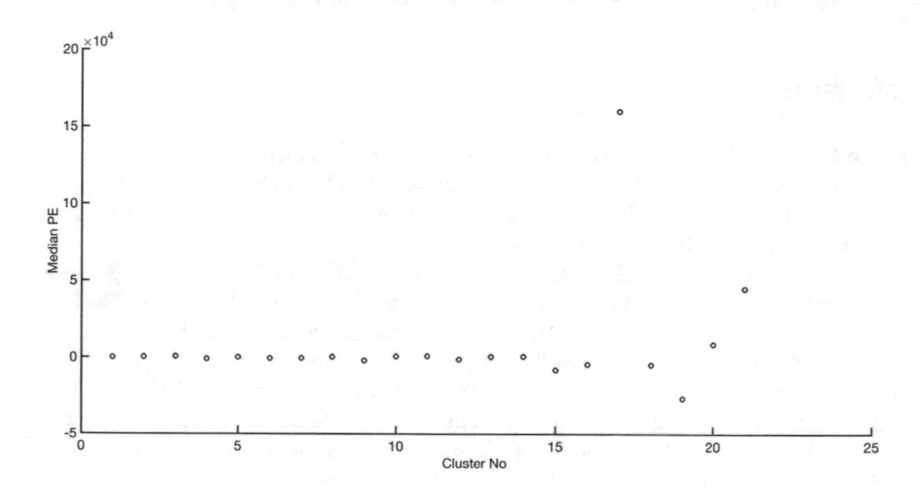

Fig. 4. Median PE for clusters

RCMLR model has found 21 clusters in total, which are not equal in performance. Total performance (Table 3) is influenced by several clusters, which median prediction error is significantly higher than others.

Regarding RQ1, we can declare that regression clustering can improve prediction ability of MLR models. As can be seen, some cluster performs reasonably well, but another no. Those bad performers influenced overall model ability. This is a reason, why H_0 can be rejected at 5% significance level.

5 Conclusion

In this paper, an RCMLR model for software development effort was proposed. The model is using a function points features for clustering. Then the stepwise linear regression is applied. The prediction model is computed for each of identified clusters. The proposed model can identify a number of clusters – count of clusters is not set a priori.

Median PE values are lower when compared to MLR model with no clustering, but overall performance when MAPE or SSE are used shows the worse performance of RCMLR.

Some clusters perform reasonable better (see Fig. 4), but rest of them has very high prediction error.

It can be concluded that RCMLR can find very similar projects which form clusters. Those clusters are naturally suitable for the regression model, but prediction capability of RCMLR is outperformed by non-clustered MLR.

In our future research, we will investigate RCMLR further, which means dealing with metric for cut-off a project from training dataset, investigation and evolution of the clustering ability, investigation and evaluation a classification approach.

References

1. Silhavy, R., Silhavy, P., Prokopova, Z.: Analysis and selection of a regression model for the use case points method using a stepwise approach. J. Syst. Softw. **125**, 1–14 (2017)
2. Silhavy, R., Silhavy, P., Prokopova, Z.: Evaluating subset selection methods for use case points estimation. Inf. Softw. Technol. **97**, 1–9 (2018)
3. Azzeh, M., Nassif, A.B., Banitaan, S.: Comparative analysis of soft computing techniques for predicting software effort based use case points. IET Softw. **12**(1), 19–29 (2018)
4. Abualkishik, A.Z., Lavazza, L.: IFPUG function points to COSMIC function points convertibility: a fine-grained statistical approach. Inf. Softw. Technol. **97**, 179–191 (2018)
5. Abualkishik, A.Z., et al.: A study on the statistical convertibility of IFPUG function point, COSMIC function point and simple function point. Inf. Softw. Technol. **86**, 1–19 (2017)
6. Albrecht, A.J.: Measuring application development productivity. In: Proceedings of IBM Application Development Joint SHARE/GUIDE Symposium, Monterey, CA, 1979 (1979)
7. Idri, A., Amazal, F.A., Abran, A.: Analogy-based software development effort estimation: a systematic mapping and review. Inf. Softw. Technol. **58**, 206–230 (2015)
8. Lokan, C., Mendes, E.: Investigating the use of duration-based moving windows to improve software effort prediction. In: 2012 19th Asia-Pacific Software Engineering Conference (Apsec), vol. 1, pp. 818–827 (2012)
9. Silhavy, P., Silhavy, R., Prokopova, Z.: Evaluation of data clustering for stepwise linear regression on use case points estimation. In: Advances in Intelligent Systems and Computing, pp. 491–496 (2017)
10. Silhavy, P., Silhavy, R., Prokopova, Z.: Evaluation of data clustering for stepwise linear regression on use case points estimation. In: Computer Science On-line Conference, Springer, Cham (2017)
11. ISBSG: ISBSG development and enhancement repository – release 13. International Software Benchmarking Standards Group (ISBSG) (2015)
12. Stevens, J.P.: Outliers and influential data points in regression analysis. Psychol. Bull. **95**(2), 334 (1984)
13. Jayakumar, D.S., Sulthan, A.: A new procedure of regression clustering based on Cook's D. Electron. J. Appl. Stat. Anal. **8**, 13–27 (2015)
14. de Myttenaere, A., et al.: Mean absolute percentage error for regression models. Neurocomputing **192**(Suppl. C), 38–48 (2016)

Comparison of Numerical Models Used for Automated Analysis of Mechanical Structures

Andrzej Tuchołka(✉), Maciej Majewski, Wojciech Kacalak, and Zbigniew Budniak

Faculty of Mechanical Engineering, Koszalin University of Technology, Raclawicka 15-17, 75-620 Koszalin, Poland
{andrzej.tucholka,maciej.majewski,wojciech.kacalak, zbigniew.budniak}@tu.koszalin.pl
http://kmp.wm.tu.koszalin.pl

Abstract. Authors present the comparison of properties observed in numerical models, that were used to classify features of mechanical constructions. We use these models to detect similarities between mechanical designs and compare them on how well they quantify the description of the mechanical elements, and how they deal with arising challenges: multi-dimensionality of feature values, multiple classes of relations, data incompleteness and variability of format. The key conclusion here, is the ability of modern numerical models (i.e. ConvNet, CapsNet) to process information describing mechanical constructions, while including meaningful structural data in the calculations. Application of these models can provide direct assistance in the mechanical design process.

Keywords: Structure analysis · ConvNet · CapsNet

1 Introduction

Simulation based quality grading models require manual and detailed definition of the design of tested structures. Still, the precision of such quality evaluation depends on the platform, detail of the created models, and scope of simulated physics. Furthermore, even highly optimized simulations, due to their inherent complexity, lack in performance and require manual configuration and interpretation of results. To review the calculations and assess the quality of a tested element, simulation results have to be validated against a reference standard or at least against each other. Additionally, for the simulation to correctly and comprehensibly reflect the final behavior of real world objects, its design has to be complete. Such obstacles exist even for trivial design errors, limiting the overall efficiency of validating mechanical designs and delaying it until the final stages of the process.

We aim to reduce the occurrence of mechanical design errors through early detection of design mistakes by looking for similarities between the designed

R. Silhavy et al. (Eds.): CoMeSySo 2018, AISC 859, pp. 341–352, 2019.
https://doi.org/10.1007/978-3-030-00211-4_30

mechanical construction and known design errors represented as antipatterns. To enable detection of such errors we perform the classification of the tested element (when possible its decomposed parts) against the library of incorrect (antipattern) designs [7]. Such comparison can be performed using various numerical models allowing for quantifying the similarity of mechanical constructions (Fig. 1). In this paper we present the observation of the key features found in some of numerical models that increase the automated ability of computers to detect similarities in machine elements and constructions [6,10].

The selection of algorithms we have used while validating the method of Intelligent Quality Assessment [3] is aiming to verify the ability to represent and process symbolic data describing mechanical constructions. The main problem these numerical transformations aim to solve is to quantify a partial and indirect similarity between compared mechanical designs. Trained human (i.e. a mechanical designer), when presented with a technical drawing is capable of quickly detecting design errors, by identifying similarities to other incorrect designs. We aim to identify the properties of numerical methods that enable computers to perform a such a classification activity.

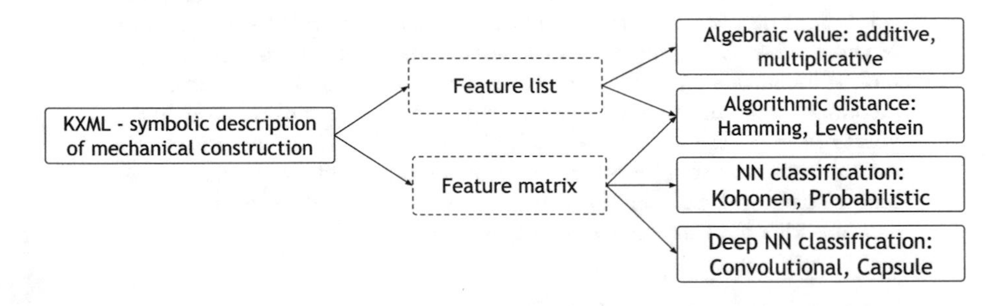

Fig. 1. Overview of the data structures used for selected numerical models

As a reference construction we use a worm gearbox, which as a mechanical construction described with a concrete structure and features, aims to provide several properties of the overall construction: casing for the oil, structural protection, heat dispersion, and worm gear positioning. The quality and predicted performance of a gearbox design is evaluated through the performance in delivering these properties to fulfill the function of the gearbox. Such properties of mechanical elements emerge from a concrete configuration of features and structure of nodes used to describe a gearbox. Considering such configurations of gearbox features, the construction can be malfunctioning (incorrectly designed) in a manner that is repeatable among many designs and can be characterized as containing at least one antipattern that is responsible for such malfunction.

The example antipattern visualized on Fig. 2, is based on the lack of stiffening ribs, which would reduce the vibrations of the gearbox, increase the stiffness of the nodes used for mounting bearings and reduce the possibility of shrinkage cracks. Compared with human interactions with technical drawings, computer software even with several neural network designs specifically created for image processing and detecting similarities in embedded objects, cannot cope with the nature of technical drawings and lack of associated context. Rotations, varying notation styles and often implied character of the notations used to describe the mechanical constructions make it a challenge for direct processing of technical drawings.

The patterns representing incorrect configuration of features and structure (i.e. antipatterns) are directly related with the structure and features of the mechanical design, so the comparison of parts of these constructions provides us with an ability to detect incorrect data patterns (i.e. antipatterns). For a mechanical design expert, searching for such mistakes is a cumbersome and lengthy process due to complexity of the designs and the large amount of potential mistakes that can be present. Due to inconsistency in human expertise levels (e.g. amount of known antipatterns to avoid) and limits of human memory, creation of an automated method capable of performing such similarity classification can result in a relatively cheap and widespread increase of quality of mechanical designs. In this paper we review the properties and the ability, of four kinds of numerical models, to highlight the properties supporting the automated analysis of mechanical constructions.

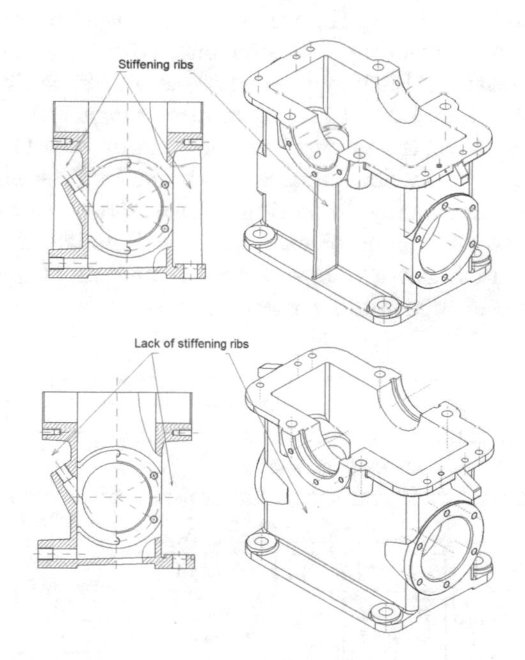

Fig. 2. Antipattern: lack of stiffening ribs on gearbox's case

Each of tested numerical models (e.g. Antipattern Matching Factor, Modified Hamming Net [11], Probabilistic Neural Network, and Convolutional Neural

Network) processes the data describing the structure in a manner that is varies in scope of analyzed data relations and a varying ability to process consistent abstractions of similar data.

The scope of the symbolic representation of antipattern features, as a reference quality measure, should allow to clearly define the pattern we want to detect and the incorrect nature of the antipattern. Among used models, only ConvNet and CapsNet directly support decomposition of the structured description, even if requiring reduced amount of dimensions to be computationally feasible.

2 Symbolic Description of Mechanical Constructions

Many of currently used 3D data formats and related structures [12], are built with abstractions and low level primitives unrelated directly to the function of the modeled object but often specific to the features of the platform defining the description.

To avoid such non-essential complexity of existing computer interpretable data formats (e.g. CAD formats, Standard for the Exchange of Product model data (ISO 10303)), we have created a symbolic language (i.e. KXML [13]) to describe and enable processing of mechanical constructions. To represent a mechanical design with KXML, we manually decompose the construction described on a technical drawing, to a symbolic, object-oriented, tree-like structure of nodes and features. Such symbolic representation can be used embedded with many programming and data representation languages.

We have defined a set of assumptions for creation of a language, such that a construction can be easily decomposed into nodes, and features can be distinguished as elements for altering dimensions, shape, structure, or other properties of processed objects. Still, none of used numerical models allow for structured object oriented input, hence we observe a universal loss of information coming from the limitations of input data structures.

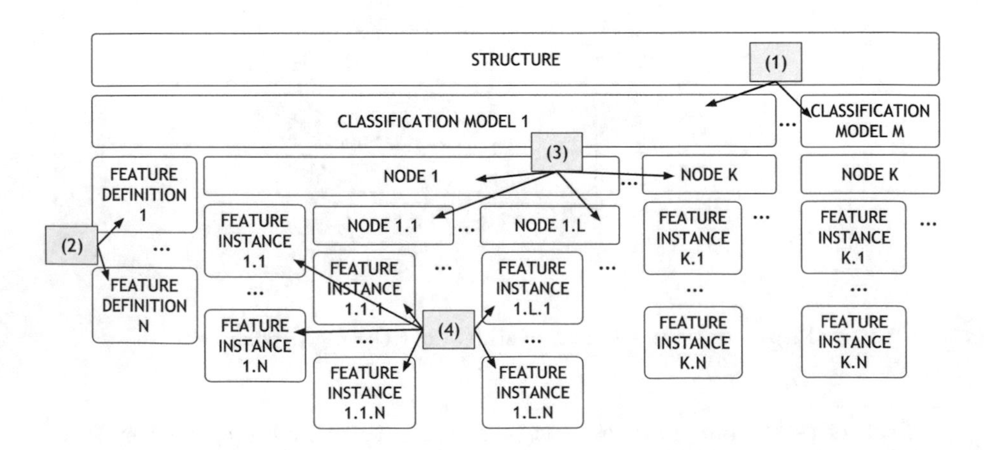

Fig. 3. Overview of KXML symbolic structure

To represent the data describing the construction in KXML (Fig. 3) we use four main elements: (1) a classification model - providing the context for the analysis and interpretation of the features and structural nodes; (2) definitions of features - structuring the data used for calculating the properties of the construction; (3) instances of nodes and their structure - allowing for representing the construction decomposed explicitly into a node structure; (4) instances of features - containing values or math functions allowing to use them in the context of the concrete classification model and its position in the structure of nodes.

Relations emerging from the structure of the language, through the properties of those four main elements, allow to represent contextual information between: (A) elementary elements of the construction - through the structure of classification models, nodes, feature definitions and instances; (B) class definitions for features and nodes enabling common reasoning on values enclosed in feature instances through class definition reuse; (C) inherited node and feature definitions allowing for common analysis of represented elements based on their ancestry and common feature scopes; (D) feature instances in different classification models - through the node identity mechanism; (E) indirect numerical relations included in mathematical functions relating feature values to each other.

Besides increased verbosity that simplifies the integration of the data, proposed KXML language (and in general a concise computer readable format) makes it easier to generate large populations of examples of a specific antipattern. Larger amounts of samples are required by more advanced numerical models (i.e. neural networks [11]) to be able to properly abstract over the features they use to classify the mechanical structure.

3 Data Format Normalization

Automated systems for designing machine components [1,2,4,5,8,9] need to reflect on at least these main properties of the symbolic description: composition of functional elements forming the construction, the values and structure of features, and geometrical relations of elements forming the construction. Traditional computer simulations lack subjective information aiming to define a function of a specific element, while emphasizing completeness of its physical description, such that the function (e.g. providing casing for oil in a gearbox) can be observed and validated by a human interpreting simulation results.

When it comes to verbal communication of mechanical designs between humans, the lexical structure and content are presented on a different level of abstraction focusing on decomposing the structure into objects, defining their function and additional character with their features. Such lexical description might seem incomplete due to missing explicit information about shape (traditionally communicated via technical drawings). Still, by describing decomposed structure, the lexical classification of an element, and description of its features with associated values, information describing the overall shape of the elements can be reasoned upon.

Following outlined above argumentation and to enable classification of mechanical constructions based on their function and features, we have created KXML - an object oriented, symbolic language to represent the structure and feature values describing the construction and avoiding pitfalls of existing computer data formats [7]. Attempting to process as much information from the KXML symbolic representation (e.g. Listing 1), we have to include the feature value and class information in the context and structural position of the node class and its nesting.

On the other hand, the requirements of the numerical models are much more rigid, mostly requiring a low-dimensional input. The additive (i.e. Antipattern matching factor [7]) and multiplicative models we've tested require one-dimensional input, and so provide a very limited ability to process tree-like datasets to detect similarities. Such ability is not allowing for differentiation between tree's nodes and hence becomes useless if the input data's structure is not static (impractical for real life structures).

```
<struct class=''shaft" id=''s01">
  <model class=''mechanical">
    <feature id=''l" unit=''mm"/>
    <feature id=''d" unit=''mm"/>
    <feature id=''R" unit=''mm"/>
    <feature id=''a" unit=''deg" vector=''(0,0,1)"/>
    <node class=''mandrel" id=''p01">
      <feature id=''l">48</feature>
      <feature id=''d">35</feature>
      <node class=''edge">
        <feature id=''l">12</feature>
        <feature id=''a">45</feature>
      </node>
      <node class=''thread">
        <feature id=''l">24</feature>
      </node>
      <node class=''undercut">
        <feature id=''R">8</feature>
        <feature id=''d">8</feature>
      </node>
    </node>
  </model>
</struct>
```

Listing 1. KXML description of a simple shaft

The application of more complex algorithms and neural networks allow for more structure in the input and so contain more of information describing the structure of the element, still due to computational complexity we have to transform the object-oriented description into some low-dimension data-set, matching the requirements of a concrete numerical model. Introducing any dimension reducing model (e.g. Principal Component Analysis) would increase the blurring of the relationship between the data, so aiming to preserve as much information as possible, we're using data matrices with lexically normalized structure of features.

3.1 Feature List

To facilitate calculation of a similarity between two mechanical constructions using algebraic and algorithmic models, we use normalization of feature values into a list. The values of features contained in such a list-based description of a construction, require to be extensively standardized within a consistent value scale (to accommodate for variations of values in scale), description scope (to accommodate for inability to decompose or process parts of he description). Also, the calculations done using these models are most effective, when performed on derived properties (i.e. weight to length ratio or difference between angles) enabling a simple comparisons between features of the construction.

Attempting to avoid adding unnecessary complexity (through deriving properties), each of the features included in the calculation is standardized to the same base (i.e. 1...100). While there can be different functions used to normalize concrete features, the base value of each has to be identical. This assures that the impact of the calculated distance between features is equally weighted and can be universally processed.

Furthermore, introduction of such normalization functions provides us with several opportunities to improve the efficiency of performed calculations. One lies in mapping the correctness bounds (of feature value) in each class of features. For example, measuring the collineation angle between the bolt and the slot, within the normalization function we can assume that all values above five degrees are incorrect, while collineation value of zero is correct. This indicates, that bounded algebraic functions can be very useful in tuning the calculation of the feature distances. Furthermore, these modifiers can be used when there is a need to demonstrate the non-proportional nature of distances between feature values. For example, a 25% difference between a collineation value in an example range (0...5) can represent a "drop in quality" which could differ between 0 and 1 compared to distance between 4 and 5.

As much as we have found several techniques improving the efficiency of this data format, the lack of structural information and the inability to contain data describing the relations, make it (and associated numerical models) hardly useful for the purpose of comparing mechanical structures.

3.2 Feature Matrix

While increasing the amount of structural information passed to the numerical model, we can see how simple adding of class name, provides us with additional analytical data (e.g. through similarity of function and features) that could be used to detect structural similarities with increased precision. Flattening out the internal tree-like data structure of KXML, the two dimensional input matrix

(1) and (2) present a map-like matrices of feature values of each node in the antipattern structure (Listing 1).

$$s01 = \begin{bmatrix} mandrel & 48 & 35 & 0 & 0 \\ edge & 12 & 0 & 0 & 45 \\ thread & 24 & 0 & 0 & 0 \\ undercut & 0 & 8 & 8 & 0 \end{bmatrix} \quad (1)$$

Including class names (1) also allow for better interpretation of feature values by placing them in a lexical context which could be further analyzed (e.g. for correctness of the description). Still, this approach limits the conveyed information with regards to the structural relations (e.g. nesting) between the nodes of the construction.

To improve upon the above format, we can use a symbolic dictionary to add the information about structure and uniqueness of the node's classes (2). This format requires additional mechanisms for parsing and processing of the symbols used in the first column, and is not providing easy means for including structural data in the calculation.

$$s01 = \begin{bmatrix} 1 & 48 & 35 & 0 & 0 \\ 1.2 & 12 & 0 & 0 & 45 \\ 1.3 & 24 & 0 & 0 & 0 \\ 1.4 & 0 & 8 & 8 & 0 \end{bmatrix} \quad (2)$$

Attempting to include more information from the KXML symbolic description, we have to include the feature value and class information in the context of the node class and its nesting. To avoid creation of additional dimensions, we can embed such information in a more complex matrix (3). Here, in addition to the feature values (columns 5–7), the matrix is expanded with class of the node (1st column) and node's nesting level (columns 1–4).

$$s01 = \begin{bmatrix} mandrel & & 48 & 35 & & & \\ mandrel & edge & 12 & & & & 45 \\ mandrel & thread & 24 & & & & \\ mandrel & undercut & & & 8 & 8 & \\ \ldots & \ldots & \ldots & \ldots & \ldots & \ldots & \ldots & \ldots \end{bmatrix} \quad (3)$$

In the final matrix (3), we show an additional challenge - the zero and non-existence of values (i.e. missing) or features not being assigned to a specific node. Correct interpretation of data with such property creates a challenge to address, as the input matrix formats (compared to object-oriented) impose the structure that through inclusion of incorrect (non-existent but assigned zero) feature values will result with incorrect classification output.

The most obvious problem observed here are the neural networks classifying the overwhelming similarity of zero values of non-existent features but found in incorrect matrix descriptions.

Compared, to rather minimal data matrices above, the full construction contains multiple nesting levels and feature definitions, both increasing the m (horizontal) dimension of the matrix. Its n (vertical) dimension is directly related to the amount of nodes in the description. Each row of such a matrix is a node, and having the antipatterns being formed by relations between features of such nodes, we have a low dimensional dataset that can be processed as input.

Event if this data format (3) representing structural data was most fit for our needs, it is worth noting that it still is missing information about structural relations, hinting on the possibility of greater improvements in the capability to meaningfully serialize data describing mechanical constructions.

4 Classification of Mechanical Structures

Detecting similarities in mechanical structures through classification will vary depending on the ability of the model to: consistently process feature values contained in the symbolic description, decompose the structure and process its parts, and avoid data driven challenges (i.e. multi-dimensionality, position variation) without excessive data loss. Furthermore, as observed while testing the Method of Intelligent Quality Assessment using Antipatterns [3] the usefulness of such classification is limited unless the classification is performed against a large population of structural, mechanical solutions. As we aim to detect design errors (and probably for most other applications), the ability of the model to detect as many design errors as possible, is crucial and has to be computationally efficient.

In addition to the transformations performed during the data format normalization, we observe an increase in relevance of the output with the change in methods used for processing of the input data. The algebraic, additive and multiplicative models process feature values that have been normalized within the scope of each of the structures and also among all structures. This results in heavy loss of information throughout the whole process, emphasized by the need to work with derived, indirect features.

Similar loss of information can be observed in the model implementing the Modified Hamming Distance. In this case, such limitation arises from the need to represent feature values with the same amount of information bits. This problem doesn't occur in a model based on Levenshtein distance as it is a numbering required changes - indirectly processing the input feature values.

Neural Networks enable a more direct processing of the symbolic feature data input (with limited support for structural data), at least on the first layer of the network.

An exception here are Kohonen maps which require scaling of the feature values (within a concrete feature dimension) among the whole population of maps representing incorrect topologies. Attempting to avoid overlapping and natural blurring of data occurring on creating maps representing unrelated classes of structures, the Kohonen classifiers (i.e. concrete maps created per structure or antipattern class) have to independently and coherently represent structural patterns by processing pre-scaled input data (feature values and scopes of represented relations between features).

Among tested models, only the features specific to convolution layers and introduced by ConvNet design (i.e. kernels and filters) provided us with the capability to adapt the design of the model to reflect on the composition and relations found in the symbolic description of the construction. Comparing that with the ability of CapsNet to abstract over specific objects (rather then hoping the sliding window will contain the local patterns), we observe an obvious improvement arising from the ability to dynamically detect and abstract over concrete structures (i.e. patterns) found in data.

Reflecting on the capabilities of models based on Deep Neural Networks to process the structure, again the CapsNet design with its unique approach to internal network connectivity (i.e. routing) demonstrates an improvement by enabling more flexibility and so, the ability to represent more of the relations found in the symbolic data (Table 1).

Table 1. Key properties of selected numerical models

Numerical model	*Values*	*Relations*	*Structure*	*Complexity*	*Output*
CapsNet	direct	multi-feature	per design	linear	pattern similarity
ConvNet	direct	per design	scoped	linear	pattern similarity
Kohonen nets	pre-scaled	scaled	scoped	exponential	limited precision
Probabilistic net	direct	indirect	indirect	linear	hard to justify
Levenshtein dist.	indirect	n/a	n/a	exponential	error prone
Hamming dist.	normalized	n/a	n/a	exponential	low precision
Additive	normalized	n/a	n/a	exponential	error prone
Multiplicative	normalized	n/a	n/a	exponential	hard to relate

The largest impact on relevancy of the output data, has the model's ability to process the feature relations and the structure of the description. In case of simple models, we have not been able to adapt them to include such information. Furthermore, their inability to dynamically adapt the internal variables, creates an exponential increase in complexity that quickly gets out of control with the increase of reference structures and features used to describe them. Neural models require an additional phase of training, still the complexity of classifying a mechanical structure is linear (as in requiring one execution of a "black box" NN classifier to produce output).

Compared with simple algebraic models, neural networks require additional configuration, data normalization to be performed adjusting the data and the model to fit each other. The benefits of applying neural networks become barely visible with ConvNet design, due to: local feature detectors (sliding window of convolutions) enabling structural analysis of the construction, and the layered design enabling varying scopes of convolutions on tested data. It is possible to further increase the efficiency of the design by introducing custom kernels and approach to data convolution.

ConvNet's two main features: the location invariance and compositionality, enable detection of patterns in any location (position) of the symbolic data and combining data patterns found in relations (supported by designed kernel and filter configurations). Comparing the availability of such features in the numerical method used for Intelligent Quality Assessment, demonstrates a clear advantage of ConvNet design (and these features) compared to other numerical models missing an ability to process the relations [14]. Through different kinds of kernels, both ConvNet and Kohonen maps require manual scoping of the calculations, limiting the benefits coming from inclusion of structural data in the analysis. This property is further improved upon in the CapsNet design by allowing to design the network to detect patterns with structure processing layers designed to fit the structure sizes common for patterns found in the data.

Comparing the outputs obtained from these models, we observe a clear correlation between the relevance and precision of results and the ability of the model to process structured data. It is important to note here, that all of the models were working with input that had already lost some of the structural data during the format normalization. Attempting to include some of structural data (e.g. as demonstrated on Eq. 3) requires designing custom kernels. Such limitation of ConvNet is partially addressed in CapsNet through its equivariance property enabling abstraction over specific patterns (i.e. objects) that are detected by the network.

5 Conclusions

Modern, Deep Neural Network models (i.e. ConvNet, CapsNet) demonstrate an increased ability to detect similarities in symbolic descriptions of mechanical constructions. Such improvement arises from application of techniques enabling these models to decompose, interpret, and dynamically process the input data. Pattern recognition capabilities of the Deep Neural Networks, allow us to overcome the limitations of algebra and simple algorithms by including computationally effective features: sliding kernels, custom designed filters, capsule routing. Still, by using pooling, fully connected, softmax layers, both ConvNet and CapsNet loose relevancy of processed data making it difficult to track, analyze or make guarantees with regards to their output. Such improvement in the ability to process mechanical constructions is directly correlated with advances in image, text, and medical data analysis. Calculation models, based on neural networks, have created a unique opportunity to widen the tool-set of mechanical designers by automated algorithms detecting common design errors. We foresee several data integration and processing challenges (e.g. automating structure decomposition, dynamic or intelligent scoping of features) that still that can further increase the precision of classifying mechanical structures.

References

1. Kacalak, W., Majewski, M.: New Intelligent interactive automated systems for design of machine elements and assemblies. In: Huang, T., Zeng, Z., Li, C., Leung, C.S. (eds.) ICONIP 2012, Part IV. LNCS, vol. 7666, pp. 115–122. Springer, Heidelberg (2012)
2. Kacalak, W., Majewski, M.: Interactive design of machine elements and assemblies. Arch. Mech. Technol. Autom. **34**(3), 13–22 (2014)
3. Kacalak, W., Majewski, M., Tuchołka, A.: Intelligent assessment of structure correctness using antipatterns. In: International Conference on Computational Science and Computational Intelligence, pp. 559-564. IEEE Xplore Digital Library. IEEE (2015)
4. Kacalak, W., Majewski, M., Budniak, Z.: Intelligent automated design of machine components using antipatterns. In: Jackowski, K., Burduk, R., Walkowiak, K., Wozniak, M., Yin, H. (eds.) Intelligent Data Engineering and Automated Learning. Lecture Notes in Computer Science, vol. 9375, pp. 248–255. Springer, Cham (2015)
5. Kacalak, W., Majewski, M., Stuart, K., Budniak, Z.: Interactive systems for designing machine elements and assemblies. Manag. Prod. Eng. Rev. **6**(3), 21–34 (2015)
6. Kacalak, W., Majewski, M., Budniak, Z.: Worm gear drives with adjustable backlash. ASME J. Mech. Robot. **8**(1), 014504 (2015). ASME Press
7. Kacalak, W., Majewski, M., Tuchołka, A.: A method of object-oriented symbolical description and evaluation of machine elements using antipatterns. J. Mach. Eng. **16**(4), 46–69 (2016)
8. Kacalak W., Majewski M.: Interactive design of machine elements in uncertainty and unrepeatability. In: Manufacturing 2014: Contemporary Problems of Manufacturing and Production Management, pp. 57–64 (2016)
9. Kacalak W., Majewski M., Budniak Z.: Analysis of similarities between structural features of designed machine elements and corresponding antipatterns. In: Manufacturing 2014: Contemporary Problems of Manufacturing and Production Management, pp. 135–142 (2016)
10. Kacalak, W., Majewski, M., Budniak, Z.: Innovative design of non-backlash worm gear drives. Arch. Civ. Mech. Eng. **18**(3), 983–999 (2018)
11. Majewski, M., Zurada, J.M.: Sentence recognition using artificial neural networks. Knowl. Based Syst. **21**(7), 629–635 (2008)
12. McHenry, K., Bajcsy, P., (2008) An Overview of 3D Data Content, File Formats and Viewers, National Center for Supercomputing Applications, University of Illinois at Urbana-Champaign, Urbana, IL, Technical report ISDA 08-002
13. Tuchołka, A., Majewski, M., Kacalak, W.: Object-oriented, symbolic notation for design features, relations and structures. Mach. Eng. **20**(1), 112–120 (2015)
14. Tuchołka, A., Majewski, M., Kacalak, W., Budniak, Z.: A method for intelligent quality assessment of a gearbox using antipatterns and convolutional neural networks. In: Silhavy, R. (ed.) Artificial Intelligence and Algorithms in Intelligent Systems. Advances in Intelligent Systems and Computing, vol. 764, pp. 57–68. Springer, Cham (2019)

Particular Analysis of Normality of Data in Applied Quantitative Research

Marek Vaclavik[1(✉)], Zuzana Sikorova[1], and Tomas Barot[2]

[1] Department of Education and Adult Education, Faculty of Education, University of Ostrava, Fr. Sramka 3, 709 00 Ostrava, Czech Republic
{Marek.Vaclavik, Zuzana.Sikorova}@osu.cz
[2] Department of Mathematics with Didactics, Faculty of Education, University of Ostrava, Fr. Sramka 3, 709 00 Ostrava, Czech Republic
Tomas.Barot@osu.cz

Abstract. Statistical methods have been widely applied in the concrete realizations of researches. A suitable procedure for providing of a significant research is generally based on a complementation of two types of approaches – a qualitative or a quantitative research. In the current researches in a field of education, the quantitative research is principally bound on computations of appropriate statistical methods; however, a modular approach of a software using of these methods can be frequently seen. Implementations and mathematical principles can be hidden and often unknown for some users of software support, because a modern software interface can return results in a comfortable form of *p*-values. In this paper, a normality testing of data is highlighted and further analyzed with focusing on relations between a normality occurrence and size of population using an example of a practical application of a pedagogical research.

Keywords: Normality of data · Test of hypotheses · p-value · Parametric tests Non-parametric tests · Statistical research

1 Introduction

Established statistical methods [1–5] or their further researches e.g. [6] can be applied in a statistical processing of data of statistical researches, e.g. [7–11], in a wide spectrum of areas of sciences. This paper is focused on area of education [12–18], where inappropriate phenomenons bound on level of mathematical background of academics can be appeared. Statistical software solutions, e.g. a free-available software PAST described or denoted in [19, 20], assume a required knowledge of a statistical theoretical background known by their users. Main problem about properties of data, especially based on normality [21], is discussed in this contribution.

Properties of data should be analyzed in a phase of their processing; however, this analysis has not been seen in all presented researches; especially, a testing of a normality of data is a significant important condition for a further testing of hypotheses. With respect to its conclusions, a particular statistical method is then used in a

R. Silhavy et al. (Eds.): CoMeSySo 2018, AISC 859, pp. 353–365, 2019.
https://doi.org/10.1007/978-3-030-00211-4_31

framework of a testing of hypothesis. These methods can be divided into opposite categories - the parametric [3–5] and non-parametric tests [1, 2].

In category of the parametric tests, generally used tests ANOVA [3] and T-test [4] are determined for applications in case of a hypotheses testing on significant dependences between one categorical and one numerical variable, in one dimensional case. A problem is that these methods are sometimes used also in cases, where the non-parametric tests should be used corresponding with conclusions of a testing of a normality of data. The non-parametric statistical methods are not based on a normal probability distributed data; therefore, these tests are developed for these specific purposes. Hypotheses are then tested with inappropriate conclusions, when data is normal distributed and the parametric tests are used.

For purposes of testing of hypotheses on significant dependences between variables, following alternatives are used: for ANOVA – the Kruskal-Wallis test [1], for T-test – the Mann-Whitney test [2]. Utilization of a concrete variant of tests depends on a specified number of items by a categorical variable. A selection between the parametric and non-parametric tests can be determined using two similar tests on a normality of data by the Shapiro-Wilk [21] and Anderson-Darling tests [21].

In favor of correct conclusions in a quantitative research, this described procedure should be always processed. This recommendation can be suitable implemented using statistical software solutions, as can be seen in a majority cases of all presented researches of academics. However, a deeper understanding of equations in each statistical method is not necessary, because the software solutions access to a testing of hypothesis by a modular approach with results in a form of p-value [15]. This representation of software results is same in application of a wide spectrum of methods – in case of methods aimed for a testing of hypotheses as well as for a testing of a normality of data.

An advantage of using of p-values is bound with their interpretation, that can be related to a significance level [15] in a quantitative research. In a pedagogical area, a significance level equal to 0.05 is frequently used. In technical and medical sciences, a stricter significance level [15] equal to 0.01 or 0.005 can be seen. In this paper is deeply analyzed a strategy of using of p-values [15] in case of their utilization depended on a number of population of a same data source. An appropriate conclusion should have a form of a similar trend of this dependency by an acceptation of stochastic fluctuations. In a tests of hypotheses and in tests of a normality of data, results of p-values should cause same conclusions in an interval of data. These aspects are detail analyzed in case of a practical pedagogical research on significance level 0.05. Conclusions of this analysis can be advantageous for a positive approach to application of a p-values strategy [15], in testing of a normality and hypotheses in pedagogical researches.

2 Statistical Methods for Purposes of Testing of Hypotheses

A frequently used tests on hypotheses in an academicals researches consist of testing on statistical significant dependences. The following conclusions are related to a significance level, which is considered in humanistic sciences as 0.05. In technical or medical sciences can be this constraint more strictly defined.

With focusing on a testing of statistical significant dependences, number r of hypotheses can be declared in a quantitative research. In case of two variables, the first variable has a categorical type with s items and the second variable has a numerical form. Settings of number s can be expressed in (1) and for variable indices (2) can be denoted also other aspects of a testing of hypotheses: a sequence of numerical variables (3), hypotheses (4) and mean values of numerical variables (5).

$$r, s \in \mathcal{N}; s \geq 2 \tag{1}$$

$$i, j \in \mathcal{N}; i \in \langle 1; r \rangle; j \in \langle 1; s \rangle \tag{2}$$

$$X_1^1, \cdots, X_j^i, \cdots, X_s^r \tag{3}$$

$$H^1, \cdots, H^i, \cdots, H^r \tag{4}$$

$$\mu_1^1, \cdots, \mu_j^i, \cdots, \mu_s^r \tag{5}$$

Each testing of hypotheses can be consisted of two opposite parts: a zero hypothesis (6) and an alternative hypothesis (7) with definitions in Table 1.

$$H_0^i : \mu_1^i = \cdots = \mu_s^i \tag{6}$$

$$H_1^i : \mu_1^i \neq \cdots \neq \mu_s^i \tag{7}$$

Table 1. Definition of hypotheses about statistical significant dependences

$H^i; i \in \mathcal{N}; i \in \langle 1; r \rangle$		Definition
H^i	$H_0^i : \mu_1^i = \cdots = \mu_s^i$	There are not significant dependences between numerical variable (3) and categorical variable with s items
	$H_1^i : \mu_1^i \neq \cdots \neq \mu_s^i$	There are significant dependences between numerical variable (3) and categorical variable with s items

In a concrete phase of the testing of hypotheses, p-value [15] can be determined by a statistical software, e.g. in [19]. Then this p-value is compared with a declared significance level α [15], which is a rational number. As can be seen in (8), two cases of relations between these values can occurred: a zero hypothesis is failed to reject or a zero hypothesis is rejected in favor of a significance level α.

$$\left. \begin{array}{c} i, j \in \mathcal{N}; i \in \langle 1; r \rangle; j \in \langle 1; s \rangle; \alpha \in \mathcal{Q}; \\ \exists! p = p(H^i) \in \mathcal{R}; p \in \langle 0; 1 \rangle : \begin{cases} p > \alpha \Rightarrow H_0^i : \mu_1^i = \cdots = \mu_s^i; \alpha \\ p \leq \alpha \Rightarrow H_1^i : \mu_1^i \neq \cdots \neq \mu_s^i; \alpha \end{cases} \end{array} \right\} \tag{8}$$

Before these tests of hypotheses should be included a significant important part - tests of normality of research data. On a significance level α, a normal probability distribution N of numerical variables (3) should be analyzed. A normal probability distribution is given by mean value μ equal to 0 and unit variation σ^2 in this cases. A probability density function of distribution N can be described as $h(x)$ (9).

$$\left.\begin{array}{c} i \in \langle 1; r \rangle; j \in \langle 1; s \rangle; \forall x \in X_j^i : \\ \left(X_j^i \sim N(\mu = 0; \sigma^2 = 1)\right) \wedge \left(h(x) = \frac{1}{\sqrt{2\pi}} e^{-\frac{x^2}{2}}\right) \end{array}\right\} \tag{9}$$

For each combination of data, which belongs to a particular item j of a categorical variable bound on a concrete i-th hypothesis in (10), definition of hypothesis for purposes of a test of a normality can be seen in (11)–(12). In description (11), zero hypotheses are displayed and are the mirror opposites to alternative hypotheses in (12).

$$\tilde{H}^{1,1}, \cdots, \tilde{H}^{i,j}, \cdots, \tilde{H}^{r,s} \tag{10}$$

$$\left.\begin{array}{ccc} \tilde{H}_0^{1,1} : X_1^1 \sim N(0; 1) & \cdots & \tilde{H}_0^{1,s} : X_s^1 \sim N(0; 1) \\ \vdots & \ddots & \vdots \\ \tilde{H}_0^{r,1} : X_1^r \sim N(0; 1) & \cdots & \tilde{H}_0^{r,s} : X_s^r \sim N(0; 1) \end{array}\right\} \tag{11}$$

$$\left.\begin{array}{ccc} \tilde{H}_1^{1,1} : X_1^1 \nsim N(0; 1) & \cdots & \tilde{H}_1^{1,s} : X_s^1 \nsim N(0; 1) \\ \vdots & \ddots & \vdots \\ \tilde{H}_1^{r,1} : X_1^r \nsim N(0; 1) & \cdots & \tilde{H}_1^{r,s} : X_s^r \nsim N(0; 1) \end{array}\right\} \tag{12}$$

Results of a test of a normality of a concrete data are evaluated using rules (13), which are similar to rules (8). Generally used methods for a providing of a procedure (13) can be the Shapiro-Wilk and Anderson-Darling tests [21].

$$\left.\begin{array}{c} i,j \in \mathcal{N}; i \in \langle 1; r \rangle; j \in \langle 1; s \rangle; \alpha \in \mathcal{Q}; \\ \exists! p = p(\tilde{H}^{i,j}) \in \mathcal{R}; p \in \langle 0; 1 \rangle : \\ \left\{\begin{array}{l} p > \alpha \Rightarrow \tilde{H}_0^{i,j} : X_j^i \sim N(\mu = 0; \sigma^2 = 1); \alpha \\ p \leq \alpha \Rightarrow \tilde{H}_1^{i,j} : X_j^i \nsim N(\mu = 0; \sigma^2 = 1); \alpha \end{array}\right. \end{array}\right\} \tag{13}$$

Corresponding with conclusions of procedure (13), a final statistical method for test of hypothesis can be selected. For a case of normal data, a parametric ANOVA test [3] (for s greater or equal to 3) or T-test [4] (for s equal to 2) can be applied. In an opposite case of application of non-parametric tests for non-normal distributed data, the Kruskal-Wallis test [1] (for s greater or equal to 3) or the Mann-Whitney test [2] (for s equal to 2) can be further used.

3 Proposal of Particular Analysis of Normality of Data

In a framework of an appropriate range (14) of data, respectively of samples of a research population with n units, results of a software processing of particular statistical methods should have a majority stable behaviour in this interval. Therefore, the approach of using of p-values should be more wide applied in the quantitative researches as a significant safe strategy. In case of a determination of a normality of data is even this approach necessary with further influence on a quality of conclusions of tests of hypotheses. An analysis for a mapping of advantages of a p-value strategy [15] is proposed in this chapter.

In expressions (14), constraints for a range of population are defined in sense of generally rules of a quantitative research. A particular natural numbered variable n_k is further used in a proposal of an analysis in this paper. This value is constrained by a minimum and maximum limits of a selected samples of data with a final number of n samples. Returnees of responds of data can be expressed as η.

$$\left.\begin{array}{c} n_k \in \mathcal{N};\ n_k \in \langle n_{\min}; n_{\max} \rangle \\ n_{\min} \leq n_k \leq n_{\max} \leq n \\ \eta = \frac{n_{\max}}{n}\,[\%] \end{array}\right\} \quad (14)$$

Functions, which can be used for an expression of dependences between p-values and variable n_k, can be denoted as f and g. Two cases are categorized: a dependence between p-value on variable n_k in case of a testing of a normality of data (15) and a dependence between p-value on variable n_k in case of a testing of hypotheses (16). Because there can be a high number of hypotheses and items of a categorical variables - a notation with indices i and j is further used.

$$i,j \in \mathcal{N};\ p = p(\tilde{H}^{i,j}) = f(n_k) \equiv p(\tilde{H}^{i,j}, n_k) \quad (15)$$

$$i \in \mathcal{N};\ p = p(H^i) = g(n_k) \equiv p(H^i, n_k) \quad (16)$$

In favor of a mapping of the described forms of dependences, a concrete set of a relation can be defined by (17). In this paper, a providing analysis can be bound on relation (18) for case of a testing of a normality of data and on relation (19) for purposes of a testing of hypotheses.

$$\Omega = \{\geq, >, =, <, \leq\} \quad (17)$$

$$i,j \in \mathcal{N}; i \in \langle 1; r \rangle; j \in \langle 1; s \rangle; \circ \in \Omega : \left(p(\tilde{H}^{i,j}, n_k) \circ \alpha\right) \quad (18)$$

$$i \in \mathcal{N}; i \in \langle 1; r \rangle;\ \circ \in \Omega : \left(p(H^i, n_k) \circ \alpha\right) \quad (19)$$

In both cases, p-value is analyzed for constrained variable n_k as can be seen in (20). Dependences can be further displayed in charts in a comparison with a significance level 0.05 using expression (20).

$$\circ \in \Omega;\ \forall n_k \in \{\varepsilon + 2\}_{\varepsilon = n_{\min}}^{n_{\max}} : \left(p(n_k) \circ \alpha\right) \tag{20}$$

4 Results

Because a potential absence of a test of a normality of processed data can have a negative influence on results of a testing of hypotheses, it is a necessary part of procedures in quantitative researches. However, this testing of a normality of data has not been appeared in all published researches in humanities sciences.

Particular research, presented in this paper as an applied example for a demonstration of aims, is focused on a pedagogical research. Random variables, in a form of research data, and their mean values are denoted by a number of each hypothesis in an upper index and by a number of a particular number of variant of a course (1-practice, 2-lecture, 3-seminar). A full statistical population was consisted of 127 students of the course based on didactics in education [22], which is taught in a form of three variants – a practice, a lecture and a seminar. In this research, following hypotheses (21) were defined and are bound on a didactical analyses of teacher's approach to education. A zero hypothesis is denoted by number 0 in a lower index and an opposite alternative hypothesis by number 1 in a lower index. The upper indices are depended on number of a particular hypothesis.

$$i,j \in \mathcal{N};\ i,j \in \langle 1; r = s = 3 \rangle : \left\{ \begin{array}{l} H_0^i : \mu_1^i = \cdots = \mu_{s=3}^i \\ H_1^i : \mu_1^i \neq \cdots \neq \mu_{s=3}^i \end{array} \right\} \tag{21}$$

Table 2. Definitions of hypotheses in particular quantitative research

Hypothesis		Definition
H^1	$H_0^1 : \mu_1^1 = \mu_2^1 = \mu_3^1$	There are not significant dependences between a level of teacher's references to results of a research in an education and variants of the course
	$H_1^1 : \mu_1^1 \neq \mu_2^1 \neq \mu_3^1$	There are significant dependences between a level of teacher's references to results of a research in an education and variants of the course
H^2	$H_0^2 : \mu_1^2 = \mu_2^2 = \mu_3^2$	There are not significant dependences between a level of teacher's recommendation of a current literature in an education and variants of the course
	$H_1^2 : \mu_1^2 \neq \mu_2^2 \neq \mu_3^2$	There are significant dependences between a level of teacher's recommendation of a current literature in an education and variants of the course
H^3	$H_0^3 : \mu_1^3 = \mu_2^3 = \mu_3^3$	There are not significant dependences between a level of teacher's utilization of information technologies in an education and variants of the course
	$H_1^3 : \mu_1^3 \neq \mu_2^3 \neq \mu_3^3$	There are significant dependences between a level of teacher's utilization of information technologies in an education and variants of the course

Sentences of definitions of hypotheses can be seen in Table 2. This course is visited by students of a present form of a bachelor studies at institution of authors of this paper. In the statistical population, each respondent got one respond in this research with respect to multiplied possibility of an attendance in all variants of the course.

The significance level was declared as 5%, respectively 0.05, corresponding with usual applications in the pedagogical science as can be seen e.g. in [15–18]. A minimal number of a selected population can be expressed by interval (22) in case of 127 respondents in the full population. A number of obtained questionnaires was 103. A ratio of returnees of responses can be expressed as approximately 81.1%.

A phase of a collecting of the responses was based on a form of an online questioner. Where data for a testing of hypotheses were supported by 20 questions related to the relevant appropriate topics. For purposes of this paper, only three questions were used and have the similar definition as hypotheses in Table 2. Data was bound on interval scales. Before these main important questions about "a level of teacher's references to results of a research in an education", "a level of teacher's utilization of information technologies in an education" and "a level of teacher's recommendation of a current literature in an education", items for a factographical analysis were included. This part was consisted of questions: a gender, a form of a study, a type of a study, a year of a study, a variant of a visited course. Only the last factographical item of them is further considered in this paper.

$$\left.\begin{array}{c}(n_{\min} = 86) \le n_k \le (n_{\max} = 103) \le (n = 127) \\ \eta = \frac{n_{\max}}{n} = 0.811 = 81.1\%\end{array}\right\} \tag{22}$$

For each defined hypothesis, the obtained responds from students were recorded into a form of categorical items (a variant of a course) and numbers. All data assigned for each hypothesis were tested on a normality. Particular results of these testings (23), with all random variables in form of obtained data, were achieved as p-values, as can be seen in Tables 3 and 4. A normal probability distribution of data was considered with mean value μ equal to 0 and with variation σ^2 equal to 1.

$$\left.\begin{array}{l}\left.\begin{array}{l}\tilde{H}_0^{1,1} : X_1^1 \sim N(0;\,1);\ \tilde{H}_0^{1,2} : X_2^1 \sim N(0;\,1);\ \tilde{H}_0^{1,3} : X_3^1 \sim N(0;\,1) \\ \tilde{H}_1^{1,1} : X_1^1 \nsim N(0;\,1);\ \tilde{H}_1^{1,2} : X_2^1 \nsim N(0;\,1);\ \tilde{H}_1^{1,3} : X_3^1 \nsim N(0;\,1)\end{array}\right\} \\ \left.\begin{array}{l}\tilde{H}_0^{2,1} : X_1^2 \sim N(0;\,1);\ \tilde{H}_0^{2,2} : X_2^2 \sim N(0;\,1);\ \tilde{H}_0^{2,3} : X_3^2 \sim N(0;\,1) \\ \tilde{H}_1^{2,1} : X_1^2 \nsim N(0;\,1);\ \tilde{H}_1^{2,2} : X_2^2 \nsim N(0;\,1);\ \tilde{H}_1^{2,3} : X_3^2 \nsim N(0;\,1)\end{array}\right\} \\ \left.\begin{array}{l}\tilde{H}_0^{3,1} : X_1^3 \sim N(0;\,1);\ \tilde{H}_0^{3,2} : X_2^3 \sim N(0;\,1);\ \tilde{H}_0^{3,3} : X_3^3 \sim N(0;\,1) \\ \tilde{H}_1^{3,1} : X_1^3 \nsim N(0;\,1);\ \tilde{H}_1^{3,2} : X_2^3 \nsim N(0;\,1);\ \tilde{H}_1^{3,3} : X_3^3 \nsim N(0;\,1)\end{array}\right\}\end{array}\right\} \tag{23}$$

Each p-value is depended on a concrete value of a number of a population n, which is increased in a frame of interval (22) in favor of purposes of a further analysis in this paper. The Shapiro-Wilk (Table 3) and Anderson-Darling (Table 4) methods were proceeded using a free-available software PAST of version 2.17 [19].

Table 3. p-values < 0.05 in testing on normality of data using shapiro-wilk test

$p(\tilde{H}^{i,j}, n_k)$, for hypotheses $\tilde{H}^{i,j}$; $i,j \in \langle 1; 3 \rangle$									
n_k	Practice ($\tilde{H}^{1,1}$)	Lecture ($\tilde{H}^{1,2}$)	Seminar ($\tilde{H}^{1,3}$)	Practice ($\tilde{H}^{2,1}$)	Lecture ($\tilde{H}^{2,2}$)	Seminar ($\tilde{H}^{2,3}$)	Practice ($\tilde{H}^{2,3}$)	Lecture ($\tilde{H}^{3,2}$)	Seminar ($\tilde{H}^{3,3}$)
87	1.8E−03	1.7E−07	9.4E−08	1.3E−05	6.5E−06	6.0E−11	8.2E−04	9.7E−06	1.4E−08
89	1.9E−03	1.7E−07	1.2E−07	2.8E−04	6.5E−06	5.9E−10	1.6E−03	9.7E−06	2.6E−08
91	1.9E−03	4.2E−08	1.2E−07	2.8E−04	1.9E−06	5.9E−10	1.6E−03	8.6E−05	2.6E−08
93	1.7E−03	4.2E−08	1.5E−07	6.9E−04	1.9E−06	1.9E−09	1.9E−03	8.6E−05	3.9E−08
95	9.9E−04	4.2E−08	1.6E−07	3.7E−04	1.9E−06	4.4E−09	1.3E−03	8.6E−05	5.4E−08
97	9.9E−04	4.2E−08	8.5E−08	3.7E−04	1.9E−06	3.9E−09	1.3E−03	8.6E−05	3.4E−08
99	9.9E−04	8.9E−07	9.4E−08	3.7E−04	1.0E−06	7.2E−09	1.3E−03	5.1E−05	4.4E−08
101	5.3E−03	8.9E−07	6.8E−08	5.7E−04	1.0E−06	6.5E−09	1.3E−03	5.1E−05	3.5E−08
103	5.3E−03	3.2E−06	7.2E−08	5.7E−04	1.9E−05	1.0E−08	1.3E−03	2.9E−05	4.2E−08

Table 4. p-values < 0.05 in testing on normality of data using Anderson-Darling test

$p(\tilde{H}^{i,j}, n_k)$, for hypothesis $\tilde{H}^{i,j}$; $i,j \in \langle 1; 3 \rangle$									
n_k	Practice ($\tilde{H}^{1,1}$)	Lecture ($\tilde{H}^{1,2}$)	Seminar ($\tilde{H}^{1,3}$)	Practice ($\tilde{H}^{2,1}$)	Lecture ($\tilde{H}^{2,2}$)	Seminar ($\tilde{H}^{2,3}$)	Practice ($\tilde{H}^{2,3}$)	Lecture ($\tilde{H}^{3,2}$)	Seminar ($\tilde{H}^{3,3}$)
87	1.19E−04	5.7E−14	6.8E−15	1.3E−08	2.6E−09	6.1E−27	3.8E−05	8.6E−09	2.3E−17
89	1.81E−04	5.7E−14	2.2E−14	2.6E−06	2.6E−09	8.3E−24	1.9E−04	8.6E−09	4.1E−16
91	1.81E−04	7.2E−16	2.2E−14	2.6E−06	8.5E−11	8.3E−24	1.9E−04	3.6E−07	4.1E−16
93	1.93E−04	7.2E−16	5.4E−14	2.2E−05	8.5E−11	1.3E−21	3.9E−04	3.6E−07	3.3E−15
95	6.13E−05	7.2E−16	1.1E−13	5.3E−06	8.5E−11	5.0E−20	1.6E−04	3.6E−07	1.5E−14
97	6.13E−05	7.2E−16	1.0E−14	5.3E−06	8.5E−11	5.2E−20	1.6E−04	3.6E−07	2.7E−15
99	6.13E−05	1.1E−12	2.0E−14	5.3E−06	1.4E−11	8.1E−19	1.6E−04	9.4E−08	9.7E−15
101	3.50E−04	1.1E−12	6.1E−15	1.8E−05	1.4E−11	7.3E−19	2.3E−04	9.4E−08	4.1E−15
103	3.50E−04	3.5E−11	1.1E−14	1.8E−05	5.2E−09	6.1E−18	2.3E−04	2.3E−08	1.2E−14

For Tables 3 and 4 with p-values for aimed for a normality, corresponding charts were displayed in Figs. 1, 2 and 3. In these figures, a significant independence of p-values on variable n_k can be seen. Although p-values have a stochastic progress on interval (22), due to a random selection of an increasing of added data, these values did not cross a constant significance level of 5% in all measured situations. Therefore, the normality of all hypotheses was consistent by each hypothesis.

Important conclusion (24) can be seen in Figs. 1, 2 and 3. The conclusions about the normality did not change and a final statistical method for a testing of hypotheses did not change in interval (22) in this research.

$$\left.\begin{array}{c} \forall n_k \in \{\varepsilon + 2\}_{\varepsilon=85}^{103};\ \forall i,j \in \langle 1; r = s = 3 \rangle : \\ (\circ \equiv <) \wedge \left(p\left(\tilde{H}^{i,j}, n_k\right) \circ \alpha\right) \Leftrightarrow \left(p\left(\tilde{H}^{i,j}, n_k\right) < \alpha\right) \end{array}\right\} \tag{24}$$

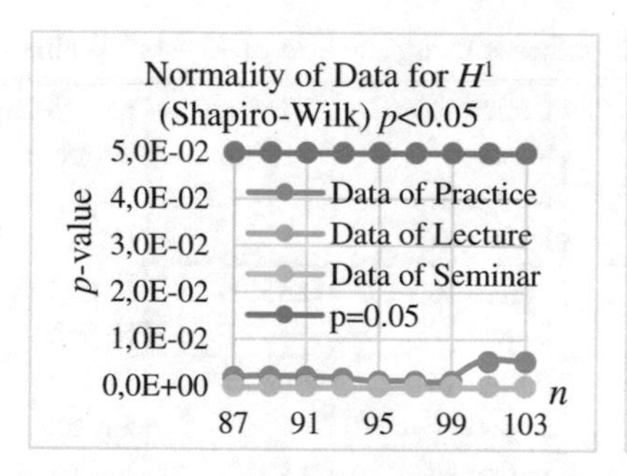

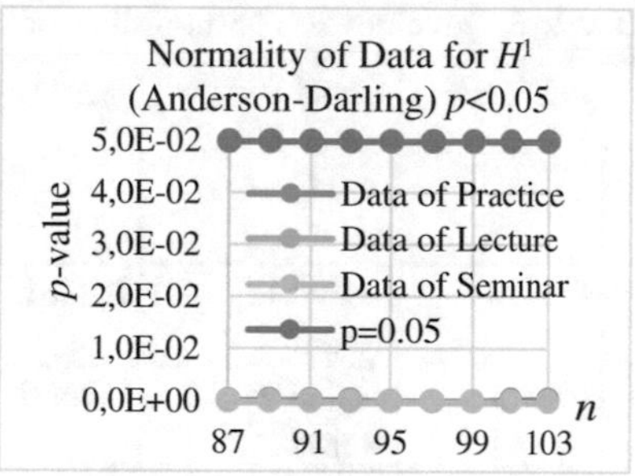

Fig. 1. Dependences of p-values of testing on normality of data on n_k for case of relevance to hypothesis H^1 Using Shapiro-Wilk and Anderson-Darling tests

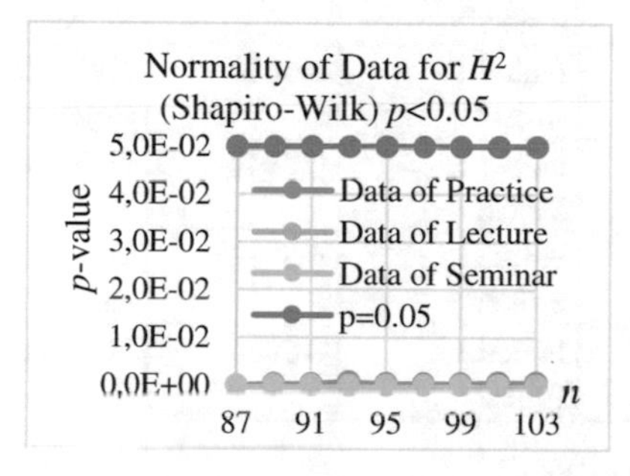

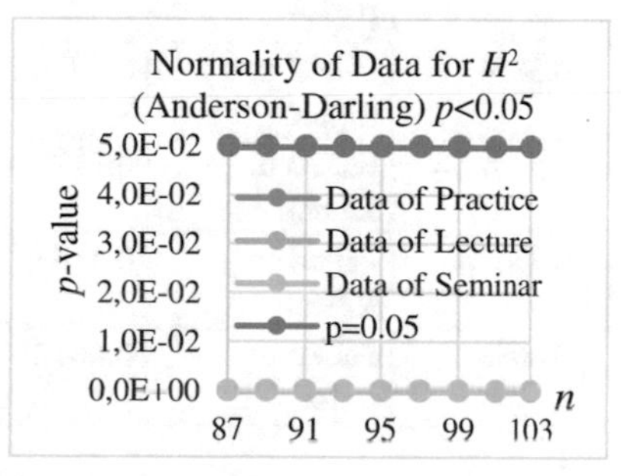

Fig. 2. Dependences of p-values on testing of normality of data on n_k for case of relevance to hypothesis H^2 using Shapiro-Wilk and Anderson-Darling tests

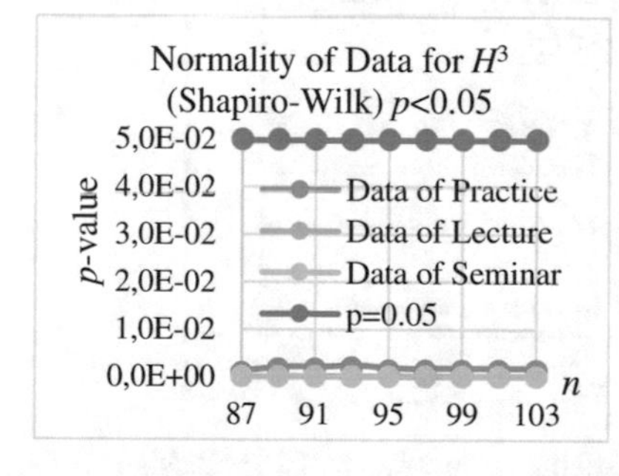

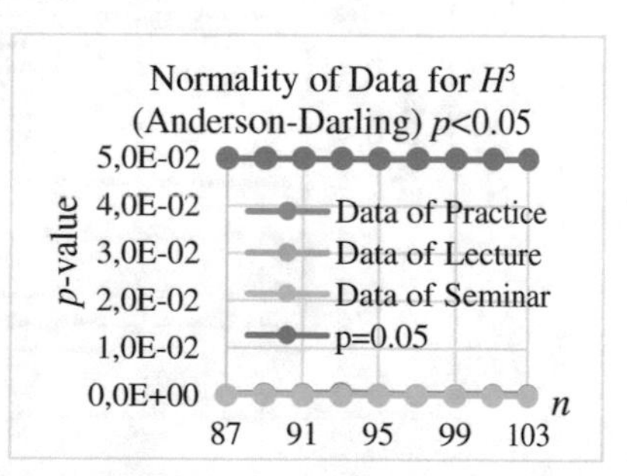

Fig. 3. Dependences of p-values on testing of normality of data on n_k for case of relevance to hypothesis H^3 using Shapiro-Wilk and Anderson-Darling tests

For all hypotheses, the Kruskal-Wallis test was applied for purposes of the testing of defined hypotheses. A reason of an application of this non-parametric test is related to unfulfilled normality of all data in the research and by identifying of 3 items of the

Table 5. p-values > 0.05 in testing of hypotheses using method of Kruskal-Wallis

n_k	$p(H^1, n_k)$	Conclusion about H^1 on signif. level 0.05	$p(H^2, n_k)$	Conclusion about H^2 on signif. level 0.05	$p(H^3, n_k)$	Conclusion about H^3 on signif. level 0.05
87	0.6390	Failed to Reject	0.8868	Failed to Reject	0.8531	Failed to Reject
89	0.4811	Failed to Reject	0.7565	Failed to Reject	0.6885	Failed to Reject
91	0.4878	Failed to Reject	0.8749	Failed to Reject	0.7312	Failed to Reject
93	0.3520	Failed to Reject	0.7001	Failed to Reject	0.6083	Failed to Reject
95	0.3899	Failed to Reject	0.6518	Failed to Reject	0.6139	Failed to Reject
97	0.3874	Failed to Reject	0.6613	Failed to Reject	0.6450	Failed to Reject
99	0.5107	Failed to Reject	0.7177	Failed to Reject	0.7144	Failed to Reject
101	0.3143	Failed to Reject	0.5143	Failed to Reject	0.5313	Failed to Reject
103	0.4093	Failed to Reject	0.6613	Failed to Reject	0.5955	Failed to Reject

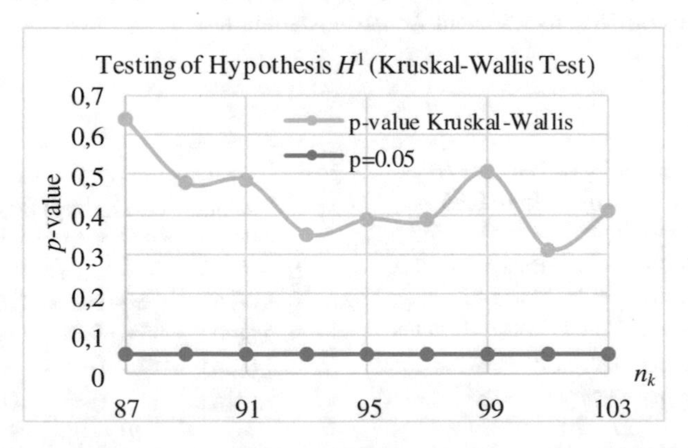

Fig. 4. Dependence of p-values on variable n_k in testing of hypothesis H^1

categorical variable. As can be seen in Figs. 4, 5, 6 and in Table 5, analysis verified still trend of p-values in this testing of hypotheses using the Kruskal-Wallis test. Displayed p-values did not underload a significance value 0.05; therefore, relation (24) is fulfilled.

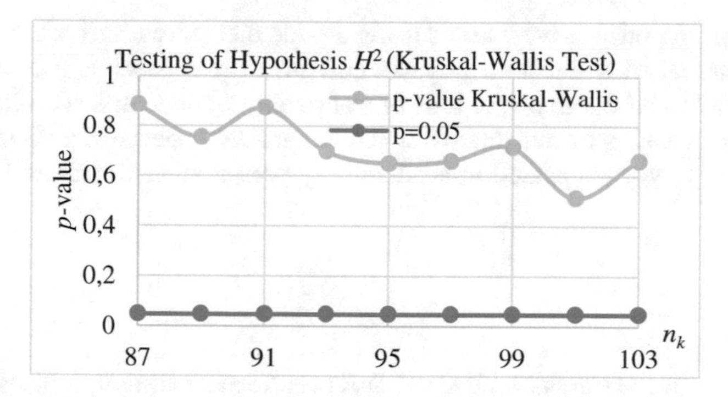

Fig. 5. Dependence of p-values on variable n_k in testing of hypothesis H^2

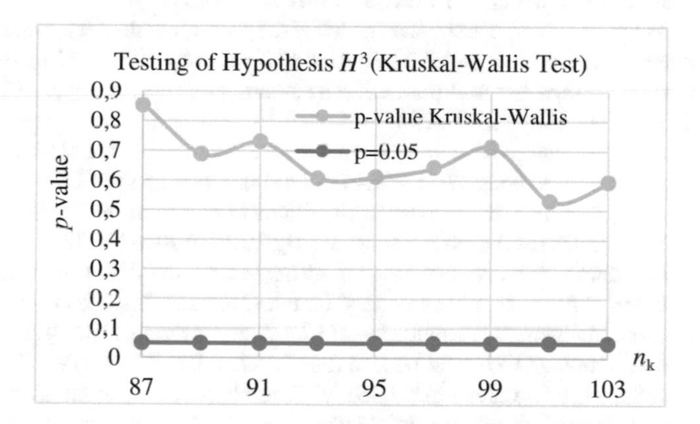

Fig. 6. Dependence of p-values on variable n_k in testing of hypothesis H^3

$$\left.\begin{array}{c} \forall n_k \in \{\varepsilon + 2\}_{\varepsilon=85}^{103};\ \forall i \in \langle 1; r = 3 \rangle : \\ (\circ \equiv\ >\ ) \wedge (p(H^i, n_k) \circ \alpha) \Leftrightarrow (p(H^i, n_k) > \alpha) \end{array}\right\} \qquad (25)$$

5 Conclusion

In this paper, a relation between results of a testing of a normality of a researched data and a variable number of respondents was analyzed in an environment of a particular pedagogical research. By three different hypotheses testing, results of testing of a normality of data, in a form of p-values, followed a stochastically fluctuation of values; however, these values have not been underload or overload a declared significance level of 0.05. Analyses were determined for allowed interval of sizes of selected population and conclusions about a normality caused an appropriate selection of own statistical methods for purposes of a testing of hypotheses on the same significance level 0.05. All hypotheses were tested using a same method with respect to a variety of a defined interval of a size of a population. Aim of this paper is focused on a recommendation of an importance of tests of a normality of data. In a complex concept of a statistical processing of quantitative based researches, aspects of a normality testing have not been always appeared in academically presentations, that can be in general improved.

References

1. Kitchenham, B., Madeyski, L., Budgen, D., et al.: Robust statistical methods for empirical software engineering. Empir. Softw. Eng. 1–52 (2016). ISSN 1573-7616
2. Fischer, D., Oja, H.: Mann-Whitney type tests for microarray experiments: the R package gMWT. J. Stat. Softw. **65**(1), 1–19 (2015). ISSN 1548-7660
3. Lazic, S.E.: Why we should use simpler models if the data allow this: relevance for ANOVA designs in experimental biology. BMC Physiol. **8**(18) (2008). ISSN 1472-6793
4. Rasch, D.: The two-sample t test: pre-testing its assumptions does not pay off. Stat. Pap. **52** (1), 219–231 (2011). ISSN 1613-9798
5. Mewhort, D.J.K.: A comparison of the randomization test with the F test when error is skewed. Behav. Res. Methods **37**(3), 426–435 (2005). ISSN 1554-3528
6. Krivy, I., Tvrdik, J., Krpec, R.: Stochastic algorithms in nonlinear regression. Comput. Stat. Data Anal. **33**(12), 277–290 (2000). https://doi.org/10.1016/s0167-9473(99)00059-6
7. Achuthan, K., Murali, S.S.: Virtual lab: an adequate multi-modality learning channel for enhancing students' perception in chemistry. In: 6th Computer Science On-line Conference: Cybernetics and Mathematics Applications in Intelligent Systems, pp. 419–433. Springer (2017). https://doi.org/10.1007/978-3-319-57264-2_42. ISBN 978-3-319-57263-5
8. Lech, P., Wlodarski, P.: Analysis of the IoT WiFi mesh network. In: 6th Computer Science On-Line Conference: Cybernetics and Mathematics Applications in Intelligent Systems, pp. 272–280. Springer (2017). https://doi.org/10.1007/978-3-319-57264-2_28. ISBN 978-3-319-57263-5
9. Meghanathan, N.: Correlation analysis of decay centrality. In: 6th Computer Science On-Line Conference: Cybernetics and Mathematics Applications in Intelligent Systems, pp. 407–418. Springer (2017). https://doi.org/10.1007/978-3-319-57264-2_41. ISBN 978-3-319-57263-5
10. Kularbphettong, K.: Analysis of students' behavior based on educational data mining. In: 1st Conference on Computational Methods in Systems and Software: Applied Computational Intelligence and Mathematical Methods, pp. 167–172. Springer (2017). https://doi.org/10.1007/978-3-319-67621-0_15. ISBN 978-331967620-3
11. Sulovska, K., Belaskova, S., Adamek, M.: Gait patterns for crime fighting: statistical evaluation. In: Proceedings of SPIE - The International Society for Optical Engineering. **8901**. SPIE (2013). https://doi.org/10.1117/12.2033323. ISBN 978-081949770-3
12. Schoftner, T., Traxler, P., Prieschl, W., Atzwanger, M.: E-learning introduction for students of the first semester in the form of an online seminar. In: Pre-Conference Workshop of the 14th E-Learning Conference for Computer Science, pp. 125–129. CEUR-WS (2016). ISSN 1613-0073
13. Kostolanyova, K.: Adaptation of personalized education in e-learning environment. In: 1st International Symposium on Emerging Technologies for Education, pp. 433–442. Springer (2017). https://doi.org/10.1007/978-3-319-52836-6_46. ISBN 978-3-319-52835-9

14. Krpec, R.: Alternative teaching of mathematics on the second stage of elementary schools using the scheme-oriented education method. In: EDULEARN 2015 Proceedings, Barcelona, pp. 7338–7344. IATED Academy (2015). ISBN 978-84-606-8243-1
15. Cortes, J., Casals, M., Langohr, M., et al.: Importance of statistical power and hypothesis in P value. Med. Clin. **146**(4), 178–181 (2016). ISSN 0025-7753
16. Barot, T.: Possibilities of process modeling in pedagogical cybernetics based on control-system-theory approaches. In: 6th Computer Science On-Line Conference: Cybernetics and Mathematics Applications in Intelligent Systems, pp. 110–119. Springer (2017). https://doi.org/10.1007/978-3-319-57264-2_11. ISBN 978-3-319-57263-5
17. Kostolanyova, K.: The evaluation of adaptive study material - the automatic feedback for content creators. In: International Conference of Numerical Analysis and Applied Mathematics 2016, vol. 1863. American Institute of Physics (2017). https://doi.org/10.1063/1.4992245. ISBN 978-073541538-6
18. Horsley, M., Sikorova, Z.: Classroom teaching and learning resources: international comparisons from TIMSS - a preliminary review. Orbis Scholae **8**(2), 43–60 (2017). https://doi.org/10.14712/23363177.2015.65. ISSN 1802-4637
19. Hammer, O., Harper, D.A.T., Ryan, P.D.: PAST: paleontological statistics software package for education and data analysis. Palaeontol. Electron. **4**(1) (2001). http://palaeo-electronica.org/2001_1/past/issue1_01.htm
20. Barot, T.: Adaptive control strategy in context with pedagogical cybernetics. Int. J. Inf. Commun. Technol. Educ. **6**(2), 5–11 (2017). ISSN 1805-3726
21. Alizadeh Noughabi, H.: Two powerful tests for normality. Ann. Data Sci. **3**(2), 225–234 (2016). ISSN 2198-5812
22. Adkins, D., Guerreiro, M.: Learning styles: considerations for technology enhanced item design. Br. J. Educ. Technol. **49**(3), 574–583 (2018). ISSN 0007-1013

Towards Smartphone-Based Navigation for Visually Impaired People

Izaz Khan, Shah Khusro(✉), Irfan Ullah, and Saeed Mahfooz

Department of Computer Science, University of Peshawar, Peshawar 25120, Pakistan
{izazcs, khusro, cs.irfan, saeedmahfooz}@uop.edu.pk

Abstract. Mobility – the ability of users to navigate from one point to another and the environmental awareness – knowing about one's surroundings (indoor and outdoor) are some of the challenging tasks for blind people. They either rely on their human companions for navigation or traditional/technological assistive solutions including a guide dog, white cane and electronic travel and orientation aids. It is the position put forward here that these solutions are either too much expensive to afford (e.g., human guide and guide dogs), limited in offering full navigational support (e.g., white cane), or also costly in adopting them as they need customizations in the targeted building. To mitigate this issue, low-cost smartphone-based solutions can be developed that exploit color and text/pattern recognition algorithms to sense objects, buildings, and important landmarks to aid blind users in mobility, navigation, and environmental awareness. This position paper uncovers such possibilities and proposes one possible solution that may open new avenues of research and development in making the life of blind users much easier especially in mobility, navigation, and obtaining information about their surroundings.

Keywords: Human-computer interaction · White cane · Smartphone Mobility · Navigation · Orientation · Blind and visually impaired people

1 Introduction

Mobility – the ability of users to navigate from one point to another and the environmental awareness – knowing about one's surroundings (indoor and outdoor) are some of the challenging tasks for blind people. Traditionally, a human (and sometimes a guide dog) aids them in navigating and keeping them updated about their surroundings, which is quite expensive and most of the time not feasible. They use a white cane, which is simple, reliable, low-cost, efficient but limited in detecting trunk and head-level obstacles as well as other over-hanging objects [1]. Also, it needs direct physical contact with the ground and obstacles and unable to inform the user about the approaching obstacles [2]. To mitigate this issue, several automatic/semi-automatic traditional and technological assistive tools have been designed from time to time to make them independent and self-confident while navigating in the surroundings. These technological aids include electronic travel aids, electronic orientation aids, and

R. Silhavy et al. (Eds.): CoMeSySo 2018, AISC 859, pp. 366–373, 2019.
https://doi.org/10.1007/978-3-030-00211-4_32

position locator devices [3]. Mostly, these are handheld devices or designed in combination with traditional aids to give good environmental perception. Most of these solutions use dedicated hardware or need customization to the structure of the building for their adoption, which is a severe limitation and makes the existing solutions less useful. Therefore, this position paper briefly highlights what has been achieved towards the mobility, orientation, and environmental awareness from the blind user perspective and what needs to be done. In Sect. 2, the author presents a brief survey of the state-of-the-art and identifies its limitations. Section 3 discusses and analyzes the position put forward in this position paper and shows some avenues of research and development. Section 4 concludes the paper. References are given at the end of the paper.

2 Survey of the Literature

Most of the research work for aiding blind people have been carried out around exploiting technologies including computers, wireless and mobile devices, and sensors including, e.g., ultrasonic, infrared, laser, RFID, Bluetooth, Wi-Fi, etc., for mobility and environmental awareness. Examples include solutions using ultrasonic [4], laser [5], infrared [6], Computer Vision [7], RFID [8], or the fusion of the sensory data from multiple sensors [9] to analyze the surroundings and help the user in mobility and navigation. In addition, the modern-day Smartphones have made most of the complicated and challenging tasks much easier and in real-time.

Smartphones come with easy-to-use input/output methods, storage space, computational and sensing capabilities, which make them ideal candidates for aiding blind users in navigation and give them much more useful information about the surroundings. Researchers have exploited smartphones in developing several latest solutions. Virtual White Cane [10] captures laser beam's reflection using the Smartphone camera. The distance between virtual white cane and the reflected beam is computed through active triangulation method. For the user awareness, personalized vibration patterns are used, vibration magnitude reflects the distance from the obstacle, which is used in avoiding collisions with the incoming obstacles. However, unlike the traditional white cane, it is limited to communicate information about objects including their shape, size, type, and landmarks, etc. The ioCane [11] uses an ultrasonic sensor unit attached to the white cane and hand-held Smartphone. The ultrasonic sensor unit is cane-mounted and easily detachable, which sends contextual data wirelessly to smartphone application and the user is informed about the object height and proximity with built-in mobile phone vibrations. The system provides increased assistance in mobility with minimum learning efforts to avoid obstacles. However, it is costly due to an extra set of hardware with a smartphone.

An obstacle detection system proposed in [12] uses smartphone orientation and short distance sensors for detecting objects. It is attached to the back of smartphone having a pair of ultrasonic sensors and Bluetooth connectivity module to connect with a smartphone. The aim is to increase sensing range and detect obstacles to make the user mobility easier. The extra hardware and inability to precisely detect obstacles and accurately identify essential objects and landmarks are some of the noteworthy limitations.

The pAth Recognition for Indoor Assisted NavigatioN with Augmented perception (ARIANNA) [13] is a low-cost navigation system for blind users. It uses the path made with stickers or painted on the floor of the building to identify the point of interest using barcodes for particular landmarks. The smartphone is used for path detection and user guidance through vibration, however, it relies on the dedicated infrastructure (requires customization of the building, e.g., paintings/stickers, etc.) and the barcode requires line of sight, which makes it less applicable.

A smartphone-based guidance system [14] is designed for blind people. This system detects obstacles coming in front of the user even in unfamiliar environments and aids user in obstacle avoidance. It combines image recognition using Computer and smartphone. It works in online and offline modes based on the availability of the Internet. It detects obstacles and sends the sensory data (images, etc.) to the backend server to be interpreted for obstacle recognition. The output is reported back to the user using text-to-speech (TTS) on the Smartphone. According to the authors, the system shows 60% accuracy, however, it lacks in instant response, requires extensive image processing, battery power, network usage.

The solution in [15] estimates depth-data for real-time obstacle detection and avoidance in indoor environments. This study uses Google Tango device as the development platform to create a cost-effective, portable and lightweight solution with the freedom from network connectivity for image recognition, etc. To provide a user-centric solution, the authors involved the visually impaired individuals in the design and testing phase through interviews and field tests. This solution also offers multi-model feedbacks to the user to understand the generated information easily and capable of differentiating floor from walls with obstacle specifications. However, it is costly because of the dependency on Google Tango device and therefore, not affordable for most of the blind users.

By looking at the state-of-the-art solutions, it can be judged that most of these solutions are expensive, use dedicated hardware or require modifications to the existing infrastructure of the buildings, which make their use and utility limited. Although smartphone-based solutions have been proposed, they are unable to exploit the potential of the smartphone technology fully. The most related research work, in this regard, is the ARIANNA [13] but that also requires some modifications in the existing surroundings through painting and stickers. Others [14, 15] need high processing and battery power and use expensive devices, etc. Therefore, a solution is required that is cheap, feasible and free from such modifications and dependencies. This can be achieved using pattern and color recognition software (preferably open-source applications) to use the existing infrastructure, information about the objects such their color and already drawn text/images on these objects, etc., to detect and guide the blind user in navigation and in providing accurate information about the surrounding environment. This can be achieved using smartphone camera and accelerometer sensors to identify objects, tagged doors, infrastructural data (shapes, etc.) of the corridors, walls and other landmarks by sensing and analyzing their colors and geometrical shapes etc., whereas the accelerometer can be used in changing the application sensing mode. Such an application can be trained by a normally sighted person (or persons in the community) for a blind user to collect all the data about the surroundings in which the user lives and performs most of their daily life activities. Once trained, the application can be used by the blind user to gain assistance in navigation and obtaining meaningful information about their surroundings. One possible solution is discussed and analyzed in the next Section, which may bring new avenues of research and development.

3 Discussion and Analysis

Looking at the survey of the literature in Sect. 2, a solution is required that is low-cost, feasible and easy-to-adopt in any indoor/outdoor environment. One possible solution can be using pattern and color recognition software (preferably open-source applications) to use the existing infrastructural information including the color, color patterns, geometrical shape and the already available tags (e.g., Room 202, Class Room, Computer Lab, Professor XYZ, Use Me (on dustbins), Committee Room, and navigation symbols for lift, corridor, stairs etc.) to detect, analyze and then assist blind users in navigation and in accessing accurate information about the surrounding environment. One such possible solution can be achieved using smartphone camera and accelerometer sensors to identify objects, tagged doors, infrastructural data (shapes, etc.) of the corridors, walls and other important landmarks by sensing and analyzing their colors and geometrical shapes, etc., whereas the accelerometer can be used in changing the application sensing mode. This is graphically visualized in Fig. 1, which summarizes the general flow of the process involving several components including user interfaces for blind and sighted users, sensors, Internet, information about the building (captured by sighted users, who can be friends, relatives, volunteers, etc., who upload such information), and server for storage and access to such information.

The solution, which may be in the form of a smartphone app, offers two separate user interfaces for the sighted and blind users so that the former can capture and train the app for their companion blind user or upload the captured information to the server, and the later can use the trained app directly or download trained data from the server about a specific place or building. The application logic behind these interfaces, as depicted in Fig. 2, will enable the app to be operated in two modes i.e., sighted user mode and blind user mode, where modes can be changed using sensors such as an accelerometer. The former enables the sighted users to train the app for the blind user as well as upload the captured data about certain places or buildings to the server, where it is stored as a configuration file in any format including e.g., XML (eXtensible Markup Language), etc., so that it can be available in downloadable and usable form to other users. The user after successful registration or login (mandatory for blind users as well) can add and modify data about objects, structure and landmarks of the building captured through the camera and analyzed and processed by the application logic. For this purpose, several pattern recognition and visual analytics algorithms can be used. The sighted user mode gets benefited from three modules including Sensing module, Business Intelligence module and Uploader/trainer module as shown in Fig. 2.

The blind user interface enables the blind users to efficiently use the application in navigation and obtaining information about the surroundings from the current point of reference using the information stored in the app or downloaded from the server. Note that Internet is required only when certain data about a place or building is required, for which their sighted companion has not collected data themselves. In addition, any

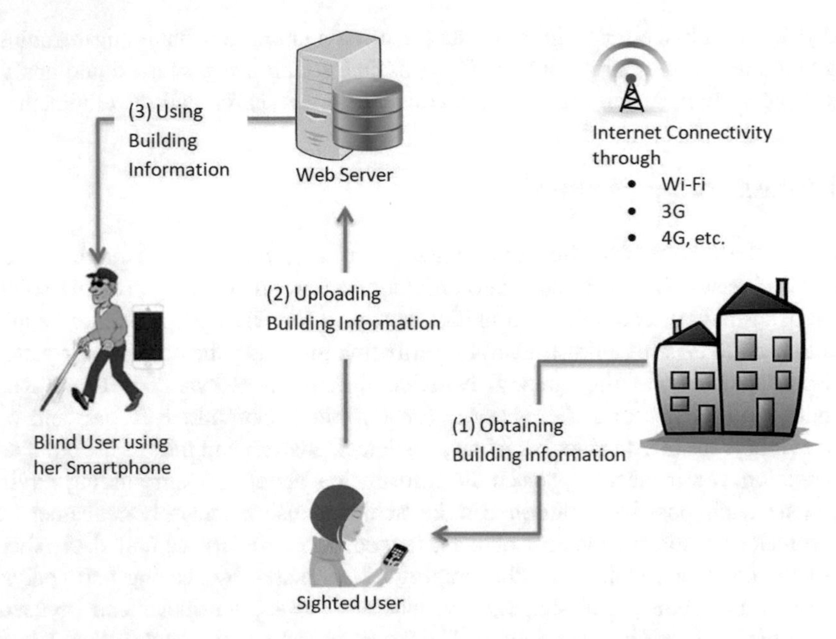

Fig. 1. Smartphone-based indoor navigation system for blind and visually impaired people – block diagram

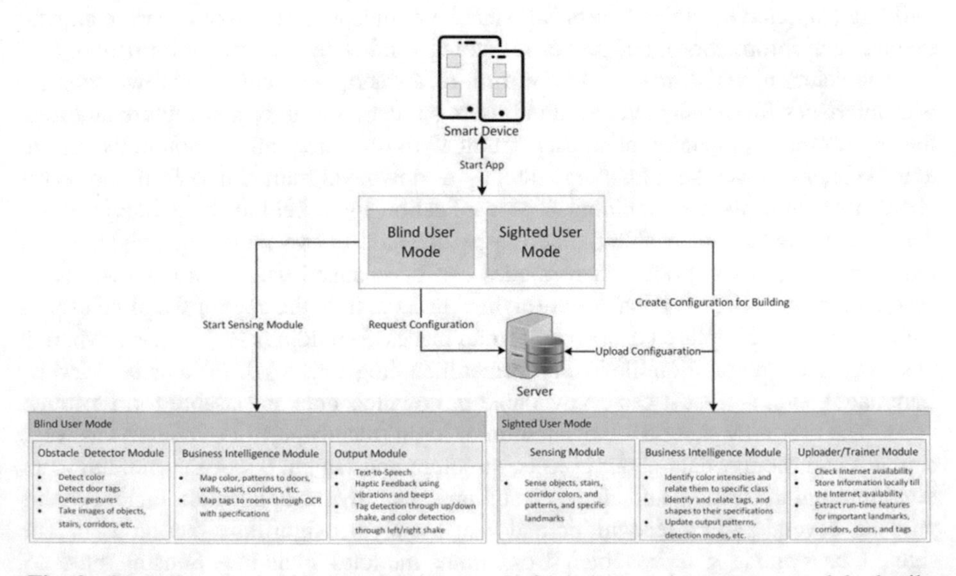

Fig. 2. Smartphone-based blind navigation system for indoor environments – module details

suitable input/output method, e.g., TTS, vibration, etc., can be used for communicating with the app so that the information about the destination can be inputted along with the current location (current location can also be given using GPS sensor if connected).

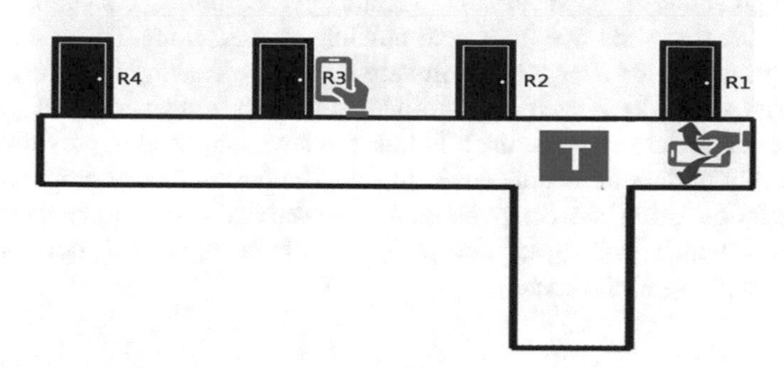

Fig. 3. Using the smartphone-based solution in the corridor and for detecting doors with tags

The blind user mode may first search its local repository for such information and if not available then request the server to send the details necessary for navigation. Three modules are responsible for capturing the necessary information and assisting the user in navigation. These include Obstacle Detector, which using the camera captures the surrounding objects (e.g., doors, important landmarks etc.) coming in the pathway while navigation (or standing at a certain point) to sense its colors or take pictures, which are then passed to the Business Intelligence module to understand the color combinations (or patterns) and match it against the stored data (training data) to detect the object, landmarks, etc., and guide the user accordingly. The Output module guides the user during navigation and keeps them aware of the surrounding. In its most basic form, the solution uses color sensing and analytics so that no phone calibration or specified angling of the camera is required. The blind user needs to hold the smartphone in portrait or landscape mode to sense the environment for colors and tags. Further, the blind user can move the phone from left-to-right or right-to-left and may point vertically to left-side and right-side for scanning walls, doors, and capturing landmarks, etc., this is shown in Fig. 3.

4 Conclusion

Mobility (navigation) and the awareness about the surroundings (environmental awareness) are some of the daunting tasks in the life of blind people. Typically, a human (and sometimes a guide dog) assists them in navigation and keeping them updated about their surroundings, which is quite expensive and most of the time not feasible. To mitigate this issue, several automatic/semi-automatic traditional and technological assistive tools have also been designed from time to time to make them independent and self-confident while navigating in the surroundings.

It was the position put forward that these solutions are either costly (e.g., human guide, guide dog), less productive (white cane) or suffer from limited adaptability such as requiring modifications to the existing infrastructure of the buildings, costly hardware, continuous Internet connectivity, high-processing and battery costs, etc. Although smartphone-based solutions have been proposed, they are unable to fully exploit the potential of the smartphone technology. Therefore, a solution is required that is cheap, feasible and free from such building modifications. This can be achieved using pattern and color recognition software to use the existing infrastructure, information about the objects such their color and already drawn text/images on these objects, etc., to detect and guide the blind user in navigation and in providing accurate information about the surrounding environment. The position paper proposed one such possible solution using the computational, connectivity and sensory capabilities of smartphones, which will open new avenues of research and development for researchers working in this area.

References

1. Pyun, R., Kim, Y., Wespe, P., Gassert, R., Schneller, S.: Advanced augmented white cane with obstacle height and distance feedback. In: IEEE International Conference on Rehabilitation Robotics (ICORR), pp. 1–6. IEEE (2013)
2. Wang, Y., Kuchenbecker, K.J.: HALO: haptic alerts for low-hanging obstacles in white cane navigation. In: Haptics Symposium (HAPTICS), pp. 527–532. IEEE (2012)
3. Nieto, L., Padilla, C., Barrios, M.: Design and implementation of electronic aid to blind's cane. In: 2014 III International Congress of Engineering Mechatronics and Automation (CIIMA), pp. 1–4 (2014)
4. Kamaludina, M.H., Mahmooda, N.H., Ahmada, A.H., Omara, C., Yusofb, M.A.M.: Sonar assistive device for visually impaired people. TC **500**, 1 (2015)
5. Mekhalfi, M.L., Melgani, F., Zeggada, A., De Natale, F.G., Salem, M.A.-M., Khamis, A.: Recovering the sight to blind people in indoor environments with smart technologies. Expert Syst. Appl. **46**, 129–138 (2016)
6. Kher Chaitrali, S., Dabhade Yogita, A., Kadam Snehal, K., Dhamdhere Swati, D., Deshpande Aarti, V.: An intelligent walking stick for the blind. Int. J. Eng. Res. General Sci. **3**, 1057–1062 (2015)
7. Bonetto, M., Carrato, S., Fenu, G., Medvet, E., Mumolo, E., Pellegrino, F.A., et al.: Image processing issues in a social assistive system for the blind. In: 2015 9th International Symposium on Image and Signal Processing and Analysis (ISPA), pp. 216–221 (2015)
8. Yelamarthi, K., Haas, D., Nielsen, D., Mothersell, S.: RFID and GPS integrated navigation system for the visually impaired. In: 2010 53rd IEEE International Midwest Symposium on Circuits and Systems (MWSCAS), pp. 1149–1152 (2010)
9. Guerrero, L.A., Vasquez, F., Ochoa, S.F.: An indoor navigation system for the visually impaired. Sensors **12**, 8236–8258 (2012)
10. Vera, P., Zenteno, D., Salas, J.: A smartphone-based virtual white cane. Pattern Anal. Appl. **17**, 623–632 (2014)
11. Leduc-Mills, B., Profita, H., Bharadwaj, S., Cromer, P.: ioCane: a smart-phone and sensor-augmented mobility aid for the blind (2013)
12. Tange, Y., Takeno, S., Hori, J.: Development of the obstacle detection system combining orientation sensor of smartphone and distance sensor. In: 2015 37th Annual International Conference of the IEEE Engineering in Medicine and Biology Society (EMBC), pp. 6696–6699 (2015)
13. Croce, D., Gallo, P., Garlisi, D., Giarré, L., Mangione, S., Tinnirello, I.: ARIANNA: a smartphone-based navigation system with human in the loop. In: 2014 22nd Mediterranean Conference of Control and Automation (MED), pp. 8–13 (2014)

14. Lin, B.-S., Lee, C.-C., Chiang, P.-Y.: Simple smartphone-based guiding system for visually impaired people. Sensors **17**(6), 1371 (2017)
15. Jafri, R., Khan, M.M.: User-centered design of a depth data based obstacle detection and avoidance system for the visually impaired. Hum. Centric Comput. Inf. Sci. **8**(1), 14 (2018). https://doi.org/10.1186/s13673-018-0134-9

Ordering Assertive Strategies for Corporate Travel Agencies: Verbal Decision Analysis Model

Maria Luciana Santiago Leite[1], Plácido Rogério Pinheiro[2](✉), Marum Simão Filho[3], and Maria Lianeide Souto Araújo[4]

[1] Fortaleza, Brazil
[2] Graduate Program in Applied Informatics, University of Fortaleza, Fortaleza, Brazil
placido@unifor.br
[3] 7 de Setembro College, Fortaleza, Brazil
marum@edu.unifor.com
[4] Federal Institute of Education, Science and Technology of Ceará (IFCE), Fortaleza, Brazil
lianeide.araujo@yahoo.com.br

Abstract. The internet has brought revolutionary changes to tourism, especially to the distribution channels of tourism products and services, enabling the final consumer to get what they need directly from suppliers in order to exclude the role of the intermediary. In the corporate travel market, travel agencies called Travel Management Companies (TMCs) are under strong pressure to find ways to reinvent themselves in the face of the imminent threat of disintermediation. In order to know the assertive strategies that the Travel Management Companies (TMC's) could adopt, in the perspective of their contracting clients, a Multi-criteria model was applied in the Verbal Decision Analysis using the ZAPROS III-i method. The method proposes to list, in order of relevance, the strategies that could interfere in the decision analysis of the travel manager when hiring the services of a TMC.

Keywords: Corporate travel · TMC · Disintermediation · ZAPROS III-i Verbal Decision Analysis

1 Introduction

Tourism, like any other sector of the economy, has undergone several transformations with the future of the internet. The travel agency, in performing its intermediary function between the final customer and the supplier, had to remodel to remain attractive in the market, both in the leisure segment and in the business segment. In the corporate travel industry, in the business segment, the Travel Management Companies (TMC's), as corporate travel agents are known, excel as service specialists for small and large companies who need a service focused on a demanding profile and specific.

The phenomenon of disintermediation is presented in order to show an imminent threat to the existence of the TMC's, considering that the Federal Government already

R. Silhavy et al. (Eds.): CoMeSySo 2018, AISC 859, pp. 374–384, 2019.
https://doi.org/10.1007/978-3-030-00211-4_33

buys the national airline tickets from its servers directly with the respective airlines. Following the same trend in the public sector, some companies, such as Natura® and Riachuelo®, obtained their own online reservations license, and started to have their own agency, with the autonomy to negotiate and buy directly from the supplier.

Propose strategic assertive actions that can be adopted by the TMC's, and therefore, the proposal of this article, in order to minimize or delay the tendency to disintermediation with suppliers of corporate travel services.

The ZAPROS III-i multicriteria analysis method was chosen for this study because it is based on obtaining preferences, allowing levels for the established criteria and comparing the proposed alternatives [12].

2 Justifications

The disintermediation of the travel agent damages the sector, with the potential to unbalance a sector of the national economy because it may represent the extinction of corporate travel agencies.

The indications for the tendency towards disintermediation to gain strength and consolidation are many, including:

- The conversion of former TMC's partners into competitors such as Online Booking Tools (OBT's) Reserve® and Argo it®;
- The identification of fraud by the Public Prosecutor's Office in air tickets sent by travel agents to public agencies, indicating over-billing of amounts;
- The growing autonomy of employees of private companies to purchase their own corporate tickets, naturally leading them to more practical internet access through their smartphones or tablets to the websites of the virtual agencies known as OTAs, Online Travel Agencies.

According to [9], the biggest OTA's of the moment are: Decolar.com®, Booking.com®, Expedia®, Peixe Urbano®, Tripadvisor®, Submarino Viagens®, CVC Viagens® and Viajanet®.

3 Theoretical Reference

According to [3], the development of information technology has inevitably had a major effect on the operation, structure and strategies of tourism organizations worldwide.

According to [8], the emergence of the Internet allowed faster and less costly access for travel agents, but it was not long before the global distribution system (GDS), which until then was exclusively used by operators and agents possible to have online access to information for any destination and its respective costs.

From then on, the trend began that led to the beginning of the worldwide social phenomenon of disintermediation of travel agency services.

The popularization of the internet has allowed to take the systems previously exclusive to the agencies, to the final consumer, leading to a significant reduction in distribution costs and causing suppliers to compete with their own travel agencies. Thus, the agent has become an unnecessary intermediary in many situations [4].

For [6], the problem of disintermediation is already in place, and it is therefore important to understand how service providers in general, who are threatened by the transparency of the online market, struggle not to be relegated to the role of simple intermediaries information or transaction parasites.

Disintermediation occurred quickly in the leisure segment. Then some private sector companies started to negotiate directly with hotel chains and car rental companies. Finally, in 2015 the Ministry of Planning, Budget and Management of the Federal Government launched the Direct Purchase project, now buying directly with the main Brazilian airlines. Currently, more than 430 out of 627 travel agent contracts have been canceled and agencies have started to buy direct according to project guidelines [10].

In the two segments of tourism, leisure and business, the disintermediation happened initially in the product air tickets, either through collective purchasing sites or even on the websites of the airlines. This fact is critical and worrisome when it comes to a TMC, since the product of air tickets, specifically the national, is a considerable part of the revenues of these companies.

According to data from the Brazilian Association of Corporate Agencies (ABRACORP) [1], any change that involves the aerial product will significantly compromise the survival of those agencies that have the specialty in the business traveler.

4 Methodology

In decision-making processes that involve various points of view and analysis of alternatives, experts suggest the use of methodologies that favor reflection on the decision object, increasing the decision-maker's confidence through the generated knowledge, and consequently facilitating the choices of according to experience and preferences.

The Multicriteria Methodology supports the decision making process by offering tools that allow defining existing problems, considering the risks, impacts and available resources [11, 13].

From these approaches are derived several methods and techniques that can be used in the different types of problems.

According to [12], verbal decision analysis (VDA) comprises a set of several methods for classifying and ordering alternatives which consider multiple criteria in problem solving.

The structure of the methods known as the VDA Framework is composed of methods that have great applicability in problems that present many alternatives.

According to [14], among the classification methods, we can mention ORCLASS, SAC, DIFCLASS and CYCLE. Some ordering methods are PACOM, ARACE and those of the ZAPROS family (ZAPROS-LM, STEPZAPROS, ZAPROS III and III-i), according to Fig. 1.

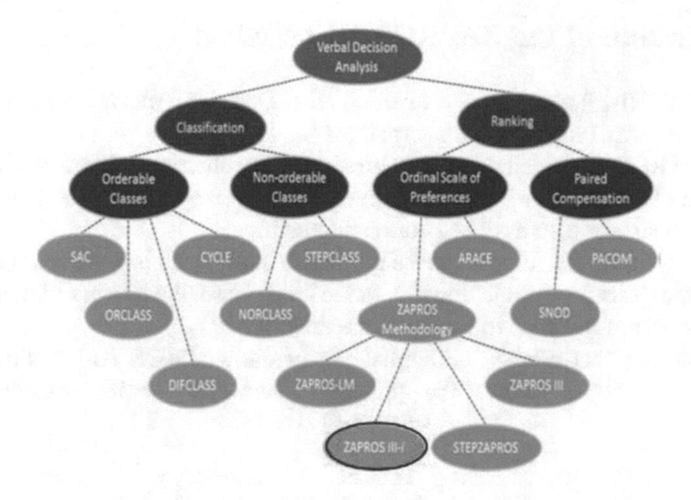

Fig. 1. Some methods of Verbal Decision Analysis [17]

The Verbal Decision Analysis is presented as support in the decision-making process for unstructured problems. Some characteristics of unstructured problems are cited by [7]:

- They are unique and new for decision making;
- These are problems that have a qualitative nature formulated in natural language;
- Evaluations of alternatives can be acquired only by those who know the problem;
- The quality of the evaluation of the alternatives can be acquired through subjective preferences of the decision maker;

In unstructured problems, the central source of information is human judgment. For this reason, the proposed methods should consider the validity of the data and possible inconsistencies. According to [5], the ZAPROS III-i method, an evolution of the ZAPROS LM method, brought in its new version, more efficiency and more precision regarding the inconsistencies.

Therefore, ZAPROS III-i was adequate for the problem studied because besides identifying the assertive strategies, it still allowed a ranking in order of preference of the decision maker.

Some benefits identified for choosing the ZAPROS III-i method were:

- It presents questions about the elicitation of the preference process that are understandable to the decision maker, based on the criteria values;
- It presents considerable resistance to the contradictory inputs of the decision maker, being able to detect them and ask for a solution to those problems;
- Specifies all qualitative comparison information in a language that is understandable to the decision maker [15].

5 Application of the ZAPROS III-i Method

The ZAPROS III-i method is one of the Verbal Decision Analysis methods and was developed to order multi-criteria alternatives [2, 19].

The ZAPROS III-i method is structured in three phases. In the first phase, of the formulation of the problem, the alternatives and their respective criteria are obtained. Levels of values are also assigned for each criterion.

In the second phase, a more or less preferable order scale is generated based on the choices of the decision makers. Finally, in the third phase, the process of comparing the alternatives occurs in order to establish the ranking.

The software that allows the application of the ZAPROS III-i method is called ARANAÚ. In Fig. 2 shows the process framework for the specific software application [20]:

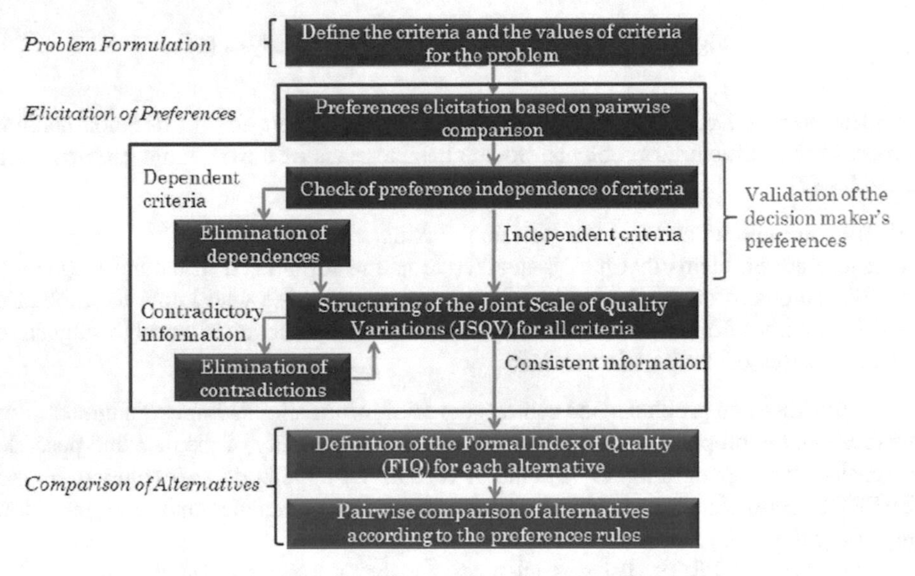

Fig. 2. Procedure for applying the ZAPROS III-i method

Seven alternatives were identified that could be used in the routine of corporate agencies as strategic actions with the objective of obtaining a competitive position differentiated from the competition, in the search to maintain contracts or conquer new partnerships. Figure 3 shows the 7 strategies that supported the applied questionnaire.

Estratégia	Alternativas
1	Oferecer **PROGRAMA DE VANTAGENS** para funcionários como descontos para viagens particulares/a lazer e em eventos (famtours)
2	Disponibilizar serviço de **CONSULTORIA**, trazendo novidades do mercado
3	Ter **SUPORTE EM EMERGÊNCIAS**, principalmente em viagens internacionais, em feriados e fins de seman
4	Dispor de tarifas diferenciadas em hospedagens e locações em virtudes de **PARCERIAS** firmadas
5	Apresentar **TRANSPARÊNCIA** nas transações comerciais e financeiras"
6	Possuir um **SISTEMA DE GESTÃO PRÓPRIO**, flexível e moldado à necessidade do cliente
7	Ser atuante e **FOMENTAR O PROFISSIONALISMO** e o networking no mercado e em eventos organizacionais

Fig. 3. Strategies as alternatives for decision-maker analysis

6 Definition of Criteria

The definition of the criteria is one of the most delicate parts in the formulation of decision problems and they are assigned ordering values according to a structure of preference of the decision maker [18, 21, 22].

The criteria (A, B, and C) were naturally and logically defined to the question of the problem involving the disintermediation phenomenon of the corporate travel agencies, that is, in the decision to hire or direct purchase of corporate travel services. Therefore, were thus defined:

A - possibility of increase of direct purchase;
B - possibility of increasing the intermediation of CMD;
C - possibility of maintaining direct purchase and CMD concomitantly;

Thus, the rules of preference are summarized as: A is preferable to B, B is preferable to A and/or A and B are likewise preferable.

For each criterion, three scales were considered in order to measure the preferences of the decision maker for the proposed alternatives: high, moderate and low.

7 Results

A questionnaire was prepared and sent by e-mail to 9 travel managers working in Fortaleza, capital of the State of Ceará - Brazil, based on the 7 alternatives defined as strategic. This questionnaire contained 21 questions. The first four had the function of delineating the profile of the manager, and the other four to delineate the relevance of the alternatives in the decision process. The alternatives were assigned criteria of choice and their respective levels of possibilities. Figure 4 presents a sample question template used in the questionnaire applied to travel managers.

Perguntas de pesquisa acadêmica
*Obrigatório

1 - De acordo com a sua experiência, considerando os serviços prestados por uma TMC, a atividade: ***"Realizar reuniões quinzenais com o (a) gestor (a) para o nivelamento de novidades e tendências do mercado"***, você acredita que: *

A – Possibilidade de aumento da compra direta	B–Possibilidade de aumento da intermediação da TMC	C – Possibilidade de manter compra direta e TMC concomitantemente
( )Alta ( ) Moderada ( )Baixa	( )Alta ( ) Moderada ( )Baixa	( )Alta ( ) Moderada ( )Baixa

2 - De acordo com a sua experiência, considerando os serviços prestados por uma TMC, a atividade: ***"Avaliar oportunidades de economias, assim como trazer soluções para redução de custos"***, você acredita que: *

A – Possibilidade de aumento da compra direta	B –Possibilidade de aumento da intermediação da TMC	C – Possibilidade de manter compra direta e TMC concomitantemente
( )Alta ( ) Moderada ( )Baixa	( )Alta ( ) Moderada ( )Baixa	( )Alta ( ) Moderada ( )Baixa

3 - De acordo com a sua experiência, considerando os serviços prestados por uma TMC, a atividade: ***"Ser comprometido com o planejamento estratégico e orçamentário de viagens da empresa, traçando metas e elaborando indicadores de desempenho"***, você acredita que: *

A – Possibilidade de aumento da compra direta	B –Possibilidade de aumento da intermediação da TMC	C – Possibilidade de manter compra direta e TMC concomitantemente
( )Alta ( ) Moderada ( )Baixa	( )Alta ( ) Moderada ( )Baixa	( )Alta ( ) Moderada ( )Baixa

Fig. 4. Sample of applied questionnaire questions

With answers to the questions applied exclusively to the travel managers, it was possible to organize the answers in a table, according to Fig. 5, allowing to visualize the calculation of the choices that each criterion received and the level of agreement with the proposed alternatives.

The answers were submitted to the application of the ARANAÚ tool, starting with the inclusion of the alternatives, as shown in Fig. 6, after the criteria and results that prevailed for each level of applied criteria.

The following parameters were followed to define the criteria:

- A1, B1 and C1 correspond to a "HIGH"
- A2, B2 and C2 correspond to a "MODERATE"
- A3, B3 and C3 correspond to a "LOW OR NULL"

In the next stage, called preference elicitation, a set of answers considered ideal for the problem of ensuring the existence of the role of Travel Management Company as an intermediary in the provision of corporate travel services was assigned. In the evaluation and preference of the decision maker, an ideal object, [A3 B1 C3], was assumed as the reference parameter.

The preference elicitation consists in comparing pairs of all the scales of the criteria A, B and C, from questions posed to the decision maker, for example: "It is preferable

	A - Possibilidade de aumento da compra direta			B - Possibilidade de aumento da intermediação da TMC			C - Possibilidade de manter compra direta e TMC concomitantemente					
Alternativas	A1	A2	A3	B1	B2	B3	C1	C2	C3	Resultados		
Disponibilizar serviço de consultoria, trazendo novidades do mercado;	8	5	14	13	7	7	4	12	11	A3	B1	C2
Oferecer diferencial para funcionários como descontos para viagens particulares/a lazer e em eventos (famtours);	2	4	12	11	6	1	8	4	6	A3	B1	C1
Apresentar TRANSPARÊNCIA nas transações comerciais e financeiras;	4	9	14	13	10	4	4	11	12	A3	B2	C2
Possuir um Sistema de Gestão próprio, flexível e moldado à necessidade do cliente;	3	6	9	7	10	1	6	9	3	A3	B2	C2
Ter suporte em emergências, principalmente em viagens internacionais, em feriados e fins de semana;	3	5	10	14	3	1	5	9	4	A3	B1	C2
Dispor de tarifas diferenciadas em hospedagens e locações em virtudes de parcerias firmadas;	2	2	14	11	6	1	5	9	4	A3	B1	C2
Ser atuante e fomentar o profissionalismo e networking no mercado e em eventos organizacionais;	1	7	19	10	11	6	4	12	11	A3	B2	C2

Fig. 5. Characterization of the alternatives according to the applied questionnaire

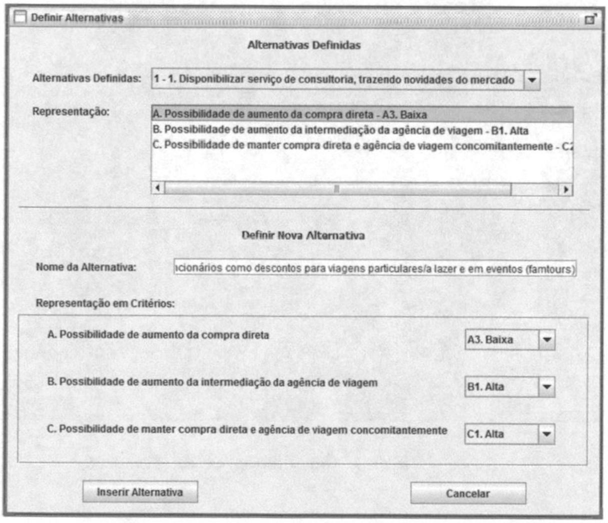

Definir Alternativas
Alternativas Definidas
Alternativas Definidas: 1 - 1. Disponibilizar serviço de consultoria, trazendo novidades do mercado
Representação:
A. Possibilidade de aumento da compra direta - A3. Baixa
B. Possibilidade de aumento da intermediação da agência de viagem - B1. Alta
C. Possibilidade de manter compra direta e agência de viagem concomitantemente - C
Definir Nova Alternativa
Nome da Alternativa: ıcionários como descontos para viagens particulares/a lazer e em eventos (famtours)
Representação em Critérios:
A. Possibilidade de aumento da compra direta A3. Baixa
B. Possibilidade de aumento da intermediação da agência de viagem B1. Alta
C. Possibilidade de manter compra direta e agência de viagem concomitantemente C1. Alta
Inserir Alternativa Cancelar

Fig. 6. Introduction of the alternatives in the tool ARANAÚ

to have a LOW possibility to hire a TMC or have an AVERAGE chance to have the two modes of shopping channels applied concomitantly?"

In the preference elicitation phase, it was possible to create a scale of values, defined from the differences between the results of each level of a criterion and the ideal object assumed as the most preferable for the solution of the study problem. For this variant identified in the comparison is given the name of Quality Variation (VQ).

Alternativas	Representação	FIQ
2. Oferecer diferencial para funcionários como descontos para viagens particulares/a lazer e em eventos (famtours)	A3. Baixa B1. Alta C1. Alta	3
1. Disponibilizar serviço de consultoria, trazendo novidades do mercado	A3. Baixa B1. Alta C2. Moderada	7
5. Ter suporte em emergências, principalmente em viagens internacionais, em feriados e fins de semana	A3. Baixa B1. Alta C2. Moderada	7
6. Dispor de tarifas diferenciadas em hospedagens e locações em virtudes de parcerias firmadas	A3. Baixa B1. Alta C2. Moderada	7
3. Apresentar TRANSPARÊNCIA nas transações comerciais e financeiras	A3. Baixa B1. Alta C3. Baixa	9
4. Possuir um Sistema de Gestão próprio", flexível e moldado à necessidade do cliente	A3. Baixa B2. Moderada C2. Moderada	14
7. Ser atuante e fomentar o profissionalismo e networking no mercado e em eventos organizacionais	A3. Baixa B2. Moderada C2. Moderada	14

Fig. 7. Setting the formal quality index (IFQ's)

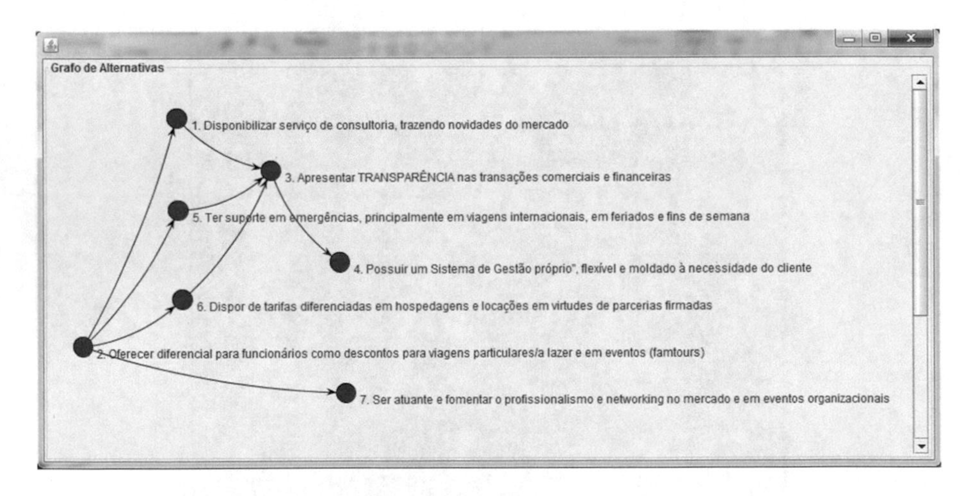

Fig. 8. Graph of the ranking of alternatives

Therefore, the alternative that obtained the lowest formal quality index (FIQ), that is, the one that most approached the ideal object [A3 B1 C3], revealed the best alternative, according to the participants of the research. Figure 7 indicates the ranking of the smallest to largest FIQ.

The ARANAÚ tool also provides the result on a dominance chart, as shown in Fig. 8.

Therefore, the result obtained, which had as decision makers, professionals who act as travel managers in the Brazilian capital of Fortaleza, showed that the offer of an incentive program that can be extended to the family and employees' aggregates, is important and influences positively in the decision to hire the services of a Travel Management Company (TMC).

8 Conclusion and Future Work

The result of the applied method evidences the realization that the applied management practices reflect the behavior of the commercial relations of a certain place. Therefore, the strategies for achieving an objective may be identical, but the order of relevance will be determined by the decision makers of the problem presented.

The ZAPROS III-i method proved to be efficient in meeting the proposal of presenting strategies to travel managers that could minimize or delay the tendency to disintermediation with suppliers of corporate travel services.

However, it is important to note that all seven alternatives have a decisive influence on the contracting of a TMC. However, with the applied method, it was possible to order a strategic selection based on the preferences of the decision maker.

As work and future studies it is suggested:

- The use of the same methodology and method based on information from a greater number of respondents and from different regions of Brazil, for a wider and heterogeneous evaluation.
- Application of research in organizations that have adopted the direct purchase with the dispensability of the travel agency. Considering that, for all the alternatives evaluated, the C2 level (moderate degree) prevailed for the criterion of maintaining the contraction of the MCT and direct purchase concomitantly. Therefore, although the intermediation of the agency is preferable to the decision makers of this research, the openness to the practice of direct purchase shows a considerable tendency.

Acknowledgment. The author's thanks to the National Council for Technological and Scientific Development (CNPq) through grants #305805/2017-7, Edson Queiroz Foundation/University of Fortaleza and FINEP (Funding Agency for Studies and Projects) for all the support provide.

References

1. ABRACORP; Números: Estatística. Net. 2012–2017 Disponível em <shttp://abracorp.org.br/numeros/> Acesso em 14 nov. 2017
2. Castro, L., Gomes, L.: Escolhas Estratégicas a partir da linguagem natural - a análise verbal de decisões. XXXV SBPO, Natal -RN (2003)
3. Cooper, C.: Turismo: princípios e práticas. Bookman, Porto Alegre (2001)
4. Lago, R., Cancellier, L.P.L.: Agências de viagens: desafios de um mercado em reestruturação. Turismo: Visão e Ação **7**(3), 507–514 (2005)
5. Larichev, O., Moshkovich, H.: Verbal Decision Analysis for Unstructured Problems. Kluwer Academic Publishers, Boston (1997)
6. Lévy, P.: O que é o virtual?. Ed.34, São Paulo (1996)

7. Moshkovich, H.M., Mechitov, A.I., Olson, D.L.: Verbal decision analysis. In: Figueira, J., Greco, S., Ehrgott, M. (eds.) State of the Art of Multiple Criteria Decision Analysis. International Series in Operations Research and Management Science. Springer, New York (2005)
8. O'Connor, P.: Distribuição da informação eletrônica em turismo e hotelaria. Bookman, São Paulo (2001)
9. Otto, G.: Tudo o que você precisa saber sobre as OTA's (Online Travel Agencies). Disponível em: <http://gabrielaotto.com.br/blog/tudo-o-que-voce-precisa-saber-sobre-as-otas-online-travel-agencies/>. Acesso em 17 set. 2017
10. PORTAL BRASIL. Governo amplia economia com compra direta de passagens aéreas. Disponível em: <http://www.brasil.gov.br/economia-e-emprego/2016/03/governo-amplia-economia-com-compra-direta-de-passagens-aereas>. Acesso em: 17 set. 2017
11. Silva, C.F.G., Pinheiro, P.R., Barreira, O.: Multicriteria problem structuring for the prioritization of information technology infrastructure problems. In: Silhavy, R., Silhavy, P., Prokopova, Z. (eds.) Applied Computational Intelligence and Mathematical Methods, vol. 662, 1st edn, pp. 326–337. Springer, Berlin (2017). 151f
12. Simão Filho, M., Pinheiro, P.R., Albuquerque, A.B.: Analysis of task allocation in distributed software development through a hybrid methodology of verbal decision analysis. J. Softw. Evol. Process **Único**, e1867 (2017). https://doi.org/10.1002/smr.1867
13. Ferreira, C., Nery, A., Pinheiro, P.R.: A multi-criteria model in information technology infrastructure problems. Proc. Comput. Sci. **91**, 642–651 (2016)
14. Gomes, U.R.P., Simao Filho, M., Pinheiro, P.R.: Project portfolio prioritization aided by verbal decision analysis. In: Proceedings of the 13th Iberian Conference on Information Systems and Technologies (2018, accept to publication)
15. Ustinovich, L., Kochin, D.: Verbal decision analysis methods for determining efficiency of investments in construction. In: Simão Filho, M. (ed.) An approach structured on verbal decision analysis to support the allocation of tasks in projects of distributed software development (2017). 212 f. Tese de Doutorado. Programa de Doutorado em Informática Aplicada, Universidade de Fortaleza (UNIFOR). Fortaleza (2017)
16. Tamanini, I., Pinheiro, P.R.: Reducing incomparability in multicriteria decision analysis: an extension of the ZAPROS method. Pesquisa Operacional **31**(2), 251–270 (2011). https://doi.org/10.1590/S0101-74382011000200004
17. Vasconcelos, M.F., Pinheiro, P.R., Simao Filho, M.: A multicriteria model applied to the choice of a competitive strategy for the printed newspaper. In: Silhavy, R., Silhavy, P., Prokopova, Z. (eds.) Cybernetics Approaches in Intelligent Systems, vol. 1, 1st edn, pp. 206–215. Springer, Berlin (2017)
18. Vincke, P.: Multicriteria Decision-Aid. Wiley, Chichester (1992)
19. Simão Filho, M., Pinheiro, P., Albuquerque, A.: Task assignment to distributed teams aided by a hybrid methodology of verbal decision analysis. IET Softw. **1**, 1–23 (2017)
20. Pinheiro, P.R., Tamanini, I., Pinheiro, M.C.D., Albuquerque, V.H.C.: Evaluation of the Alzheimer's disease clinical stages under the optics of hybrid approaches in verbal decision analysis. Telemat. Inf. **1**, 1 (2017)
21. Barbosa, P.A.M., Pinheiro, P.R., de Vasconcelos Silveira, F.R., Simão Filho, M.: Applying verbal analysis of decision to prioritize software requirement considering the stability of the requirement. In: Advances in Intelligent Systems and Computing, vol. 575, 1st edn, pp. 416–426. Springer (2017)
22. Simão Filho, M., Pinheiro, P.R., Albuquerque, A.B.: Task allocation in distributed software development aided by verbal decision analysis. In: Silhavy, R., Senkerik, R., Oplatkova, Z. K., Silhavy, P., Prokopova, Z. (eds.) Advances in Intelligent Systems and Computing, vol. 465, 1st edn, pp. 127–137. Springer, Berlin (2016)

Author Index

R. Silhavy et al. (Eds.): CoMeSySo 2018, AISC 859, pp. 385–386, 2019.
https://doi.org/10.1007/978-3-030-00211-4

Printed in the United States
By Bookmasters